EVC 1999

Critical Limb Ischemia

CRITICAL LIMB ISCHEMIA

Edited by

ALAIN BRANCHEREAU, MD

University Hospital, Marseille, France

&

MICHAEL JACOBS, MD

Academic Medical Center, Amsterdam, The Netherlands

FUTURA PUBLISHING COMPANY, INC.

ARMONK, NY

Library of Congress Cataloging-in-Publication Data

Critical limb ischemia / edited by Alain Branchereau & Michael Jacobs.
p. cm.
European Vascular Course 99 - Contents p.
Includes bibliographical references.
ISBN 0-87993-412-3
1. Peripheral vascular disease. 2. Ischemia. 3. Extremities (anatomy)--Blood-vessels--Diseases.
I. Branchereau Alain, MD. II. Jacobs Michael, MD. III. European Vascular Course (1999)
[DNLM: 1. Extremities--blood supply Congresses. 2. Ischemia--diagnosis Congresses.
3. Ischemia--therapy Congresses. WG 510 C93307 1999]
RC694.C755 1999
616.1'31--dc21
DNLM/DLC
for Library of Congress 99-14083
CIP

Published by
Futura Publishing Company, Inc.
135 Bedford Road
Armonk, NY 10504-0418

LC #: 986275
ISBN #: 0-87993-412-3

Printed in France on acid-free paper.

CONTRIBUTORS

Samuel S. AHN
Division of Vascular Surgery
200 UCLA Medical Plaza, Suite 626
Los Angeles, CA 90095, USA

Jean-Noël ALBERTINI
Hôpital La Timone
Service de Chirurgie Vasculaire
264, rue Saint Pierre
13385 Marseille Cedex 05, France

Xavier BARRAL
Centre Hospitalier Universitaire Nord
Avenue Albert Raimond
42055 Saint-Etienne Cedex 2, France

Jonathan D. BEARD
Sheffield Vascular Institute
Herries Road
Sheffield S5 7AU, United-Kingdom

Peter R.F. BELL
Department of Surgery
Robert Kilpatrick Building
Leicester Royal infirmary
Leicester LE2 7LX, United-Kingdom

Pauwel BERNAERTS
Department of Vascular Diseases
University Clinic Gasthuisberg
Leuven, Belgium

Fausto BIANCARI
Division of Vascular Surgery
Helsinki University Central Hospital
PO Box 262, 00029 Hyks, Finland

Jan D. BLANKENSTEIJN
University Hospital Utrecht
Department of Surgery
PO Box 85500
3508 GA Utrecht, The Netherlands

Christian BOISSIER
Service d'Angiologie. Centre Hospitalier
Universitaire - Hôpital Nord
42055 Saint-Etienne Cedex 2, France

Amman BOLIA
Department of Radiology
Leicester Royal Infirmary NHS Trust
Leicester LE1 5WW, United-Kingdom

Alain BRANCHEREAU
Hôpital La Timone
Service de Chirurgie Vasculaire
264, rue Saint Pierre
13385 Marseille Cedex 05, France

John BUDD
Royal United Hospital
Bath, United-Kingdom

Jaap BUTH
Department of Surgery
Catharina Hospital
Michelangelolaan 2
5623 EJ Eindhoven, The Netherlands

Christophe CAMIADE
Service de Chirurgie Vasculaire
Hôpital Jean Bernard
Centre Hospitalier Universitaire
86021 Poitiers Cedex, France

Stefano CAMPARINI
Department of Vascular Surgery
San Agostino Hospital
Modena, Italy

Adhisabandh CHULAKADABBA
Department of Vascular Surgery
Southampton General Hospital
Southampton SO16 6YD, United-Kingdom

Gioacchino COPPI
Department of Vascular Surgery
San Agostino Hospital
Modena, Italy

Lorenzo DI GIULIO
Hôpital Saint-Joseph
26, boulevard de Louvain
13285 Marseille Cedex 8, France

Bertrand EDE
Hôpital La Timone
Service de Chirurgie Vasculaire
264, rue Saint Pierre
13385 Marseille Cedex 05, France

Gianfranco FADDA
Hôpital Saint-Joseph
26, boulevard de Louvain
13285 Marseille Cedex 8, France

Bengt FAGRELL
Karolinska Institute
Department of Medicine
Karolinska Hospital
S- 171 76 Stockholm, Sweden

Jean-Pierre FAVRE
Centre Hospitalier Universitaire Nord
Avenue Albert Raimond
42055 Saint-Etienne Cedex 2, France

Michel FERDANI
Hôpital Saint-Joseph
26, boulevard de Louvain
13285 Marseille Cedex 8, France

Robert K. FISHER
Regional Vascular Unit
Royal Liverpool University Hospital
Prescott Street, Liverpool, United-Kingdom

Jean-François GARBE
Clinique Esquirol
15, rue Pontarique, BP 159
47005 Agen Cedex, France

Andrew H. GARNHAM
Department of Vascular Surgery
Selly Oak Hospital
University Hospital NHS Trust
Birmingham B29 6JD, United-Kingdom

Stefano GENNAI
Department of Vascular Surgery
San Agostino Hospital
Modena, Italy

Jose A. GONZALEZ-FAJARDO
Division of Vascular Surgery
Hospital Clínico Universitario
47011 Valladolid, Spain

Henri GUIDICELLI
Centre Hospitalier Universitaire
BP 127
38043 Grenoble Cedex 09, France

Jean-Louis GUILMOT
Médecine Interne B
Centre Hospitalier Universitaire Bretonneau
37044 Tours Cedex, France

Peter L. HARRIS
Regional Vascular Unit
Royal Liverpool University Hospital
Prescott Street, Liverpool, United-Kingdom

Mirza M. IDU
Department of Surgery
Catharina Hospital
Michelangelolaan 2
5623 EJ Eindhoven, The Netherlands

Michael JACOBS
Department of Vascular Surgery
Academic Medical Center
PO Box 22700
1100 DE Amsterdam, The Netherlands

Jean-Michel JAUSSERAN
Hôpital Saint-Joseph
26, boulevard de Louvain
13285 Marseille Cedex 8, France

Hélène JULLIAN-PAPOUIN
Hôpital La Timone
Service de Chirurgie Vasculaire
264, rue Saint Pierre
13385 Marseille Cedex 05, France

Mark KOELEMAY
Department of Vascular Surgery
Academic Medical Center
PO Box 22700
1100 DE Amsterdam, The Netherlands

Hendrik LACROIX
Department of Vascular Diseases
University Clinic Gasthuisberg
Leuven, Belgium

Anthony LAMBERT
Derriford Hospital
Plymouth, United-Kingdom

Dink LEGEMATE
Department of Vascular Surgery
Academic Medical Center
PO Box 22700
1100 DE Amsterdam, The Netherlands

Mauri LEPÄNTALO
Division of Vascular Surgery
Helsinki University Central Hospital
PO Box 262, 00029 Hyks, Finland

Henning LOEPRECHT
I. Surgical Klinik
Zentralklinikum Augsburg
86009 Augsburg, Germany

Frank W. LOGERFO
Division of Vascular Surgery
Beth Israel Deaconess Medical Center
Boston, Massachusetts, USA

Pierre-Edouard MAGNAN
Hôpital La Timone
Service de Chirurgie Vasculaire
264, rue Saint Pierre
13385 Marseille Cedex 05, France

Jean-Luc MAGNE
Centre Hospitalier Universitaire
Service de Chirurgie Vasculaire
BP 127
38043 Grenoble Cedex 09, France

Nella MANGIACOTTI
Hôpital La Timone
Service de Chirurgie Vasculaire
264, rue Saint Pierre
13385 Marseille Cedex 05, France

Roberto MORATTO
Department of Vascular Surgery
San Agostino Hospital
Modena, Italy

André NEVELSTEEN
Department of Vascular Diseases
University Clinic Gasthuisberg
Leuven, Belgium

Lars NORGREN
Department of Surgery
University Hospital
S- 221 85 Lund, Sweden

William P. PAASKE
Department of Cardiothoracic
and Vascular Surgery
Aarhus University Hospital, Skejby Sygehus
Dk- 8200 Aarhus N, Denmark

Roberto PACCHIONI
Department of Vascular Surgery
San Agostino Hospital
Modena, Italy

Didier PLISSONNIER
Chirurgie Vasculaire
Hôpital Charles Nicolle
1, rue Germont
76031 Rouen Cedex, France

Giovanni RAGAZZI
Department of Vascular Surgery
San Agostino Hospital
Modena, Italy

Jim REEKERS
Department of Radiology
Academic Medical Center
PO Box 22700
1100 DE Amsterdam, The Netherlands

Jean-Baptiste RICCO
Service de Chirurgie Vasculaire
Hôpital Jean Bernard
Centre Hospitalier Universitaire
86021 Poitiers Cedex, France

Kyung M. RO
Department of Surgery
Division of Vascular Surgery
200 UCLA Medical Plaza
Los Angeles, CA 90095, USA

J. ROBINSON
Sheffield Vascular Institute
Herries Road
Sheffield S5 7AU, United-Kingdom

Serge ROUCHET
Polyclinique du Maine
4, avenue des Français Libres
53000 Laval, France

Robert B. RUTHERFORD
0146 Springbeauty Dr.
Mesa Cortina, PO Box 23159
Silverthorne, CO 80498, USA

Clifford P. SHEARMAN
Department of Vascular Surgery
Southampton General Hospital
Southampton SO16 6YD, United-Kingdom

Roberto SILINGARDI
Department of Vascular Surgery
San Agostino Hospital
Modena, Italy

Malcolm H. SIMMS
Department of Vascular Surgery
Selly Oak Hospital
University Hospital NHS Trust
Birmingham B29 6JD, United-Kingdom

Luc STOCKX
Department of Vascular Diseases
University Clinic Gasthuisberg
Leuven, Belgium

Jacques TESTART
Chirurgie Générale et Vasculaire
Hôpital Charles Nicolle
1, rue Germont
76031 Rouen Cedex, France

Thomas TROËNG
Department of Surgery
Blekinge Hospital
371 85 Karlskrona, Sweden

Erkki TUKIAINEN
Division of Plastic Surgery
Helsinki University Central Hospital
PO Box 262, 00029 Hyks, Finland

Dirk T. UBBINK
Department of Vascular Surgery
Academic Medical Center
PO Box 22700
1100 DE Amsterdam, The Netherlands

Carlos VAQUERO
Division of Vascular Surgery
Hospital Clínico Universitario
47011 Valladolid, Spain

Jacques WATELET
Chirurgie Vasculaire
Hôpital Charles Nicolle
1, rue Germont
76031 Rouen Cedex, France

FOREWORD

As far as history goes back, limb-threatening ischemia has always existed, initially challenging barbers to safely amputate the leg and keep the patient alive. In all of history, the twentieth century has known the most impressive developments regarding surgical methods and lesser invasive, innovative techniques.

The incidence of leg ischemia has progressively increased and currently constitutes one of the major features of generalized arterial atherosclerotic disease, being the most common cause of death in the Western world for many years.

During the last decades it has become apparent that different degrees of limb ischemia with subsequent different management strategies should be distinguished. The term "critical limb ischemia" (CLI) was introduced and several working parties on both sides of the Atlantic have developed criteria to define this entity.

This year's European Vascular Course addresses the subject of CLI and this book contains twenty-eight chapters covering the wide spectrum of definition and criteria, pathophysiology, epidemiology, assessment, and therapy. Objective criteria associated with the definition of CLI have been the source of many debates and therefore their disadvantages and limitations are discussed and alternatives and recommendations are proposed.

Another decisive role in the management of CLI is the pathophysiology, which not only involves large artery disease, but specifically affects changes in the microcirculation. Hence, diagnostic assessment includes modalities that allow evaluation of skin nutritional perfusion in patients with critical ischemia.

Angiography is the traditional gold standard for the assessment of the arterial in- and outflow and the main technique on which surgical and endovascular strategies are based. During the past decade the role of non-invasive techniques has become more important and color-duplex scanning presently replaces diagnostic angiography in a considerable number of patients and sets the stage for endovascular and surgical interventions. Endovascular management is currently not limited anymore to balloon dilatation, but involves minimally invasive revascularization of crural arteries, applying a wide range of techniques, such as atherectomy, thrombolysis and suction embolectomy as an adjunct to PTA.

Present-day medical treatment of CLI comprises spinal cord stimulation, antiplatelet drugs, defibrinating agents, vasoactive substances and prostanoids.

Distal bypass surgery with autogenous vein remains the therapy of choice for extensive arterial disease. The long-term results of distal popliteal, crural, and pedal vein grafts justify an aggressive surgical approach in patients with CLI, even in diabetics. Alternatives in the absence of saphenous vein are discussed in detail: other venous conduits, prosthetic grafts with adjunctive procedures, and arterial allografts. Arterial reconstruction associated with vascularized skin flaps is described as an effective procedure for the management of major tissue loss due to critical ischemia.

Intraoperative assessment and postoperative surveillance of endovascular and surgical reconstructions with subsequent intervention appears to be crucial for an adequate and improved outcome. Apart from technical considerations several pivotal issues which determine the life of patients with CLI are addressed: the general medical condition, quality of life, long-term morbidity and mortality, and the role of primary amputation.

Our goal was to publish a reference book, gathering the most appealing subjects related to CLI, written by key opinion vascular specialists. We would like to thank the authors for their contribution, which is obviously vital for the realization of this book. Major effort and editorial assistance was provided by Dirk Ubbink in Amsterdam, and Pierre-Edouard Magnan, Eugenio Rosset and Bertrand Ede in Marseille. We are greatly indebted to the secretaries and collaborators from our departments, who invested tremendous energy in the preparation of this book and the organization of the European Vascular Course: Annie Barral, Dorine van Ringelestijn and Caroline Lorenzati.

The continuous drive and eternal devotion of Marie-France Damia and the team of ODIM guaranteed the high quality of the book, both in the English and French versions. Futura Publishing Company contributed significantly with editorial support of Kirstin Bellhouse and strategic advises of Jacques Strauss.

This high-quality textbook would not have been possible without the substantial support of the biomedical industry. We are very grateful to our major sponsors for their contribution and confidence in this project.

Maryse and Florence are specifically to be acknowledged for their infinite participation during all the months of this mission.

Amsterdam - Marseille, March 1999

Michael Jacobs

Alain Branchereau

CONTENTS

Contributors VII
Foreword XI

1 The definition of critical limb ischemia: advantages and limitations
Robert B. Rutherford 1

2 The pathophysiology of critical limb ischemia
Clifford P. Shearman
Adhisabandh Chulakadabba 11

3 Epidemiological and economic considerations in the critically ischemic limb
Robert K. Fisher, Peter L. Harris 19

4 The diabetic foot: clinical and pathophysiological aspects
Frank W. LoGerfo 27

5 The importance of the medical condition of the patient with critical limb ischemia
Hélène Jullian-Papouin
Pierre-Edouard Magnan 33

6 Assessment of the microcirculation in critical limb ischemia
Bengt Fagrell 39

7 Arterial mapping by duplex scanning in patients with critical limb ischemia
Peter Bell 49

8 Duplex scanning and angiography of the lower leg and foot arteries
Mark Koelemay, Dink Legemate
Jim Reekers, Michael J.H.M. Jacobs 55

9 Medical treatment with prostanoids in patients with critical ischemia
Jean-Louis Guilmot, Christian Boissier 63

10 Non-prostanoid medical therapies in critical limb ischemia
Lars Norgren 69

11 Spinal cord stimulation in critical limb ischemia
Dirk Th. Ubbink, Michael J.H.M. Jacobs 75

12 Percutaneous transluminal therapy for critical limb ischemia
Amman Bolia 85

13 Changing patterns in the treatment of lower limb ischemia
Samuel S. Ahn, Kyung M. Ro 95

14 Adjunctive techniques to PTA: thrombolysis, suction embolectomy and rotablator
Gioacchino Coppi, Roberto Silingardi, Roberto Moratto, Stefano Camparini, Giovanni Ragazzi, Stefano Gennai, Roberto Pacchioni 101

15 Transluminal angioplasty combined with surgical reconstruction for lower limb ischemia
Jean-Baptiste Ricco, Christophe Camiade
Nella Mangiacotti, Bertrand Ede
Alain Branchereau 119

16 Critical limb ischemia due to popliteal artery aneurysm: endovascular approach
André Nevelsteen, Hendrik Lacroix, Luc Stockx, Pauwel Bernaerts 135

17 Femoropopliteal reconstruction
William P. Paaske 147

18 Femorocrural bypass for limb salvage: real indications and results
Jose A. Gonzalez-Fajardo, Carlos Vaquero 165

19 Bypass to the ankle and foot arteries
Malcolm H. Simms, Andrew H. Garnham 183

20 Technical improvement of in-situ venous grafts
H. Loeprecht 193

21 Prosthetic femorocrural bypass and adjunctive procedures
Jean-Michel Jausseran, John Budd Serge Rouchet, Jean-François Garbe Michel Ferdani, Anthony Lambert Gianfranco Fadda, Lorenzo Di Giulio 197

22 Infrainguinal bypass in the absence of saphenous vein: alternative venous conduits
Jacques Watelet, Didier Plissonnier Jacques Testart 207

23 Infrainguinal arterial allografts for limb salvage
Jean-Noël Albertini, Xavier Barral Alain Branchereau, Jean-Pierre Favre Henri Guidicelli, Pierre-Edouard Magnan Jean-Luc Magne 215

24 Distal arterial reconstruction combined with free muscle flaps
Mauri Lepäntalo, Fausto Biancari Erkki Tukiainen 225

25 Is there a place for primary amputation?
Joy Robinson, Jonathan D. Beard 239

26 Intraoperative control of endovascular and surgical reconstructions for critical limb ischemia
Jan D. Blankensteijn 247

27 The cause of mid-term and long-term failures of arterial reconstructions: the role of surveillance with subsequent action
Jaap Buth, Mirza M. Idu 253

28 Long-term morbidity and mortality of patients treated for critical limb ischemia
Thomas Troëng 261

1

THE DEFINITION OF CRITICAL LIMB ISCHEMIA: ADVANTAGES AND LIMITATIONS

ROBERT B. RUTHERFORD

Before definitions of critical limb ischemia were developed, there were serious problems in interpreting reports on the management of degrees of limb ischemia in which the threat to viability was deemed great, i.e. those characterized by ischemic rest pain and/or foot lesions with actual tissue loss. Published reports of the natural history or success of treatment of such patients (e.g. claimed limb salvage rates) varied widely and were much greater than patency rates [1]. Once the need for a uniform definition and objective criteria for this level of disease (ultimately named chronic critical ischemia [CCI]) was widely recognized [2], different modifications of the originally suggested criteria were advanced on both sides of the Atlantic, subsequently adopted, and put to the test in clinical studies and trials. Even though the concept of CCI is now widely accepted, and studies employing its specific criteria are routinely reported, with time the limitations of certain of these criteria were recognized, and expressed in published negative opinions of its value. However the reasons behind this dissatisfaction, i.e. limitations in the specific criteria for CCI, are readily understandable, even though not widely appreciated. With little modification they can continue to serve as a necessary and valuable standard, and although some of their limitations are not totally correctable; if understood properly, these should not interfere with appropriate application. Therefore, this chapter will discuss the definitions of CCI and their development, identify and explain the limitations of existing criteria, and suggest how these criteria can be standardized, and our use of them modified, so them may better serve their intent.

■ REASONS FOR A DEFINITION OF, AND OBJECTIVE CRITERIA FOR CCI

It is obviously important, when presenting experiences with the management of chronic leg ischemia, that levels of ischemia with very different outlooks, and thus requiring different management strategies (i.e. claudication versus ischemic rest pain and/or non healing foot lesions), should either not be mixed or should be clearly stratified in published reports. The Fontaine classification identifies these levels clearly (class III for rest pain and class IV for tissue necrosis, i.e. ulcers or gangrene), but assigning patients to them is subject to clinician interpretation in the absence of objective (hemodynamic) criteria. In other words, it is up to each physician or associate to say, based on clinical examination, that a patient's leg pain is actual ischemic rest pain (and not diabetic neuritis for example), or to say that a foot ulcer or an amputation of this gangrenous toe won't heal because of the level of ischemia in the forefoot. Very experienced vascular specialists may do this with reasonable accuracy. However, one only has to look at the significant differences between studies of the natural history of claudicants based on clinical examination versus clinical examination supported by non-invasive testing (as discussed below) to realize the importance of objective hemodynamic data on diagnostic accuracy in large group studies. Similarly, when patients classified as Fontaine III or IV (primarily ischemic ulcers) were entered into large randomized pharmacotherapy trials, it was not unusual to see that 25%-40% of the placebo controls actually improved (rest pain controlled, ulcers healed or healing) at the completion of the study [3]. This is too high for truly threatened limbs and can partly be explained by a significant degree of misdiagnosis of rest pain and ulcers as being on an ischemic basis. Furthermore, others have observed that conservative therapy results in *limb salvage* in a significant number of cases of *limb threatening ischemia* [4], making the point that trials of sympathectomy and other non-reconstructive therapies must employ conservatively treated controls to claim efficacy.

The implications of this are of obvious importance, because the proper diagnosis of CCI should carry with it a very high, if not inevitable, risk of major amputation if limb perfusion is not significantly improved, and this in turn constitutes a compelling indication for intervention. Such interventions are not only assessed by patency rates, but by limb salvage rates. These salvage rates, and mortality (i.e., loss of life or limb), are the traditional criteria by which the relative effectiveness of different forms of treatment for CCI, as well as different interventionalists and their institutions, are compared.

It is human nature to wish to report one's results in the best possible light. Although quite understandable, this leads to inevitable biases in selecting patients for treatment, and in interpreting and reporting the results of such treatments. Those misleading practices must be resisted and controlled if we are to learn from reported experiences and improve our management strategies. Not only can the results of competitive therapies (and of competitive care providers) not be properly compared unless they are applied to patients with similar disease severity, but, since including patients with lesser degrees of ischemia than those at genuine risk of limb loss very significantly improves overall mortality, patency, and limb salvage rates, unreal expectations may be fostered by such *contaminated* reports, misleading those with decision-making responsibilities. Without uniform standardized definitions of critical limb ischemia, with objective criteria, how is one to know that the reported treatment was effectively limited to those with impending threat of limb loss?

■ THE DEVELOPMENT OF CRITERIA FOR CRITICAL LIMB ISCHEMIA AND HOW THEY DIFFER

It was annoyance at *unbelievable* limb salvage rates being reported by many authors as much as anything which finally precipitated a meeting of concerned investigators, arranged by C.W. Jamieson at an International Vascular Symposium in London on September 16, 1981 [2]. This working party felt the term *limb salvage* should be an outcome measure and not be applied prospectively to identify a group of patients at risk of inevitable limb loss. *Critical ischemia* was felt to be preferable to *limb salvage status*, and they developed *a definition of critical ischemia of a limb* whose essential characteristics were 1) the pain must be persistent (more than four weeks) and severe (requiring strong analgesics), 2) *non-healing* ulcers must be due to ischemia (not neuropathy, infection, or other factors) and 3) gangrene must be due to *diffuse* forefoot ischemia (in contrast to most cases of atheroembolism). Recognizing the importance of objectively diagnosing such degrees of ischemia, hemodynamic criteria were proposed for ischemic rest pain (ankle pressure < 40 mmHg) and non-healing foot lesions (ankle pressure < 60 mmHg).

The author, a member of this working party, eventually incorporated similar criteria, plus equivalent toe pressure levels (< 30 mmHg and < 40 mmHg, respectively) and absent *pulsatility* on transmetatarsal or digital plethysmography to deal with diabetics with incompressible vessels, into reporting standards, which were approved and promulgated by the Society for Vascular Surgery and the International Society for Cardiovascular Surgery/North American Chapter (SVS/ISCVS) in 1986 [5]. These criteria are included in a *clinical categorization of chronic limb ischemia* which, with little modification, has been preserved in the recently (1997) revised version of these standards [6] (Table). It will be noted in this table that provisions are made not only for 6 *categories* for clinical trials and investigations, and which form the basis for a claim of *categorical improvement,* but for four clinical *grades* which are roughly equivalent to the four Fontaine levels, but numbered differently, assigning 0 to asymptomatic cases. However, in Europe the definition and criteria for CCI developed in a slightly different direction. In the document, which culminated from the second European Consensus on CCI in 1991 [7], the *same* hemodynamic criteria (ankle pressure < 50 mmHg or toe pressure < 30 mmHg) were proposed for *both* ischemic rest pain and non-healing foot lesions, under the common heading of CCI. Although the

Table CLINICAL CATEGORIES OF CHRONIC LIMB ISCHEMIA *

Grade	*Category*	*Clinical description*	*Objective creteria*
0	0	Asymptomatic - no hemodynamically significant occlusive disease	Normal treadmill or reactive hyperemia test
	1	Mild claudication	Completes tradmill exercise** and AP after exercise > 50 torr but at least 20 torr lower than resting value
I	2	Moderate claudication	Between category 1 and 3
	3	Severe claudication	Cannot complet standard tradmill exercise** and AP after exercise < 50 torr
II	4	Ischemic rest pain	Resting AP < 40 torr; flat or barely pulsatile ankle or metatarsal PVR; TP < 30 torr
III	5	Minor tissue loss - nonhealing ulcer, focal gangrene with diffuse pedal ischemia	Resting AP < 60 torr; ankle or metatarsal PVR flat or barely pulsatile; TP < 40 torr
	6	Major tissue loss - extending above TM level, functional foot no longer salvageable	Same as categorie 5

* Grades II and III, categories 4, 5 and 6 are embraced by the term chronic *critical* ischemia.
** Five minutes at 2 kmph on a 12% incline.
AP: ankle pressure PVR: pulse volume recording TP: toe pressure TM: transmetartasal

concepts and principles of this document are sound and widely accepted, these particular criteria did not separately delineate the two levels of ischemia CCI represents (i.e. ischemic rest pain and non-healing foot lesions). This was to cause problems in subsequent clinical investigations and trials, in that the common pressure criteria were too high for ischemic rest pain and too low for foot lesion healing. As a result, continuing criticisms of these criteria have been voiced by a number of European investigators [8-13]. These and other problems with existing criteria will now be discussed.

■ PROBLEMS WITH EXISTING CRITERIA OF CCI

1 - Consider the problem of linking rest pain and non-healing foot lesions with common hemodynamic criteria, as proposed by the European Consensus on CCI [7]. This differed from the original criteria proposed by the International Working Party of 1981 [2] and adopted by the SVS/ISCVS [5,6] and apparently failed to fully appreciate the rationale behind recommending two different pressure levels. Healing (of foot lesions) is associated with an inflammatory response and requires a greater level of perfusion than simply maintaining the viability of intact tissue (as is the case with ischemic rest pain alone). Therefore, while true ischemic rest pain (in the absence of tissue necrosis) rarely occurs with ankle pressures > 40 mmHg, ischemic ulcers and localized gangrene rarely heal spontaneously or with debridement/amputation, respectively, at ankle pressures < 60 mmHg. While it can be legitimately debated whether these specific levels, or equivalent levels of toe systolic pressure, tissue oxygen tension ($tcpO_2$), or other indices of perfusion, are the most appropriate ones, the principle behind separate levels for these two levels of ischemia (call them Fontaine III & IV if you will) is valid and should be followed.

2 - Having different criteria for chronic limb ischemia on the two sides of the Atlantic (or elsewhere in the world for that matter) also causes problems, to the degree we read each other's literature, which fortunately is increasing steadily and will continue to do so. However, the tendency not to read widely is the reason why many North Americans were unaware of the European definition of CCI. The criteria they were using were causing only relatively minor problems (see below). Nor did many on either side realize, when comparing results of various treatments for the category of critical limb ischemia being reported from the two sides of the Atlantic, that the treatment groups were defined by different criteria.

3 - There is the problem that, no matter how well chosen, no single arbitrary level of ankle pressure, toe pressure, or $tcpO_2$ for that matter, can separate, with absolute accuracy, those with and without ischemic rest pain or those with or without the ability to heal foot ulcers or localized gangrene. These unreal expectations are the source of some of the discontent with the existing criteria, but the same would apply to any objective criteria based on diagnostic testing. The best that can be achieved is to select levels with the greatest accuracy in discriminating between groups on either side of the clinical end points of rest pain and ischemic ulcers or necrosis. Because definitive studies comparing the accuracy of different cut-off points on large numbers of patients, using ROC curves or other statistical methods, have not yet been carried out, the proper levels are still the subject of some debate. The major point of contention here is: what is the appropriate level above which healing can be expected? That level depends to some degree on the size of the tissue necrosis and the degree of secondary infection, so again, no single arbitrary level is likely to be accurately predictive. Furthermore, if one includes diabetics, even if allowances are made by including equivalent toe pressures, plethysmography or even tissue oxygen levels, there will still be difficulties in discriminating cleanly, because their lack of resistance to infection and the often uncontrolled effects of pressure and trauma in the face of neuropathy greatly compound the healing process. Some have suggested 70 mmHg rather than 60 mmHg as the upper limit, and others have suggested no pressure limit in the face of tissue necrosis, but as will be pointed out below, neither alone will solve the problem. Others have suggested that these pressure criteria are too crude or

remote and that only local microcirculatory criteria will provide accurate predictive criteria. Transcutaneous oxygen levels, laser doppler perfusion measurements and capillary microscopy all undoubtedly can help with clinical decision making, in determining which limbs are truly at risk of amputation, but to date no single study has been shown to be predictive on its own, and/or even if it were, such methods are not widely enough accepted and/or the technology widely enough available to provide the basis for universal criteria. Nevertheless, this approach may provide a future solution.

4 - There is the *catch-22* problem of what to use as the gold standard in judging *limb threatening ischemia.* It is hardly ethical to withhold anything but conservative therapy and judge the appropriateness of specific criteria by observing which patients undergo major amputation. On the other hand, adding to amputation the need to intervene to avoid amputation, as is commonly done, does not help either because this decision is subjective in the absence of specific criteria. More acceptable endpoints than major amputation might be persistence of rest pain or lack of healing of ischemic ulcers, but these too lack objectivity unless supported by measures of foot perfusion.

5 - Extremely important is an under-appreciated conflict between the *research purpose* for which these criteria were developed, (i.e. to create relatively pure groups for comparison in clinical studies and trials), and the *clinical purpose* or goal of the clinician, which is to identify *all* patients with true ischemic rest pain and with foot lesions that will not heal spontaneously or with conservative measures, who are therefore at serious risk of limb loss, so that appropriate treatment can be rapidly initiated to avoid this dreaded outcome. The former intent is to be *exclusive,* the latter intent needs to be *inclusive.* Naturally, the same criteria cannot, and should not, be expected to satisfy both needs. Those crying for more liberal criteria may include surgeons who mistakenly feel obliged to live by the specific criteria suggested in selecting candidates for bypass, or those investigators who have had many of their patients with non healing ulcers excluded from trials in which they were participating.

6 - No matter how well and objectively it is defined, the setting up of CCI as a distinct group with a grave if not absolute risk of limb loss, in counter-distinction to claudication with its benign prognosis of minimal limb loss risk, is not only an oversimplification which fails to recognize the continuum or spectrum of risk that spans all of chronic leg ischemia, but it completely excludes a group of patients at intermediate risk. These are a group of asymptomatic patients, who are so sedentary or limited by other frailties and disabilities that they do not claudicate nor do they have rest pain, yet they are at significant risk of limb loss because of decreased limb perfusion. These individuals can best be characterized as having sufficient perfusion to *not* have ischemic rest pain but insufficient perfusion to heal a foot lesion should it develop, from minor trauma such as might be accidentally inflicted by stubbing a toe, or in cutting ones nails or trimming a callous. As such, having perfusion pressures between the limits set for ischemic rest pain and healing foot lesions may identify them. One might also include in this group some claudicants with similarly reduced perfusion pressures, who are at a much higher risk of limb loss than generally ascribed to claudicants. Whether ankle pressures of 40 and 60 mmHg, respectively, are the most appropriate levels to delineate this group may properly be debated, but since these levels are used in the literature, they can be used to give this group, now called *chronic subcritical ischemia* [6] some perspective, in term of risk of limb loss. Because sedentary patients in this group are not discovered and thus not studied, this risk must be inferred from claudicants with low ankle pressures (e.g. < 60 mmHg).

In four studies of this lower stratification of claudicants, totaling 1120 subjects [14-17], the mean risk of limb loss, or progressing to CCI and requiring intervention to prevent limb loss, was 8.5% per year. This compares with 3.7% per year in studies based on history and physical examination and 5.8% per year when the diagnosis was confirmed by objective non-invasive testing [18]. This latter observation also supports the previously stated need for objective criteria. The commonly quoted risk of 1% per year does not recognize that approximately four times the number of claudicants actually suffering major amputation require intervention to prevent this fate. Two other studies of claudicants with pressures in the CCI range, but without rest pain (e.g. ankle pressure $<$ 40 mmHg or toe pressure $<$ 30 mmHg) showed an even higher risk, averaging 13% per year [19,20]. While identification of this previously unrecognized subgroup does not justify invasive intervention, it does justify heightened surveillance, preventative measures and possibly pharmacotherapy. For this reason, this group should be considered in any *inclusive* definition of severe chronic leg ischemia applied to clinical practice. In doing so, it must be realized that while broader definitions may serve to identify anyone potentially at risk of limb loss, the indications for intervention to prevent this outcome, must, by necessity, be more selectively applied within this larger group.

This definition of chronic subcritical ischemia has been introduced in North America in the newly revised SVS/ISCVS standards [6], but it grew out of more recent discussions held with other members of the original *International Vascular Symposium Working Party* of 1981 [2]. It should be pointed out that one of them has promulgated this term in Europe under a somewhat different concept [21]. Mr. Wolfe has suggested including those with ankle pressures under 40 mmHg or any one with ulcers or gangrene in a high risk group (called critical ischemia), although more recently this has been raised to 50 mmHg. Those with rest pain who do not fulfill the above criteria for critical ischemia, i.e. those with pressures above 50 mmHg, would constitute the group with subcritical ischemia. The author's problems with this interpretation are, firstly, that rest pain is unlikely with ankle pressures above 50 mmHg, or even 40 mmHg for that matter, unless they are diabetics with falsely elevated pressures in whom toe pressures or other criteria should be used. Secondly, this scheme offers no objective criteria for ischemic ulcers or gangrene (again creating problems with diabetics in particular) and thirdly, it ignores those with low perfusion pressures who are at intermediate risk whether they are asymptomatic or claudicate. Clearly, this intermediate group deserves the same definition on both sides of the Atlantic Ocean.

7 - Finally, some criticisms of criteria for determining limb viability in *acute* limb ischemia have been voiced which unfortunately undermined the credibility of pressure criteria for chronic limb ischemia. Some of these have come from experiences by interventional radiologists with thrombolytic therapy for acute native arterial or graft thrombosis which challenge the concept that the designation of limb threatening ischemia by the criteria suggested did not preclude successful lytic therapy. One appears to have tested the original 1981 Working Party criteria [2], which were meant to apply to chronic ischemia, to the acute setting [22], but came to an appropriate conclusion that both pressure level and presence or absence of neurologic deficit were predictive. Another pointed out that an ankle pressure level of 30 mmHg or an ABI of 0.30 are not discriminating levels by which to judge *immediate* limb threat, based on the success of lytic therapy for those with lesser or no recordable pressures [23]. Unfortunately, some have apparently missed the point that any recordable ankle pressure, e.g. 30 mmHg or more, is felt to be valuable confirmatory evidence of limb viability in the absence of neurologic deficits but their absence alone was not meant to imply an immediate threat to limb viability (i.e. the presence of doppler arterial signals are a specific but not sensitive finding). One author mistakenly translated this into an ABI of 0.30 [23], which obviously represents an even higher level of perfusion. This, and the unbelievably high limb salvage rates reported for category III (irreversible ischemia) in this article, suggests that many of these patients in this experience were not truly acutely ischemic. These misinterpretations were unfortunate, coming at the same time as others had challenged the pressure criteria for CCI. The fortunate part is that recognition of these has led to a clarification and modification of the classification of acute limb ischemia, which is presented in the revised SVS/ISCVS standards [6] which now divides the *threatened limb* category in two subgroups, primarily on neurological grounds, to acknowledge that there may be time for catheter directed thrombolytic therapy in the marginally threatened group (IIa). This fits with the contention of Earnshaw [24] and has led, at the same time, to a similar division of category II, independently proposed by European investigators and included in the Consensus Guidelines for thrombolytic therapy [25].

■ WHAT ARE THE OPTIONS?

A number of suggestions have been made and other possibilities are apparent and will be mentioned for completeness, along with some qualifying remarks.

1 - Adopt the SVS/ISCVS standards, since the use of different perfusion levels for intact feet versus those with tissue necrosis is based on sound principle and they have been well accepted, in North America at least, and modify them as necessary in the light of future investigations.

2 - Abandon objective criteria entirely and revert to clinical diagnosis (i.e., the Fontaine system), since no pressure or other objective test criteria can ever be expected to serve as sharp and discriminating cut-off points.

3 - Modify the current objective criteria now, without waiting for further investigations to establish the best cut-off points, in the light of reported experiences. In this regard two basic suggestions have been made: a) raise the ankle pressure cut-off point for ulcers and gangrene to 70 mmHg, b) abandon pressure criteria for ulcers and gangrene altogether, but maintain them (e.g. 40 or 50 mmHg ankle pressure) for rest pain [21].

4 - Abandon ankle pressures entirely in favor of toe pressures, because the former measure perfusion pressures too far above the level of the worst ischemia and can not be applied to a significant portion of the at risk population, diabetics with calcified arteries. Unfortunately many diabetics will already have lost their first or second toes.

5 - Abandon pressure criteria altogether in favor of microcirculatory tests. A future consideration if and when such tests have been shown to be predictive and are widely available.

6 - Combine toe pressures with indices of microcirculation [26]. A possible compromise between 4 and 5 above.

7 - Establish equivalent cut-off levels for all of the established perfusion measures and allow the most appropriate and available criteria to be used as a local option. This allows greater flexibility but would require additional comparative studies and an international consensus to agree on equivalent cut-off points.

■ RECOMMENDATIONS

It is obvious from the above discussion that while the basic concept and definition of chronic leg ischemia may be relatively easily agreed upon, the objective criteria associated with this definition have been, and will likely continue to be, the source of much debate, at least until definitive comparative studies are done. However, our inability to come together and agree on universally accepted criteria, rather than attack the ones that have been proposed has only worsened things. Nevertheless, based on the above observations, some recommendations can be made regarding the definition and criteria for chronic critical ischemia. In making them the author recognizes that he has a bias in favor of the SVS/ISCVS criteria he helped develop and bringing this to a European meeting may meet with opposition. However, the current dissatisfaction with the European consensus criteria will hopefully create willingness among European investigators to consider the proposed alternatives.

Definition: *chronic critical ischemia* is a limb threatening degree of ischemia characterized by either persistent ischemic rest pain or areas of tissue necrosis (ulcers or gangrene) on the foot which will not heal spontaneously or with the help of local measures because of inadequate tissue perfusion.

For clinical studies and trials of CCI, in which it is necessary to be *exclusive* in order to have pure and comparable treatment groups, these two categories should be stratified and the following objective inclusion criteria are recommended for the present. For ischemic rest pain: an ankle pressure < 40 mmHg, a toe pressure < 30 mmHg, or $tcpO_2$ < 30 mmHg. For non-healing foot lesions: an ankle pressure of < 60 mmHg, a toe pressure < 40 mm Hgm or a $tcpO_2$ < 40 mmHg. It is recognized that these stricter criteria may make patient recruitment for clinical trials more stringent, but that is preferable to a high placebo effect among controls in pharmacotherapy trials, or limb salvage rates greatly in excess of patency rates in trials comparing revascularization methods, which would result is more liberal criteria were adopted.

While the above ankle and toe pressure criteria have survived in North America for more than a decade without serious challenge, they have not been put to the same critical testing that has been applied to the European consensus criteria. Therefore, it is recommended that they be used only until definitive prospective studies determine which are the most appropriate pressure levels to serve as discriminating cut-off points, using ROC curves as discussed above. To avoid the ethical problems created by the high risk of limb loss in natural history studies, it is suggested that the end points of persistence of rest pain and lack of ulcer healing be used rather than actual limb loss rates. The additional end point of requiring intervention to avoid amputation may also be acceptable if strict criteria for this decision are incorporated into the study protocol. The use of placebo groups from pharmacotherapy trials may provide an appropriate opportunity to study this issue.

For routine practice purposes, where the goal is to be *inclusive* and subject all potentially at risk patients to further diagnostic evaluation and appropriate therapy, the clinical diagnosis of ischemic rest pain or ischemic ulcers or gangrene should suffice (i.e. the Fontaine III-IV classification). If objective criteria are used for confirming that pain or non-healing ulcers are truly on an ischemic basis, they certainly can be relaxed somewhat (e.g. by 10 mmHg.). However, it is important in such patients, because of their higher anesthetic risk and more limited longevity, to avoid operative intervention where possible and consider the least morbid treatments first. Thus, even in practice it is well to obtain non-invasive tests to be sure the diagnosis of CCI and implied limb threat are certain, and that intervention is mandatory.

Those with ankle pressures in the 40-60 mmHg range regardless of symptoms (i.e., those with chronic subcritical ischemia) should be considered to be at intermediate risk of limb loss, between that of claudicants and those with CCI. They should be considered to have severe leg ischemia, deserving of thorough diagnostic evaluation, heightened surveillance and intensified preventative care. However, they should not be included in clinical trials or investigations of the management of chronic critical ischemia.

It is important to gain universal acceptance of the ultimate criteria for CLI and to use the same criteria in both Europe and North America. In this regard, the recommendations made in items 2, 4 and 5 above are in concert with those currently being incorporated into the TASC document *(Trans Atlantic Inter Societal Consensus on the management of PAOD)*, which should be published by the millenium.

REFERENCES

1 Mikulin T, Hopkinson BR, Makin GS. Major amputation compared with graft occlusion as the end point for assessing results of bypass surgery in lower limb ischaemia. *Br J Surg* 1986; 73: 200-203

2 Jamieson CW. The definition of critical ischaemia of a limb. *Br J Surg* 1982; 69 (suppl S1).

3 Kim YW, Taylor LM, Porter JM. Circulation-Enhancing Drugs. In: *Vascular Surgery*, 4th ed., Rutherford RB (ed.), WB Saunders Co., Philadelphia, 1995, pp 324-333.

4 Rivers SP, Veith FJ, Ascer E, Gupta SK. Successful conservative therapy of severe limb threatening ischemia: the value of nonsympathectomy. *Surgery* 1986; 99: 759-762.

5 Anonymous. Suggested standards for reports dealing with lower extremity ischemia. Prepared by the Ad Hoc Committee on Reporting Standards. SVS/North American Chapter, ISCVS. *J Vasc Surg* 1986; 4: 80-94.

6 Rutherford RB, Baker JD, Ernst C et al. Recommended standards for reports dealing with lower extremity ischemia: revised version. *J Vasc Surg* 1997; 26: 517-538.

7 Anonymous. Second European Consensus Document on Chronic Critical Leg Ischemia. Working group on critical chronic ischemia. *Circulation* 1991; 84 (suppl 4): 1-26.

8 Kroese AJ, Stranden E. How critical is chronic critical leg ischemia? *Ann Chir Gynaecol* 1998; 87: 141-144.

9 Chant AD. Is critical ischemia a muddle? *Eur J Vasc Surg* 1991; 5: 107.

10 McCollum PT, Spence VA, Walker WF. Arterial systolic pressures in critical ischemia. *Ann Vasc Surg* 1886; 1: 351-356.

11 Thompson MM, Sayers RD, Varty K et al. Chronic critical leg ischaemia must be redefined. *Eur J Vasc Surg* 1993; 7: 420-426.

12 Tyrrel MR, Wolfe JH. Critical leg ischaemia: an appraisal of clinical definitions. Joint Vascular Research Group. *Br J Surg* 1993; 80: 177-180.

13 Matzke S, Ollgren J, Lepäntalo M. Predictive value of distal pressure measurements in critical leg ischemia. *Ann Chir Gynaecol* 1996; 85: 316-321.

14 Jelnes R, Gaardsting O, Hougaard Jensen K et al. Fate in intermittent claudication: outcome and risk factors. *Br Med J Clin Res Ed* 1986; 293: 1137-1140.

15 Naschitz JE, Ambrosio DA, Chang JB. Intermittent claudication: predictors and outcome. *Angiology* 1988; 39: 16-22.
16 Rosenbloom MS, Flanigan DP, Schuler JJ et al. Risk factors affecting the natural history of intermittent claudication. *Arch Surg* 1988; 123: 867-870.
17 Jonason T, Ringqvist I. Changes in peripheral blood pressures after five years of follow up in non-operated patients with intermittent claudication. *Acta Med Scand* 1986; 20: 127-132.
18 McDaniel MD, Cronenwett JL. Natural history of intermittent claudication. In: *Basic Data Underlying Clinical Decision Making in Vascular Surgery*. Porter JM and Taylor LM (eds.). Quality Medical Publishing Inc., St. Louis, 1994, pp 129-133.
19 Bowers BL, Valentine RJ, Myers SI et al. The natural history of patients with claudication and toe pressures of 40 mmHg or less. *J Vasc Surg* 1993; 18: 506-511.
20 Fowl RJ, Gerwirtz RJ, Love MC, Kempczinski RF. Natural history of claudicants with critical hemodynamic indices. *Ann Vasc Surg* 1992; 6: 31-33.
21 Wolfe JH, Wyatt MG. Critical and subcritical ischaemia. *Eur J Vasc Endovasc Surg* 1997; 13: 578-582.
22 Earnshaw JJ, Hopkinson BR, Makin GS. Acute critical ischaemia of the limb: a prospective evaluation. *Eur J Vasc Surg* 1990; 4: 365-368.
23 McNamara TO, Bomberger RA, Merchant RF. Intra-arterial urokinase as the initial therapy for acutely ischemic lower limbs. *Circulation* 1991; 83 (suppl 1): 1106-1119.
24 Earnshaw JJ. Neurological deficit more reliable than doppler. *Eur J Vasc Surg* 1991; 5: 106-107.
25 Anonymous. Thrombolysis in the management of lower extremity (arterial) occlusion: a consensus Document. Working party on thrombolysis in the management of limb ischemia. *Am J Cardiol* 1998; 81: 207-218.
26 Carter SA. The challenge and importance of defining critical limb ischemia. *Vasc Med* 1997; 2: 126-131.

2

THE PATHOPHYSIOLOGY OF CRITICAL LIMB ISCHEMIA

CLIFFORD P. SHEARMAN, ADHISABANDH CHULAKADABBA

Critical limb ischemia (CLI) is ischemia to a degree which endangers the limb or part of a limb [1]. Attempts have been made to achieve an uniform definition of the situation using clinical and hemodynamic factors, but in practical terms it occurs when the nutritional blood supply falls below the level which will sustain tissue viability and tissue repair. In practice the patient's presentation is often precipitated by a minor injury, such as an abrasion from an ill fitting shoe, which due to the poor blood supply to the limb cannot heal and may be complicated by secondary infection.

■ PATHOPHYSIOLOGY

Vascular surgeons tend to direct their attention to the large arteries of the limb and the disease process within them. This is obviously important as the primary cause of CLI is most often reduced blood flow to the limb, as a result of occlusion or stenosis of the main limb vessels with atherosclerosis, which may be amenable to angioplasty or surgical bypass. The situation, however, is not as straightforward as it may seem. Atherosclerosis is a dynamic condition reflecting both hemodynamic and biological factors. The disease does not slowly progress, but tends to undergo periods of increased activity which are associated with clinical events such as stroke, myocardial infarction, or worsening lower limb ischemia. Equally the plaque can heal and in some circumstances may even regress. The blood vessel itself can undergo adaptive changes in response to the atheromatous lesion and other non-diseased blood vessels may enlarge to increase collateral flow. The effects are not limited to large blood vessels. The reduction in perfusion pressure of the blood supply to the limb has dramatic effects on the microcirculation, which can respond and adapt to various degrees. The ability of the microcirculation to respond, which can be modified by other diseases such as diabetes, ultimately determines how the limb will be affected.

Understanding of the complex interaction between the macro- and microcirculation in the presence of atheroma and other associated disease processes is important, as it may suggest ways in which the nutritional blood supply in the critically ischemic limb can be optimized, as well as methods to augment the outcome after revascularization procedures such as surgery or angioplasty.

■ THE MACROCIRCULATION

The origins of atherosclerosis appear early in life and fatty streaks, which are accumulations of lipid laden foam cells in the intima, have been identified in young children. The significance of these lesions is controversial as many appear to heal and they do not cause any luminal obstruction although experimentally it has been shown that fatty streaks may progress to a mature plaque [2]. By the second decade of life, however, few individuals will not have some evidence of atherosclerosis. This was clearly found in young Americans killed in Vietnam, of whom a high proportion had severe coronary artery atherosclerosis at post-mortem [3]. Atherosclerosis is considered to be a systemic disease, associated with risk factors such as smoking, hypercholesterolemia, hypertension, diabetes mellitus, and genetic predisposition that can affect arteries in any part of the body. However, atherosclerosis does not occur uniformly in all vascular beds and it has a predilection for certain sites such as the coronary arteries, the carotid bifurcation, the distal aorta and in the lower limb the common iliac bifurcation, the superficial femoral artery in the adductor canal, and the popliteal trifurcation. This appears to be due to the influence of hemodynamic factors. Areas of low wall shear stress, such as occur in the outer wall of arterial branches (opposite the flow *splitter*) where atherosclerosis often appears, results in slower movement of circulating *atherogenic* factors over the vessel surface resulting in more time for interaction with the endothelial cells [4]. Exercise will alter wall shear stress in an oscillatory fashion, which appears to have a beneficial effect on vascular endothelial function and as well as increasing surface flow [5].

Atherosclerotic plaques are complex in composition, but tend to have a number of characteristic features including a fibrous cap, lipid core, smooth muscle cells, macrophages, collagen and elastin. The associated vasa vasorum increase in number but tend to be thin-walled and fragile. The composition of the plaque may change. Thinning of the fibrous cap and increased lipid core are characteristics of an unstable lesion and necrosis and hemorrhage may occur within the plaque (Fig. 1). If the fibrous cap ruptures or fissures then the highly thrombogenic core is exposed to the blood stream resulting in accumulation and activation of platelets resulting in the formation of a thrombotic plug. If the injury is small the plaque may heal by reendothelialization resulting in a larger plaque and subsequent worsening of the distal perfusion. Plaque rupture may result in complete thrombotic occlusion of the vessel or distal embolization of thrombus (Fig. 2). Depending on the clinical outcome of this event the vessel may re-canalize by lysis and some degree of flow be reestablished, but often organization of the thrombus occurs and the vessel becomes permanently occluded. On questioning, many patients will report a step-wise deterioration in their condition and may finally present with an episode of acute limb ischemia presumably reflecting the activity of their atherosclerosis. Increasing evidence suggests that the risk of clinical cardiovascular events is associated with a number of plaques at risk of rupture such as those with thin fibrous caps and large lipid cores [6]. Plaque rupture is also associated with high serum HDL-cholesterol [7]. Appreciation of the pathophysiology of atherosclerosis and in particular the importance of plaque rupture and thrombosis has prompted the term atherothrombosis to describe the situation. This fits well with clinical observations of the very powerful benefit of antiplatelet drugs such as aspirin in reducing cardiovascular mortality by 25% in patients with established peripheral vascular disease [8].

Macrocirculation Compensation Mechanisms

The macrocirculation has a number of compensatory mechanisms, which may limit the effects of atherosclerosis. For unexplained reasons some blood vessels seem to be spared from the disease. The peroneal artery, especially in diabetics, is frequently found to be suitable for bypass when the other crural vessels are heavily diseased. Even in diseased arteries considerable narrowing of the lumen usually has to occur to produce a significant reduction in distal pressure. In the iliac artery, for example, most lesions under 80% do not cause a pressure gradient [9]. Increased flow during exercise in the leg may unmask a stenosis, but for a patient to develop CLI extensive arterial disease usually has to be present before perfusion falls below the critical level. Chronic increased blood flow in a normal artery leads to a compensatory dilation of the vessel mediated by increased wall shear stress. This phenomenon is clearly

FIG. 1 Complex atherosclerotic plaque showing areas of new and old intraplaque hemorrhage, lipid and a thick fibrous cap. *(Courtesy of Dr. P. Gallagher, Department of Pathology, Southampton General Hospital).*

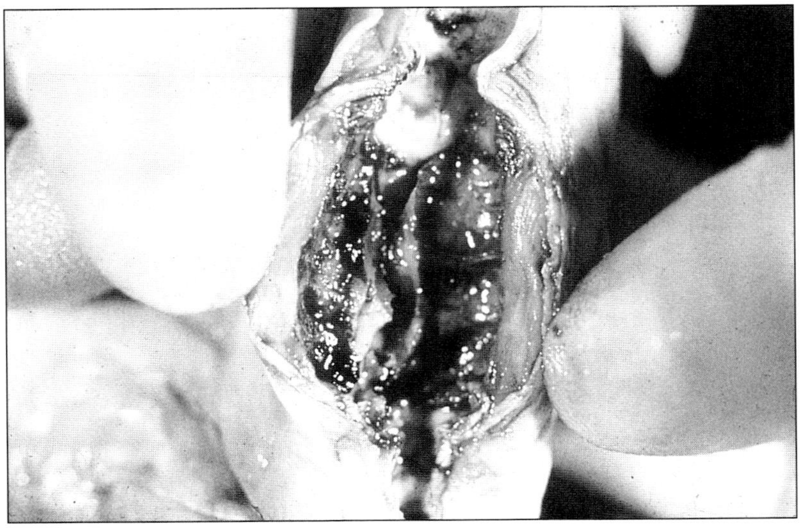

FIG. 2 Plaque rupture in coronary artery resulting in fatal myocardial infarction. Thrombotic occlusion of vessel is seen. *(Courtesy of Dr. P. Gallagher, Department of Pathology, Southampton General Hospital).*

demonstrated in arterio-venous fistula. In vessels affected by atherosclerosis there also tends to be enlargement of the vessel. This may be due to the increased flow rate through the stenosis, but more probably is due to destruction of the mural connective tissue beneath the plaque. Either way, despite encroachment of the plaque on the lumen of the vessel there tends to be a compensatory dilation. This varies considerably depending on the vessel involved and also takes time to develop and therefore this will not protect against sudden increases in plaque volume due to rupture or hemorrhage.

Probably the most important compensatory mechanism in chronic ischemia is the development of collateral circulation. These are pre-existing vessels which form alternative pathways for blood flow when the main limb vessels are diseased. There are several well-recognized collateral pathways involving vessels, which seem to be spared from the atherosclerotic process such as the connections between the profunda and geniculate arteries. The main stimulus for collateral development seems to be a pressure gradient that forms across the collateral bed when the main arteries occlude. This increases collateral blood flow, raising wall shear stress and resulting in dilation of the vessel. Collateral development is variable and takes time to establish so the outcome depends to some degree on the speed of progression of the atherosclerosis. Generally, though, it is very rare to see a patient with a single level of arterial disease who has CLI. Usually collateral development will be sufficient to prevent this although the patient may have intermittent claudication. Most patients with CLI have at least two level disease, for example aortoiliac and superficial femoral or superficial femoral and crural vessel atherosclerosis. In some patients with CLI in whom intervention has not been possible spontaneous improvement is observed presumably due to collateral development. Recently there has been interest in factors promoting the growth of new blood vessels: angioneogenesis. In experimental models of muscle ischemia new vessel growth can be observed and in tumors it frequently occurs. Early uncontrolled reports of the administration of growth factors to promote angioneogenesis have been encouraging, but much more work is needed before any firm conclusions can be drawn.

■ THE MICROCIRCULATION

The Microcirculation in Health

The microcirculation consists of arterioles, capillaries, venules and the associated interstitial spaces and capillary lymphatics. Different vascular beds have different arrangements, but generally the arterioles undergo repeated, dichotomous branching which is mirrored by the venules rejoining to form collecting veins. Vascular endothelial cells form a non-thrombotic barrier between the blood and tissue and are the final arbiter of tissue nutrition. Blood flow is determined by the arterial-venous pressure gradient and blood viscosity, but complex regulatory mechanisms exist in the microcirculation consisting of local factors and neural and biochemical regulators.

Increased demand for nutrients in tissue may itself be a direct stimulus to increase microcirculatory blood flow, but the build up of metabolic substrates such as hydrogen ions and carbon dioxide has a marked vasodilatory effect, thereby increasing flow. The smooth muscle in the arterioles may itself respond directly to pressure changes and is possibly responsible for autoregulation in some vascular beds. If the metabolic demands of the tissue chronically increases then dramatic changes occur in the vasomotion of the vascular bed.

Normally precapillary sphincters relax and constrict so that at any one time not all capillaries are transmitting blood. This is called intermittency of flow. When demands increase it can be shown that more capillaries at any one time are open. With increased demand over time new capillaries can be seen to develop and in skeletal muscle the capillary to muscle fiber ratio can increase [10]. Sympathetic vasoconstrictor fibers from the vasomotor center maintain the precapillary sphincters in a state of partial constriction, so called vascular tone. The vasomotor center can be influenced by a number of higher centers in the central nervous system, which inhibit the sympathetic tone. A complex array of biochemical compounds, with opposing biological actions, either produced in the microcirculation or transported to it in the blood stream, maintain a close balance between vasoconstriction and dilation. For example, prostacyclin produced by the endothelium is a vasodilator and inhibitor of platelet aggregation whereas thromboxane is a potent vasoconstrictor and platelet activator. One of the most important agents in the maintenance of the microcirculation appears to be nitric oxide (NO). NO is a free radical produced by the oxidation of arginine catalyzed by NO synthetase and under normal circumstances plays a key role in maintaining vasomotor tone [11].

The Microcirculation in CLI (Fig. 3)

The pivotal stimulus to changes in the microcirculation in CLI is a fall in perfusion pressure. Even with the compensatory measures outlined above there will be adverse effects and the degree to which these occur will determine the outcome for the limb. Reduced perfusion pressure results in compensatory vasodilation and abnormal vasomotion and there is significant maldistribution of nutritional blood flow making the skin particularly vulnerable to minor injuries [12]. This increases the transit times of cells passing through the microcirculation allowing greater interaction to occur with the endothelium. Under normal conditions the blood viscosity falls in the smallest vessels of the microcirculation due to streaming and flow patterns. With reduced pressure this does not happen and the viscosity in the small branches is relatively increased. Fibrinogen is also increased in CLI patients and this will also increase plasma viscosity. Red blood cells in patients with CLI become more rigid and this, together with the above, tends to encourage them to clump and block capillaries. Damaged RBC's release ADP, which promotes platelet activation, which tend to be activated more readily in patients with CLI anyway. Activated platelets interact with the endothelium and release thromboxane A2 and 5HT, which promote vasoconstriction and further platelet activation.

Neutrophils in particular seem to be a major factor in the mediation of damage to the microcirculation. Neutrophil counts are elevated in patients with arterial disease and this in itself seems to

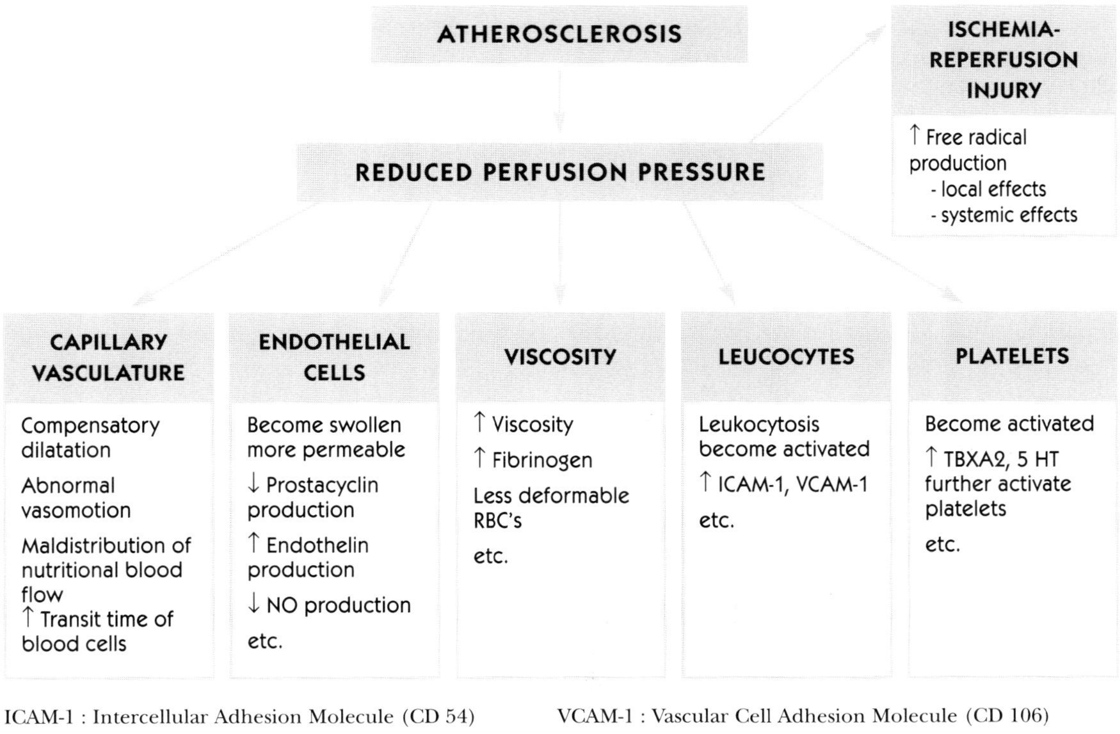

FIG. 3 A conceptual schematic of the microcirculation in CLI.

correlate with the risk of cardiovascular events and the outcome [13]. Normally the neutrophils have to squeeze through the microcirculation and interact with the vascular endothelium as they do so via adhesion molecules, predominately of the selectin group. These bonds are easily broken so the neutrophil *rolls* along the vessel surface. With reduced shear forces due to the lower perfusion pressure and increased viscosity stronger bonds between neutrophil and endothelium occur, which will arrest the neutrophil on the endothelium. The vessel may be completely blocked by the cell, which becomes more rigid due to activation or migration of the activated cell through the endothelium. If this happens then the activated neutrophil may release injurious agents such as proteolytic enzymes like elastase, collagenase and gelatinase, which damage the vascular endothelium making it more permeable. Free radicals are generated which are very destructive, especially to lipid membranes.

The target of most of this damage is the vascular endothelial cell which becomes swollen and more permeable. The swelling further reduces the lumen of the vessel and makes it more likely to become blocked by activated neutrophils or clumps of platelets and red blood cells. Increased permeability results in increased fluid in the interstitial spaces and this further impairs diffusion of nutrients to the tissues. Damaged endothelial cells are less able to produce prostacyclin and so thromboxane A_2 release by the activated platelets is unbalanced. Endothelin, a potent vasoconstrictor is released by damaged endothelial cells under conditions of reduced flow [14], and NO production is reduced [15].

As can be seen from the above, a vicious cycle of damage is set up with cellular damage resulting in worsening ischemia. To make matters worse, if flow is restored to ischemic tissue further damage occurs due to the phenomenon of ischemia-reperfusion injury. During periods of ischemic metabolism various substrates build up which on the re-introduction of oxygenated blood are metabolized to highly damaging free radicals which overwhelm the cellular protective mechanisms and cause further damage. It appears that this chain of events not only causes local tissue injury, but damage to remote endothelial surfaces such as the lung and kidney [16].

The Microcirculation in Diabetes

Diabetic patients with CLI seem to have exaggerated changes in their microcirculation including increased platelet adhesiveness and aggregation, decreased fibrinolytic activity, increased leukocyte free radical production, decreased red blood cell deformability and increased plasma viscosity. The origin of diabetic microangiopathy appears to be related to endothelial cell dysfunction associated with basement membrane thickening and increased vascular permeability. These changes are related to the duration of the disease and metabolic control as chronic elevation of blood glucose results in formation of non-enzymatic advanced glycosylation end products (AGE) such as HBA1c. These AGE products have a direct effect on the extracellular matrix, vascular permeability and nitric oxide production [17].

A number of pharmacological agents have been tried to reduce the damage inflicted on the microcirculation by CLI. Prostacyclin analogues have been used with promising results in patients with ischemic ulceration. Early results with prostacyclin during bypass surgery were encouraging, but larger studies failed to show clinical benefit. Epidural anesthesia improves bypass graft blood flow presumably by altering the balance of vasomotor tone.

■ CONCLUSION

The response of the circulation to the effects of atheroma is wide ranging and involves both the macro- and microcirculation. If the reduced blood supply to the limb occurs slowly then these compensatory mechanisms may be sufficient to avoid the development of CLI. Once established, however, a cycle of ischemic tissue damage is set up which uncorrected will result in the loss of the limb. Often the limb will require a revascularization procedure such as angioplasty or bypass surgery, but even then the damage to the microcirculation may continue. It may ultimately be possible to use pharmacological agents to attenuate this damage, but at present the most impressive results for patients with arterial disease are with the use of antiplatelet agents.

REFERENCES

1 Anonymous. Second European Consensus Document on Chronic Critical Leg Ischaemia. European Working Group on Critical Limb Ischaemia. *Eur J Vasc Surg* 1992; 6 (Suppl A): 1-32.
2 Faggiotto A, Ross R, Harker L. Studies of hypercholesterolemia in the nonhuman primate. II Fatty streak conversion to fibrous plaque. *Arteriosclerosis* 1984; 4: 341-356.
3 McNamara JJ, Molot MA, Stremple JF, Cutting RT. Coronary artery disease in combat casualties in Vietnam. *JAMA* 1971; 216: 1185-1187.
4 Zarins CK, Giddens DP, Bharadvaj BK et al. Carotid bifurcation atherosclerosis: quantitative correlation of plaque localization with flow velocity profiles and wall shear stress. *Circ Res* 1983; 53: 502-514.
5 Fry DL. Hemodynamic forces in atherosclerosis. In Scheinburg P (ed.) *Cerebrovascular Disease.* New York, Raven Press 1976; pp 77-95.
6 Davies MJ. The composition of coronary-artery plaques (ed. comment). *N Eng J Med* 1997; 336: 1312-1314.
7 Burke AP, Farb A, Malcom GT et al. Coronary risk factors and plaque morphology in men with coronary disease who died suddenly. *N Eng J Med* 1997; 336: 1276-1282.
8 Anonymous. Collaborative overview of randomised trials of antiplatelet therapy - 1. Prevention of death, myocardial infarction and stroke by prolonged antiplatelet therapy in various catergories of patients. Antiplatelet Trialists Collaboration. *Br Med J* 1994; 308: 81-106.
9 Haimovici H, Escher DJW. Aortoiliac stenosis. Diagnostic significance of vascular hemodynamics. *Arch Surg* 1956; 72: 107-117.
10 Hudlicka O, Price S, Hoppleler H et al. The long-term effect of electrical stimulation on fast muscles with limited blood supply-degeneration and regeneration. In Gordon T (ed.) *Current status of peripheral nerve regeneration.* New York: Alan Liss, 1988; pp 307-316.
11 Gimbrone MA. Vascular endothelium in health and disease. In Haber E (ed.) *Molecular cardiovascular medicine.* Scientific American, New York, 1995; pp 49-61
12 Fagrell B, Lundberg G. A simplified evaluation of vital capillary microscopy for predicting skin viability in patients with severe arterial insufficiency. *Clin Physiol* 1984; 4: 403-411.
13 Ernst E, Hammerschmidt DE, Bagge U et al. Leukocytes and the risk of ischemic diseases. *JAMA* 1987; 257: 2318-2324.
14 Yanagisawa M, Kurihara H, Kimura S et al. A novel potent vasoconstrictor peptide produced by vascular endothelial cells. *Nature* 1988; 332: 411-415
15 Luscher TF. Imbalance of endothelium-derived relaxing and contracting factors: a new concept in hypertension? *Am J Hypertens* 1990; 3: 317-330
16 Tisi PV, Shearman CP. Systemic consequences of reperfusion. In Grace PA, Mathie RT. (eds.) *Ischaemia-reperfusion injury.* Blackwell Science Ltd, Oxford, 1999; pp 20-30.
17 Jones BA, Sidaway AN, LoGerfo F. Diabetic vascular disease. In Sidaway AN, Sumpio BE, DePalma RG (eds.). *The basic science of vascular disease.* New York, Futura Publishing Company, 1997; pp 441-456.

3

EPIDEMIOLOGICAL AND ECONOMIC CONSIDERATIONS IN THE CRITICALLY ISCHEMIC LIMB

ROBERT K. FISHER, PETER L. HARRIS

There is an inherent difficulty in obtaining accurate information on the epidemiological and economic impact of chronic critical limb ischemia. One reason is that patients with arterial disease are often managed under a variety of specialists, including vascular surgeons, diabetologists, geriatricians, etc., which detracts from the formation of a comprehensive database. Another, more important factor, is the problem of definition.

■ DEFINITION

The pathophysiology of critical limb ischemia (CLI) is known to involve profound changes in both the macro- and microcirculatory systems of the lower limb, however, establishing a widely accepted definition has proved a contentious issue. After the concept of CLI was first introduced in 1982 by the *Working Party of the International Vascular Symposium* [1], it was not until 1993 that the European Consensus Meeting (ECM) standardized the definition as: persistent recurring ischemic pain requiring regular analgesia for more than 2 weeks, with an ankle systolic pressure ≤ 50 mmHg and/or a toe pressure of ≤ 30 mmHg; or ulceration or gangrene of the foot or toes, with an ankle systolic pressure of ≤ 50 mmHg or a toe systolic pressure of ≤ 30 mmHg [2].

The ECM definition, while the most objective, can also be deemed the most restrictive in its clinical application, in that there remains a difficulty in establishing the outcome of patients with CLI. Although the concept of CLI is one of imminent limb loss without intervention, under the ECM criteria, it has been found that at least 25% of limbs deemed critically ischemic survive, despite only conservative management [3]. Some surgeons have emphasized the importance of ischemic necrosis in the definition, as it appears to be a more accurate predictor of outcome, especially limb loss, and its presence, combined with the surgeon's own clinical experience, should determine the management decision [4]. In the light of this it has been suggested that the definition of CLI should probably be subdivided into patients with rest pain and those with tissue necrosis [5]. Nevertheless, the value of the ECM criteria lies in the fact that it has allowed us to standardize the classification of patients with peripheral vascular disease, enabling the formation of more precise and relevant databases.

The reality is that of a multifactorial definition encompassing all of these criteria, and one to which we refer throughout this chapter as we draw from the experience and findings of a number of centers.

■ INCIDENCE AND PREVALENCE

Symptomatic leg ischemia is increasing in its incidence, with the literature quoting figures between 1.5 and 5% in men over 50 years [6]. The European Consensus Document reports an incidence of CLI of 50-100/100 000 per year [2], based on comparisons of an assumed amputation rate of 25% for all patients presenting with CLI in USA and Denmark. However, a true amputation rate of 21.5% was found in a prospective survey by the *Vascular Surgical Society of Great Britain and Ireland* in 1993, equating to a UK incidence for CLI of 40/100 000 per year [7]. This represents 20 000 patients per year, with a prevalence of 1: 2 500 of the population annually.

Men, more than women, exhibit a propensity for critical limb ischemia, with a ratio of approximately 1.5: 1. However, this predominance diminishes with advancing age, and in those over 80 years old equal numbers of men and women are affected [8]. Age is recognized as an important parameter when considering CLI, with the peak incidence occurring between 70 and 79 years, and a four-fold increase in prevalence noted in the group older than 80 years [7]. Those presenting with intermittent claudication at the younger end of the distribution curve often have a more aggressive form of peripheral arterial disease, and may progress more rapidly towards a critically ischemic limb [9].

■ RISK FACTORS

The pathological components influencing the development of atherosclerosis and critical ischemia are well recognized, and include cigarette smoking, hypertension, obesity, hyperlipidemia, and diabetes. In all cases prevention is the ultimate aim and has attracted a great deal of attention in the context of primary health plans. It is apparent to the vascular surgeon, however, that the success of this approach is limited, with the incidence of CLI and its demands on clinical resources ever increasing.

The issue of smoking is a topical one evoking much medical publicity. A Danish study by Jürgens et al. over 30 years ago suggested smoking did not effectively predict outcome among patients with peripheral arterial disease. However, more recent results from a Swedish center indicated that in patients with intermittent claudication, cessation of smoking diminishes the risk of developing critical limb ischemia [10]. This has given credence to the adage *Stop smoking and keep walking*. It is widely accepted, however, that patients are somewhat less than open regarding their smoking status, and objective markers are often necessary to establish the non-smokers, including whole blood carboxyhemoglobin concentrations and plasma cotinine levels. Such was the case in a study by Smith et al., who looked objectively at the affect of cessation of smoking and other risk factors for peripheral arterial disease (diabetes, hypertension, and dyslipidemia) on the progression of chronic limb ischemia. They, as others, found that there was no short-term (1-3 years) benefit of smoking cessation in the progression of the disease or the symptoms; however there were clear trends for smoking to be associated with the failure of vascular reconstruction. A cessation period of over 7 years must be observed before significantly diminished rates of development of critical ischemia can be shown, and it is this surprisingly long latent period of unfelt benefits that drains the patients resolve, resulting in a return to smoking [9].

The same study did reveal that hypertension and dyslipidemia are associated with a progression in arterial disease. Interestingly, it is found that the progression of the disease is specifically related to elevated levels of triglycerids, while no association was demonstrated with hypercholesterolemia.

DIABETES

When regarding the risk factors of critical limb ischemia, both type I and type II diabetes warrant unique consideration. Approximately 30% of patients presenting with CLI are found to have some form of diabetes, and its presence constitutes a poor prognostic factor [11-13]. Patients with diabetes mellitus are five times more likely to develop CLI than non-diabetic patients, and the ischemia has a tendency to be associated with gangrene, resulting in amputation. This is often due to the atypical distribution of arterial disease observed in diabetics, superimposed on a background of neuropathic trauma and a propensity for infection. These three characteristic features act synergistically to promote further ischemic necrosis and limb loss.

The characteristic pattern of atherosclerosis in the non-diabetic patient involves the more proximal vessels of the limb, with relative preservation of the crural arteries. The opposite scenario, however, is found in the diabetic patient, in whom the superficial femoral and popliteal arteries remain comparatively disease free, while atherosclerosis occurring within the tibial and peroneal vessels leads to occlusions of varying length, and critical ischemia. In addition, the pedal arteries are found to remain patent, and it is this unusual picture of atherosclerotic distribution, which has led to recent speculation regarding more aggressive intervention, including femoropedal bypass, when managing the diabetic foot [13].

■ CLINICAL ASPECTS AND OUTCOMES OF CLI

The majority of patients who develop chronic critical limb ischemia present with symptoms for several days or weeks consisting of rest pain, with or without associated ulceration and gangrene. Symptoms in the diabetic, however, may be diminished, consisting of painless ulceration only, due to a concurrent neuropathy. Results from the British national audit showed that one-third of patients reported having symptoms for less than 1 week at presentation, while half reported symptoms extending over 3 weeks or more [7]. Ulceration and gangrene are each represented in one-third of presenting patients, with the majority, approximately two-thirds, complaining of rest pain [8]. In addition to significant morbidity from their symptoms, patients with CLI have a high risk of dying, with a mortality rate in excess of 10% per annum. Death is usually associated with additional cardiovascular manifestations, such as myocardial infarction or cerebrovascular accident, with women demonstrating a higher mortality rate than men [12]. Bilateral critical ischemia carries with it a worse survival rate, as does the presence of diabetes [3]. These observations emphasize the fact that CLI can be used as a reliable indicator of global arterial disease, which must be taken into consideration in the assessment and management of the patient.

■ THE MANAGEMENT OF CLI

The fundamental objectives in the management of critical ischemia are limb salvage, where possible, and preservation of the patient's quality of life. There is a range of treatments, including conservative management in those deemed unfit or unwilling to undergo any procedure, amputation in the unsalvageable limb, interventional radiology, and aggressive reconstructive surgery. The dilemma facing the surgeon, however, is in deciding which of these treatment options is the correct one for each individual ischemic limb. One must differentiate between those limbs that are unsalvageable, those that are amenable to some form of surgical intervention, and those that can be managed conservatively. An important compounding factor in this assessment is the lack of scientific data regarding outcome, although recently some research has been published. Lepäntalo et al. studied the outcome of CLI in patients unsuitable for reconstructive surgery, in terms of mortality and leg salvage, with particular emphasis on the definition of CLI. Despite no revascularization being performed in over 130 legs deemed critically ischemic under the European criteria, he found greater than 80% were preserved at 1 month, 70% at 3 months, about 60% at 6 months, and more than half of the limbs survived beyond 1 year [3]. As mentioned earlier, studies such as this reiterate the fact that CLI does not inevitably lead to major amputation, and that the definition, although useful, does not always correlate with the clinical situation. The presence of tissue necrosis has already been emphasized as an important prognostic indicator, and in conjunction with the experience of the surgeon is crucial when assessing the critically ischemic limb, and formulating the management plan.

The national audit [7] reveals that in Britain approximately two-thirds of patients admitted to the hospital with CLI undergo revascularization, while 16% require primary major amputation, 10% receive medical or minor surgical treatment, and 8% have symptomatic relief only. Two-thirds of revascularizations are bypass grafts and one-third are percutaneous transluminal angioplasty (PTA) or thrombolysis. Limb salvage was achieved in two-thirds of cases, with an overall amputation rate of 22% (primary amputation and failed reconstruction). One quarter of bypasses were to the below-knee popliteal artery, and nearly a third to the tibial and peroneal vessels.

Due to recent advances in distal bypass grafting, limb salvage rates have improved sufficiently to justify these long and technically difficult operations. There is considered to be enough evidence to support a more aggressive policy regarding femorocrural and pedal bypasses in CLI patients, especially in those with diabetes [14]. The materials and techniques used for bypass vary with availability and the surgeon's preference. Evidence identifies autologous vein as the gold standard in achieving optimum graft patency, but if unavailable, expanded polytetrafluoroethylene (ePTFE) as a prosthetic alternative attains acceptable results [15].

The patient's gender is an important parameter in the graft patency rate, with women exhibiting a lower rate than men [11]. In addition, bilateral disease acts as a poor prognostic indicator, while neither heart disease nor diabetes were conclusively shown to adversely affect the patency [3,8,11]. On the issue of surgical experience, it has been reported that the amount of work undertaken by the surgeon may influence the outcome. The national audit revealed that legs are more likely to be salvaged, often with the aid of radiological intervention, in centers where the surgeon performs more than ten femorodistal bypasses per annum. Based on this, the recommendation was made that surgeons looking after critical limb ischemia should either be full time vascular surgeons, or at least general surgeons with a major vascular interest [7,8].

■ ECONOMIC ISSUES IN CRITICAL LIMB ISCHEMIA

The Cost of Management

With worldwide advances in medicine, and the ever-increasing public demands and expectations of the profession, the financial considerations have become more important. On a global scale they concern the governments, while at the doctor-patient interface, the individual clinician needs to be aware of the cost-effectiveness of any treatment implemented. The *National Health Service,* a tax funded institution, is responsible for more than 90% of health care delivered to the United Kingdom, involving approximately 6% of the gross national product. This degree of government expenditure compares to figures of 9% in Germany, and 12% in the United States [15]. Within the UK, the government is responsible for the direction and limitation of finances on a nation-wide scale. However, individual Hospital Trusts manage their allocated budget, in order to best serve the local population, with each Directorate having autonomy for the distribution of money within their department.

The concepts of a national health service are shared by several European countries, who all face the same difficulties; but even within the private system of North America, cost effectiveness has become more of an issue with the advent of the *Diagnosis Related Group* (DRG), who predetermine the hospital's reimbursement depending on the disease, rather than the procedures undertaken [16,17]. Under this system an hospital can incur significant losses, especially when dealing with difficult cases of CLI, requiring several procedures which may ultimately fail.

Awareness of the financial issues has prompted consideration into the comparative expense of different management options. This is particularly important to vascular surgery, a historically expensive specialty by the very nature of the disease and its management, and one whose demands on the medical resources of the western world is set to increase as the elderly population expands, and with the advent of new treatments. Chronic limb ischemia accounts for a significant proportion of patients seen by the vascular surgeon, and has profound financial implications due to its frequently protracted history, and difficult management alternatives. A more aggressive policy is now adopted, with only 16% or less of patients presenting with CLI undergoing primary amputation, and at least two-thirds having some form of revascularization [8,16,17]. Arterial reconstruction involving the infrapopliteal vessels is frequently indicated; however, such an approach has attracted criticism, as treatment is often considered only palliative in nature, with both the patient and the revascularization procedure ultimately succumbing to the underlying disease. It is in view of this and the ever-increasing pressure on financial resources, that each of the treatment options has been studied in an attempt to identify the most cost-effective management. Mackey et al. presented the early figures for the cost of revascularization compared to amputation in critical limb ischemia, showing no statistically significant difference between the two, although amputees were found to require longer periods of hospitalization [16]. These findings were supported by other studies at that time [17].

With developments in surgical knowledge and techniques surrounding CLI, however, the balance has shifted somewhat, and subsequent results from both Europe and North America have demonstrated the mean cost of primary arterial reconstruction is substantially cheaper than amputation. In 1992 Cheshire et al. published figures regarding the economics of femorocrural reconstruction, showing that distal arterial reconstruction, including secondary operations, was 44% cheaper than primary amputation when utilizing an autologous vein, and 6% cheaper when using PTFE [15]. An even greater difference, with primary amputation being nearly three times more expensive than a successful distal bypass, was demonstrated by Panayiotopoulos et al. in 1997, whose results have raised several interesting issues when considering the cost effectiveness of bypass surgery in CLI [18].

FACTORS PREDETERMINING THE SUCCESS AND COST-EFFECTIVENESS OF MANAGEMENT

The assessment of the costs related to the various procedures is multifactorial, with difficulty in defining CLI and the prediction of its outcome again causing problems when trying to make the correct management decision. The severity of the presenting disease is important, with the presence of tissue necrosis being a significant indicator of poor limb salvage rates. Ischemic necrosis, along with diabetes, incurs greater expense, while both require repeated debridement, a higher amputation rate and extended hospital stay [15,18,19]. The condition of the runoff vessels, as might be expected, has been shown to be important regarding graft patency and cost, with significance being assigned to the number of patent calf vessels (with a greater difference between one and two vessels), the presence of healthy pedal arteries, and the number of vessels crossing the ankle. Although the state of the inflow vessels was significant, graft patency remained unaffected in cases requiring preoperative procedures to the proximal vessels, such as PTA.

Graft material is an important consideration when assessing patency rates, with autologous vein being shown to be significantly superior to PTFE, with the latter having a high rate of occlusion over the first 6 months. No difference is demonstrated between reversed and in-situ vein grafts [18]. This correlates directly with the cost-effectiveness; vein grafts, due to their greater overall success and lower reoperation rate, being the most economical treatment of CLI [16,18].

The main components dictating the cost of a procedure are the investigations, radiological procedures, theatre time, anesthetic time, synthetic grafts, intensive care and duration of hospital stay. The latter is commonly regarded as the fundamental factor governing the overall cost. Both reconstruction and amputation require lengthy periods of hospitalization, but the former is consistently shown to be significantly shorter, whereas failed revascularization and secondary amputation result in significantly longer hospitalization and greater expense [15-18]. With this in mind, one can understand the economic importance of implementing the correct management, be it arterial reconstruction with the intention of achieving high limb salvage rates, or a primary amputation in the unsalvageable limb, thereby avoiding costly graft failure and the expense of secondary amputation.

Although it is accepted that further surgery may be necessary after arterial reconstruction to maintain or restore graft patency, and that secondary intervention still constitutes a more cost-effective option than primary amputation [15], there must be a limit beyond which the limb is deemed both medically and financially nonviable. This is particularly relevant in femorocrural bypass, whose patency rates are worse than more proximal procedures, especially if autologous vein is unavailable [2,15,16,18]. The European Consensus Document (ECD) [2] suggested the patency rate for infrapopliteal grafts should be approximately 70% at 12 months for vein grafts, and 40% for PTFE with vein cuff, in order to achieve financial validity.

The Value of Distal Reconstruction over Amputation

Some surgeons speculate that rather than infrapopliteal reconstruction in critical limb ischemia, a primary amputation is indicated, as the most effective means of returning the patient, who is often frail, back to independent living. However, Cheshire et al. along with other studies have shown patency and limb salvage figures comparable to those recommended by the ECD, which would appear to justify the role of femorodistal bypass in the treatment of CLI [15,18]. More importantly, however, these studies showed that successful femorocrural reconstruction not only reduced hospital stay, but significantly improved the patients degree of mobility and independence, a fact that has both social and economic relevance. Fifty-two percent of patients undergoing reconstruction (including those who ultimately had amputation) achieved mobility and independence outside the house compared to 31% of patients undergoing primary amputation. When considering the successful reconstructions alone, nearly 80% of patients were mobile outside of the house and the remaining 20%, while retaining a viable leg, were immobile largely due to concurrent ailments such as contralateral amputation, hemiplegia, angina and rheumatoid arthritis. Of this category of patients, however, less than 1% were confined to a wheelchair. With this high degree of success without incurring additional expenditure, vascular surgeons believe they now have a case for aggressive reconstruction in critical limb ischemia [15].

In all of these studies the true cost of amputation is probably underestimated, as the continuing care of the amputee in the community has not been accurately evaluated. Although arterial reconstruction often requires repeated intervention in order to maintain graft patency, the patients' improved mobility and independence compared to those having undergone amputation has been established. This means less support is necessary on their return home, and therefore less incurred expenditure [15].

CONCLUSION

Throughout the western world, critical limb ischemia has become a progressively more common manifestation of atherosclerosis, which places the limb at imminent risk, especially in the presence of diabetes and tissue necrosis. The outcome of critical limb ischemia can be summarized by the data presented by Wolfe et al., whose 1-year estimates were that 25% of patients would have at least one leg amputated, 55% would retain both legs, and 20% would have died [4]. Its financial demands on the Health Service are profound, and have led to close appraisal of the cost-effectiveness of the management of CLI, which has been the subject of many debates. In limbs that are not irretrievably ischemic, a policy of aggressive arterial reconstruction, including distal bypass, meets with economic approval, as well as hopefully and more importantly improving the patients quality of life. In those requiring amputation, the long period of hospitalization puts significant economic strain on the system, and we should be striving to shorten this, with sufficient community support to effectively rehabilitate patients back to an acceptable lifestyle.

REFERENCES

1 Bell PRF, Charlesworth D, De Palma RG. The definition of critical limb ischaemia. Working Party of the International Vascular Symposium. *Br J Surg* 1982; 69: (suppl): S2.
2 Anonymous. Second European Consensus Document on Chronic Critical Limb Ischaemia. *Eur J Vasc Surg* 1992; 6: Suppl A: 1-32.
3 Lepäntalo M, Mätzke S. Outcome of unreconstructed chronic critical leg ischaemia. *Eur J Vasc Endovasc Surg* 1996; 11: 153-157.
4 Wolfe JHN. Defining the outcome of critical ischaemia: a one-year prospective study. *Br J Surg* 1986; 73: 321.
5 Thompson MM, Sayers RD, Varty K et al. Chronic critical limb ischaemia must be redefined. *Eur J Vasc Surg* 1993; 7: 420-426.
6 Dormandy J, Mahir M, Ascady G et al. Fate of the patient with chronic leg ischaemia. A review article. *J Cardiovasc Surg* 1989; 30: 50-57.
7 Anonymous. Critical limb ischaemia: management and outcome. Report of a National Survey. The Vascular Surgical Society of Great Britain and Ireland. *Eur J Vasc Endovasc Surg* 1995; 10: 108-113.
8 Anonymous. Recommendations for the management of chronic critical lower limb ischaemia. Audit Committee of the Vascular Society of Great Britain and Ireland. *Eur J Vasc Endovasc Surg* 1996; 12: 131-135.
9 Smith I, Franks PJ, Greenhalgh RM et al. The influence of smoking cessation and hypertriglyceridaemia on the progression of peripheral arterial disease and the onset of critical ischaemia. *Eur J Vasc Endovasc Surg* 1996; 11: 402-408.
10 Jonason T, Bergstrom R. Cessation of smoking in patients with intermittent claudication. Effects on the risk of peripheral vascular complications, myocardial infarction and mortality. *Acta Med Scand* 1987; 221: 253-260.
11 Enzler MA, Ruoss M, Seifert B et al. The influence of gender on the outcome of arterial procedures in the lower extremity. *Eur J Vasc Endovasc Surg* 1996; 11: 446-452.
12 Jönsson B, Skau T. Outcome of symptomatic leg ischaemia: four-year morbidity and mortality in Vadstena, Sweden. *Eur J Vasc Endovasc Surg* 1996; 11: 315-323.
13 Da Silva AF, Desgranges P, Holdsworth J et al. The management of outcome of critical limb ischaemia in diabetic patients: results of a national survey. *Diabet Med* 1996; 13: 726-728.
14 Woelfle KD, Lange G, Mayer H et al. Distal vein graft reconstruction for isolated tibioperoneal vessel occlusive disease in diabetics with critical foot ischaemia. Does it work? *Eur J Vasc Surg* 1993; 7: 409-413.
15 Cheshire NJW, Wolfe MS, Noone MA et al. The economics of femorocrural reconstruction for critical leg ischaemia with and without autologous vein. *J Vasc Surg* 1992; 15: 167-175.
16 Mackey WC, McCullough JL, Conlon TP et al. The costs of surgery for limb-threatening ischemia. *Surgery* 1986; 99 (1): 26-34.
17 Gupta SK, Veith FJ. Is arterial reconstruction cost-effective compared with amputation? In: Greenhalgh RM et al. (eds), Philadelphia, WB Saunders. *Limb salvage and amputation for vascular disease.* 1988; pp. 447-452.
18 Panayiotopoulos YP, Tyrrell MR, Owen SE et al. Outcome and cost analysis after femorocrural and femoropedal grafting for critical limb ischaemia. *Br J Surg* 1997; 84: 207-212.
19 Schweiger H, Klein P, Lang W. Tibial bypass grafting for limb salvage with ringed polytetrafluoroethylene prostheses. Results of primary and secondary procedures. *J Vasc Surg* 1993; 18: 867-874.

4

THE DIABETIC FOOT: CLINICAL AND PATHOPHYSIOLOGICAL ASPECTS

FRANK W. LOGERFO

The etiology of ulceration in the diabetic foot is complex, involving ischemia, neuropathy, and infection. The diagnosis and correction of ischemia is the cornerstone of success in diabetic foot management. With a well-perfused foot, it is possible to perform minor amputations or foot salvage procedures with optimum success. It also greatly enhances the management of infection. Therefore, an understanding of the mechanisms leading to ischemia is of fundamental importance in this clinical problem.

■ MICROCIRCULATION

In the past, there has been some speculation that ischemia in the diabetic foot is a result of an occlusion in the microcirculation. This derives from an observational study of amputation specimens where it was thought that there was a high incidence of arteriolosclerosis in the amputation specimens from diabetic patients [1]. Strandness et al. [2] examined this in a prospective study using blinded histology, and was unable to confirm the observation. There was no evidence of arteriolosclerosis associated with diabetes. In a subsequent anatomic study, Conrad [3] used a sophisticated arterial casting technique in amputation specimens. Once again, there was no evidence of a small artery or arteriolar occlusive process that was more common in the diabetic amputation specimens than in the non-diabetic.

If diabetics had arteriolosclerosis, they would be expected to have a fixed and high peripheral resistance. Barner et al. [4] tested this hypothesis by injecting papaverine at the conclusion of femoropopliteal bypass grafting and found no difference between diabetics and non-diabetics where there was similar tibial runoff. Based on that study, he concluded that *arterial reconstruction should not be withheld (in diabetics) on the basis of arteriolar-capillary involvement.* Using non-invasive techniques, Irwin et al. [5] also was unable to demonstrate physiologic evidence of microvascular occlusive disease associated with diabetes in patients presenting with foot ulceration. Thus a great deal of evidence argues against the existence of small vessel or arteriolar occlusive disease as being more common in the diabetic as compared with the non-diabetic [6]. The term *small vessel disease* implies the existence of an occlusive lesion and is often used to explain ulceration in the diabetic foot. A rejection of this concept is essential to understand the basis for modern management of the diabetic foot. If in fact there were an occlusive lesion in the microcirculation, treatment of ischemia in the diabetic would be hopeless. In that sense, the term *small vessel disease* can be harmful and inhibit the appropriate care of diabetic foot ulcers.

There are abnormalities in the microcirculation associated with diabetes that are not occlusive in nature. The most commonly described lesion is a thickening of the capillary basement membrane [7]. Surprisingly, this is not associated with narrowing of the capillary. In fact, the luminal capillary diameter in diabetic patients is actually greater than that in non-diabetics [8]. In a study of sural nerve biopsies, it was noted that the largest capillary diameters actually occurred in the diabetic patients with the greatest degree of neuropathy [9]. While not an occlusive process, the capillary basement membrane may impair the exchange of nutrients and cells between the capillary and the interstitium. It has been demonstrated in several studies that there is an increased flux of highly charged molecules, most notably albumin, from the capillary [10]. This capillary leak has been attributed to a decreased charge on the basement membrane, due to glycosylation and replacement of sulfur groups, allowing easier movement of highly charged molecules through the membrane [11]. This may well have physiologic consequences leading to tissue that is less biologically resistant to the effects of ischemia, pressure, and infection. Looked at in another way, the compromised biology of the diabetic foot may require greater vascular perfusion in order to maintain skin integrity. It is of note that there does not appear to be any impairment of oxygen diffusion across the capillary basement membrane. In a study of diabetics and non-diabetics presenting with foot ulceration, the transcutaneous pO_2 in the diabetics was actually greater than that in the non-diabetics [12]. Thus there is no physiologic rationale to support the use of hyperbaric oxygen in the management of diabetic foot ulcers.

■ MACROCIRCULATION

When ischemia occurs in the diabetic foot, it is due to atherosclerotic occlusive disease. Histologically, this process does not appear to be different from that which occurs in non-diabetics. Symptomatic atherosclerotic occlusive disease, as manifested by claudication, is more common in diabetics [13]. In addition, most studies have supported the view that progression of atherosclerosis is more rapid in the diabetic [14]. Calcification of the media is common in diabetes, but it is not associated with arterial occlusive disease either in the periphery or the heart [15,16]. As pointed out elsewhere in this chapter, calcification can interfere with non-invasive testing and can add to the technical difficulties of arterial reconstruction.

The most distinguishing feature of atherosclerosis in diabetes is the pattern in which it occurs in the lower extremity arteries. The occlusive process tends to involve the infrageniculate arteries. This may bring up, once again, the concept of small vessel disease. However, the site of the occlusive process is not based on the size of the vessels since the smallest vessels, the foot arteries, are relatively spared. In the prospective studies by Strandness et al. [2] and by Conrad [3], apart from the absence of microvascular occlusive disease, it was also noted that the foot arteries were relatively spared of occlusive disease in the diabetic amputation specimens. In an angiographic study of the pattern of vascular disease, Menzoian et al. demonstrated that the foot vessels, especially the dorsalis pedis, were spared in the diabetic patient even though the leg arteries were occluded [17]. It is extremely important to recognize this pattern of vascular occlusion since it opens up the possibility of arterial reconstruction with bypass grafts to the inframalleolar vessels, especially the dorsalis pedis artery [18,19]. Because of this pattern of occlusion, the angiographic runoff to the foot should be determined even when the infrageniculate arteries are occluded. It is a common mistake to terminate the angiogram, upon noting that the three infrageniculate arteries are occluded, on the assumption that the distal vessels must also be occluded.

Neuropathy is almost always a contributing factor to ulceration in the diabetic foot. It leads to sensory loss, atrophy of the intrinsic muscles with the development of foot deformities, and increased pressure points as well as autonomic dysfunction [20]. When severe, this can lead to ulceration even in the presence of normal arterial circulation. In general, neuropathy makes the foot more susceptible to ulceration even at lesser degrees of ischemia that would not cause ulceration in the non-diabetic. When a diabetic patient presents with a foot ulceration, an evaluation for ischemia should always be performed even when neuropathy is also a contributing factor. Once the diagnosis of arterial occlusion is made, arteriography should be performed except in those patients with the most superficial ulcerations under circumstances where they can be closely observed with conservative management.

■ SURGICAL MANAGEMENT

When spreading or deep infection complicates the presentation in the diabetic foot, immediate debridement and drainage should be performed as noted by Gibbons [21] and detailed in this chapter. Once the infection is controlled, an effort should be made using non-invasive tests to determine whether or not ischemia is a contributing factor. If so, arteriography should be performed immediately with a view towards prompt arterial reconstruction to provide the maximum opportunity for eradication of the infection and subsequent successful wound closure.

Based on the foregoing, we have developed a management plan for diabetic foot ulcers which is based upon the following:

- a rejection of the concept that microvascular occlusive disease is the etiology of ulceration;
- deep or spreading infection must be immediately controlled by debridement or drainage;
- all diabetic foot problems should be evaluated for ischemia even in the presence of neuropathy and/or infection;
- when arteriography is performed, the status of the foot vessels should always be demonstrated even when the infrageniculate arteries are occluded;
- arterial reconstruction should be designed to restore maximum foot perfusion based upon the arterial anatomy.

Over the past several years, this management plan has been increasingly implemented at our institution as we have become more confident in the outcome. Figure 1 illustrates the marked reduction in amputation rates at all levels that we have achieved as a result of this approach [20] and emphasizes the important role of bypass grafts to the dorsalis pedis artery. It should be emphasized that optimum care of the diabetic foot is a true multi-disciplinary effort involving vascular surgeons, podiatrists, orthopedic surgeons, cardiologists, infectious disease specialists, etc. However, in recent years increased proficiency with extreme distal arterial reconstruction has been the critical advance leading to improved outcome. Surprisingly, the results of arterial reconstruction in the diabetic are as good as or better than in non-diabetics [22]. This is true even though diabetics require more distal bypasses and often present with gangrene or ulceration requiring minor amputation [23]. Again this supports the view that there is no occlusive lesion in the microcirculation with subsequent limiting runoff or impairing graft patency.

Patients with diabetic foot lesions often present complex management problems that extend beyond those of the ischemic foot in non-diabetics. However, an understanding of the pathobiology and in particular the etiology and management of ischemia will be rewarded by a greatly decreased rate of major amputation.

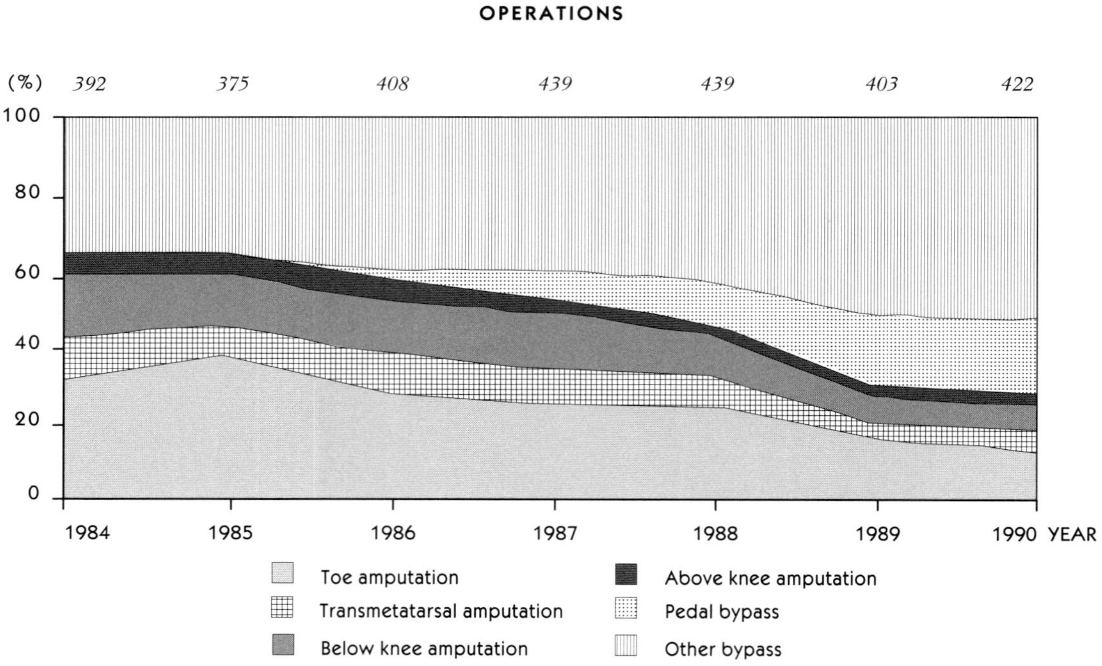

FIG. 1 This figure summarizes our operative experience at the New England Deaconess Hospital between the years 1984 to 1990. Only operations on patients with diabetes mellitus are included. The numbers at the top represent the total number of operations during a given year. Note that the incidence of amputations at all levels has decreased in recent years. This decrease in the incidence of amputations has occurred even though there has been no change in the indications for surgery, 30-day mortality, or post-bypass amputations.

REFERENCES

1 Goldenberg SG, Alex M. Joshi RA, Blumenthal HT. Nonatheromatous peripheral vascular disease of the lower extremity in diabetes mellitus. *Diabetes* 1959; 8: 261-273.

2 Strandness DE Jr, Priest RE, Gibbons GE. Combined clinical and pathologic study of diabetic and nondiabetic peripheral arterial disease. *Diabetes* 1964; 13: 366-372.

3 Conrad MC. Large and small artery occlusion in diabetics and nondiabetics with severe vascular disease. *Circulation* 1967; 36: 83-91.

4 Barner HB, Kaiser GC, Willman VL. Blood flow in the diabetic leg. *Circulation* 1971; 43: 391-394.

5 Irwin ST, Gilmore J, McGrann S et al. Blood flow in diabetics with foot lesions due to "small vessel disease". *Br J Surg* 1988; 75: 1201-1206.

6 LoGerfo FW, Coffman JD. Vascular and microvascular disease of the foot in diabetes: implications for foot care. *N Engl J Med* 1984; 311: 1615-1619.

7 Siperstein MD, Unger RH, Madison LL. Studies of muscle capillary basement membranes in normal subjects, diabetic, and prediabetic patients. *J Clin Invest* 1968; 47: 1973-1999.

8 Katz MA, McCuskey P, Beggs JL et al. Relationships between microvascular function and capillary structure in diabetic and nondiabetic human skin. *Diabetes* 1989; 38: 1245-1250.

9 Britland ST, Young RJ, Sharma AK, Clarke BF. Relationship of endoneurial capillary abnormalities to type and severity of diabetic polyneuropathy. *Diabetes* 1990; 39: 909-913.

10 Parving HH, Rasmussen SM. Transcapillary escape rate of albumin and plasma volume in short- and long-term juvenile diabetics. *Scand J Clin Lab Invest* 1973; 32: 81-87.

11 Brownlee M, Cerami A, Vlassara H. Advanced glycolysation end products in tissue and the biochemical basis of diabetic complications. *N Engl J Med* 1988; 318: 1315-1321.

12 Wyss CR, Matsen FA, Simmons CW, Burgess EM. Transcutaneous oxygen tension measurements on limbs of diabetic and nondiabetic peripheral vascular disease. *Diabetes* 1964; 13: 366.

13 Brand FN, Abbott RD, Kannell WB. Diabetes, intermittent claudication, and risk of cardiovascular events. The Framingham Study. *Diabetes* 1989; 38: 504-509.

14 Beach KW, Bedford GR, Bergelin RO et al. Progression of lower-extremity arterial occlusive disease in type II diabetes mellitus. *Diabetes Care* 1988; 11: 464-472.

15 Maser RE, Wolfson SK Jr, Ellis D et al. Cardiovascular disease and arterial calcification in insulin-dependent diabetes mellitus: interrelations and risk factor profiles. Pittsburgh epidemiology of diabetes complications study – *V. Arterioscl Throm* 1991; 11: 958-965.

16 Chantelau E, Ma XY, Herrnberger S et al. Effect of medial arterial calcification on O_2 supply to exercising diabetic feet. *Diabetes* 1990; 39: 938-941.

17 Menzoian JO, LaMorte WW, Paniszyn CC et al. Symptomatology and anatomic patterns of peripheral vascular disease: differing impact of smoking and diabetes. *Ann Vasc Surg* 1989; 3: 224-228.

18 Pomposelli FB Jr, Jepsen SJ, Gibbons GW et al. Efficacy of the dorsal pedal bypass for limb salvage in diabetic patients: short-term observations. *J Vasc Surg* 1990; 11: 745-751.

19 Pomposelli FB Jr, Jepsen SJ, Gibbons GW et al. A flexible approach to infrapopliteal vein grafts in patients with diabetes mellitus. *Arch Surg* 1991; 126: 724-729.

20 LoGerfo FW, Gibbons GW, Pomposelli FB Jr et al. Trends in the care of the diabetic foot: expanded role of arterial reconstruction. *Arch Surg* 1992; 127: 617-621.

21 Gibbons GW. The diabetic foot: amputations and drainage of infection. *J Vasc Surg* 1987; 5: 791-793.

22 Hurley JJ, Auer AI, Hershey FB et al. Distal arterial reconstruction: patency and limb salvage in diabetics. *J Vasc Surg* 1987; 5: 796-802.

23 Rosenblatt MS, Quist WC, Sidawy AN et al. Results of vein graft reconstruction of the lower extremity in diabetic and non-diabetic patients. *Surg Gynecol Obstet* 1990; 171: 331-335.

5

THE IMPORTANCE OF THE MEDICAL CONDITION OF THE PATIENT WITH CRITICAL LIMB ISCHEMIA

HÉLÈNE JULLIAN-PAPOUIN, PIERRE-EDOUARD MAGNAN

The incidence of critical limb ischemia (CLI) in Europe is estimated between 400 and 450 per million per year [1,2]. CLI occurs most frequently in patients with associated risk factors and other localizations of atherosclerotic disease. Age, smoking, diabetes, and hypertension are known risk factors of the disease. Coronary artery disease is encountered in 40% to 60% of patients suffering from CLI [3]; its frequency and severity are correlated with the ankle/brachial blood pressure index [4]. Also, cerebropetal artery disease has been found in 34% of these patients aged over 62 years [5]. The medical treatment of these risk factors and associated localizations of the disease is fundamental in order to preserve the threatened leg. This medical treatment commences before the arterial reconstruction with the evaluation and work-up of the patient and should continue during the period of hospital stay and during wound healing. It is not possible to apply a standard protocol, since every patient is a unique case for whom a compromise should be found between an ideal medical work-up and the need for quick revascularization of the threatened limb.

■ THE PREOPERATIVE PERIOD

Before the arterial reconstruction one should take into account and analyze the risk factors commonly found among patients presenting chronic critical leg ischemia [6]. In the majority of cases the patients are elderly. Even when their general condition is good, patients over 70 years of age have a reduced functional cardiopulmonary, renal and neurological reserve capacity, a more fragile mental condition and reduced defense mechanisms against infection and a tendency towards a delayed wound healing.

Coronary Artery Disease

Coronary artery disease affects 40 to 60% of this population and the perioperative risk of myocardial infarction is above 5% with a mortality of 25 to 70% [7]. The presence of coronary artery disease can be assessed easily through a medical history and electrocardiography, but it is the threshold and the extent of the ischemia that needs to be known. In most cases the patients have had previous coronary events, but they are more or less asymptomatic. Medical history and physical examination do not always allow the assessment of the progression of the disease. In this context it is wise not to loose time by additional sophisticated and expensive investigations, like stress ultrasound and myocardial scintigraphy. Initial coronary angiography should be reserved for cases with unstable or resting angina. In contrast, factors fostering coronary events, like anemia, tachycardia, hypotension and hypoxia should be avoided. Myocardial ischemia should be detected accurately, at best during the intervention by means of computerized analysis of the ST segment and after the intervention by administering troponine. Coronary disease should be treated optimally; beta blocking drugs are the most effective [8], but have a deleterious effect on the peripheral arterial perfusion. Calcium antagonists are presently preferred over nitrates when cardiac function permits.

Coronary insufficiency is the most serious risk factor, both for the survival and functional prognosis. In contrast to coronary artery disease, cardiac insufficiency has the advantage to be easily detectable from the clinical history and physical examination. Transthoracic ultrasound allows assessment of the myocardial function by measuring the ventricular ejection fraction. In addition, transesophageal ultrasound can detect valvulopathy, aortic or mitral stenosis in particular, or pulmonary hypertension, which are unfavorable conditions. This also allows assessment of the presence or absence of an intracavital thrombus in case of a nonsinusoidal rhythm. The treatment of cardiac insufficiency should be optimized preoperatively as good as possible using diuretics, digitalis, ACE inhibitors, including the prevention of infectious endocarditis in case of valvulopathy [9], especially in patients presenting with an infected skin ulcer. This treatment should be initiated as early as possible, since the maximum efficacy is obtained not until one week after the start.

Hypertension is very common: it is not a major threat if adequately controlled or when it does not aggravate coronary disease or cardiac insufficiency [10]. However, it compromises the management of the patient during the perioperative period because of the hemodynamic instability it provokes, as well as hyper- or hypotensive episodes, jeopardizing the patency of the revascularization. During the perioperative period, when the regular medication of patients is insufficient or impossible to administer, it is usually necessary to have a continuous intravenous administration of antihypertensive drugs. We use either a selective calcium antagonist like nicardipine (Loxen®) or an alpha blocking agent like uradipil (Eupressyl®), which does not increase the heart rate and does not change the cardiac function.

Arrhythmias due to conduction or excitability disturbances should be looked for systematically on the ECG. These may reveal an underlying cardiopathy. If possible they should be treated before the intervention.

Respiratory Status

Respiratory insufficiency [11] due to nicotine abuse and resulting obstructive bronchopulmonary disease increases significantly the perioperative morbidity, because the patients are exposed to bronchial obstruction, atelectasis and superimposed infections. The resulting hypoxemia compromises the energetic balance of the myocardium and increases the risk of myocardial ischemia. Functional respiratory tests do not have any interest in this context. Arterial gas analysis is sufficient to investigate blood oxygenation. Cessation of smoking is imperative, although the preoperative period is often insufficient for this to be really beneficial. Respiratory physical therapy, allowing drainage of secretions, combined with aerosols consisting of bronchial liquifiers and / or betamimetics should be given. Specific antibiotics are administered in case of bronchial superinfection.

Renal Function

Preoperative investigation of the renal function should not be neglected as it is often impaired. This is due to the atherosclerosis affecting the renal arteries and the dehydration, which is frequently observed in older patients. Two factors are found persistently in studies on the perioperative renal risk; age above 60 years and preexisting renal failure. The creatinine clearance is the major marker of the renal function; the serum creatinine level is not reliable enough. In order to improve or prevent deteriorating the preoperative renal function, one should take care of a good hydration, avoid nephrotoxic drugs or those interfering with prostaglandin synthesis, like aminoglycosides and NSAID's, and only use diuretics in case of a proper hydration status. Performing an arteriography is a problem in these patients and may increase renal insufficiency. This frequently occurring situation illustrates the importance of noninvasive investigation of these vulnerable patients by means of duplex scanning and magnetic resonance imaging, which are mentioned in other chapters of this book. At present and under these circumstances, we hydrate the patients several hours before performing the arteriography by means of a 500 ml intravenous saline solution together with 1 000 U Uricozyme® (urate oxydase) when the uric acid level is elevated.

Diabetes

Diabetes is observed frequently in this patient population. Actually, lower limb ischemia is four times higher in diabetic patients [12] and the prevalence of gangrenous lesions in diabetic patients presenting with lower limb ischemia is 20 to 30 times higher than in nondiabetics. Diabetes increases the operative risk, which is due to the often insidious character of coronary disease, cardiomyopathy, the frequency and severity of renal failure, autonomic neuropathy [13], and the higher proneness to infection. Skin lesions due to critical ischemia increase the infection risk and make regulation of the diabetes difficult. A vicious circle may evolve in which both pathological circumstances aggravate each other. Even when the diabetic is not insulin-dependent, insulin treatment is usually required. Different therapeutic schedules are possible. During the perioperative period we use a continuous infusion of regular insulin.

■ INTRAOPERATIVE PERIOD

Choosing the moment of intervention is often a compromise between the time needed to improve the patient's condition and the degree of urgency to revascularize the limb. There is no unequivocal method of anesthesia for these patients. General and locoregional anesthesia may be used and there is no difference between the two in terms of postoperative morbidity or mortality [10,14]. The choice for the type of anesthesia depends on specific criteria, taking into account the kind of surgical intervention and its estimated duration, as well as the medical condition of the patient. Epidural anesthesia is most suited for the patients who present with respiratory or cerebrovascular insufficiency [15]. The indication for this method is limited by the obligatory use of anticoagulant or antiplatelet drugs in these patients. Prevention of spinal or epidural hematomas indeed requires the placement or retraction of the catheter in a patient with a strictly normal coagulation, which is in practice hard to accomplish. On the other hand, a light sedation must often be added to reduce the poor comfort due to the duration of the intervention.

General anesthesia should have a limited effect on myocardial contractility. It should attenuate the catecholamine release provoked by the intubation and pain stimuli. We combine a short-acting morphinomimetic drug (remifentanyl), a hypnotic drug with little effect on the baroreflex (hypnomidate) and a curare derivative (atracurium, vercuronium) to obtain this result.

Irrespective of the technique chosen, the maintenance of the hemodynamic stability is the constant goal to achieve, even if the means to succeed are not always easy, given the pathology of certain patients. This stability is estimated by simple monitoring of the arterial pressure, heart frequency, hourly diuresis, and central venous pressure, when a central venous catheter is used. Correction of fluid and blood loss, the alternating use of inotropic betamimetic and vasoconstrictive alphamimetic drugs are the medical policies usually applied perioperatively. All means to correct hypothermia should be used: an ambient theatre temperature of 20°C, warming of infusion fluids, and the use of heated air blankets. A moderate hypothermia may cause cardiac arrhythmia and increase peripheral vascular resistance, being detrimental to the distal revascularization [17].

To obtain a waking as quick as possible one should use short-acting drugs. This should happen in a patient who has attained a normal central temperature (36.5°C). The disappearance of the anesthetic effects in a hypothermic patient induces protective mechanisms, vasoconstriction and shiver, which are especially harmful to fragile subjects. Analgesic drugs are to be administered from the moment of waking up.

■ POSTOPERATIVE PERIOD

During the directly postoperative period the numerous risk factors in these patients explain why a considerable number of complications may occur, among which the most important are myocardial infarction, severe cardiac insufficiency, and infectious problems. Prevention of these complications is done by the permanent checking of many parameters and the correction as quickly as possible of anemia, tachycardia, unstable blood pressure, oliguria, dyspnea, hypoxia, and hyperthermia. In France the consensus is to transfuse all patients showing anemia with a hemoglobin less than 7 g/l. Between 7 g/l and 10 g/l the indication is at the discretion of the clinician. We feel that these patients who present with critical lower limb ischemia and, thus, a high probability to have coronary lesions, should be transfused when the hemoglobin concentration is below 10 g/l. One should be particularly watchful about the clinical tolerance of additional treatments applied to improve the result of the revascularization, especially prostacyclins.

Analgesic treatment is tailored, using drugs with increasing strength. We usually start with paracetamol and then, when this is insufficient, opioid drugs, and ultimately morphine derivatives, keeping in mind that in older patients these may induce respiratory suppression or psychic decompensation.

Apart from revascularizations for ischemia or hemorrhage, which may have serious systemic consequences, these patients are sometimes reoperated for wound debridement or excision of necrotic lesions. The systemic impact of these minor interventions should not be underestimated, as they add another trauma to an area, which has reduced defense mechanisms. From this point of view it is important to plan the arterial revascularization to be as complete as possible.

The patient's rehabilitation phase should start immediately. The intensive care or recovery period should be limited, since it fosters the appearance of psychic problems in older subjects, who should be returned quickly to the ward, which facilitates the family to be present and reduces the risk of disorientation. The patient should be fed orally from the day after surgery on. Venous lines and drains should be removed as soon as possible to be able to mobilize the patient easily with the aid of physical therapists.

The rehabilitation and/or wound healing period is often long, which is cumbersome for the patient and family, and sometimes frustrating for the medical team. Problems concerning wound healing, superinfection, and malnutrition are paramount, even when other problems are not yet resolved. The nursing care plays an important role. As soon as possible one should prevent any cause for infection, stimulating the patient to eat, by using an adapted and attractive diet. Psychological support is decisive. Every positive sign should be appreciated in this development, which is sometimes of a depressing length. Analgesic treatment should be continued as long as necessary. One may add anxiolytic or psychotropic drugs if wisely prescribed, since undesirable effects are frequently observed in this population.

Dismissal as soon as possible is advocated during this phase of treatment. This allows the patient to return to his environment and family in particular, provides a better nutrition, and improves the quality of life and psychological stability. Continuation of the care at home is fundamental, consisting of local wound care on the one hand and drug treatment of the risk factors and associated defects on the other. Like during the hospital stay a balance should be found between suboptimal or insufficient drug treatment, which may promote complications, and an excess of medication, which may lead to patient discouragement and noncompliance.

Home care and surveillance facilities with repeated follow-up visits are certainly to be imagined and developed to facilitate the return of the patients to their local environment, which is also one of the essential goals of this kind of surgery.

CONCLUSION

The care for patients suffering from critical limb ischemia is teamwork, involving the surgeon, anesthesiologist, cardiologist, endocrinologist, nurses and physical therapists. The aim of this enterprise is to prevent cardiac and neurological events and to preserve the patency of the revascularization by maintaining adequate hemodynamic conditions, promoting wound healing of skin ulcers by means of suitable nutrition, counteracting infection, and correcting metabolic derailments. Finally, one should ascertain the physical rehabilitation required to obtain an acceptable functional result, without which mere limb salvage is an insufficient result.

REFERENCES

1 Catalano M. Epidemiology of critical limb ischaemia: North Italian data. *Eur J Med* 1993; 2: 1-4.
2 Anonymous. The Vascular Surgical Society of Great Britain and Ireland. Critical limb ischaemia: management and outcome. Report of a national survey. *Eur J Vasc Endovasc Surg* 1995; 10: 108-113.
3 Dormandy J, Mahir M, Ascady G et al. Fate of the patient with chronic leg ischaemia. J *Cardiovasc Surg* 1989; 30: 13-18.
4 Hertzer NR. The natural history of peripheral vascular disease. Implication for its management. *Circulation* 1991; 83 (Suppl I): 12-19.
5 Aronov WS, Ahn C. Prevalence of coexistence of coronary artery disease, peripheral arterial disease, and atherothrombotic brain infarction in men and women 62 years of age. *Am J Cardiol* 1994; 74: 64-65.
6 Norman JC, Theodore MS. Anesthesia in vascular surgery In: Miller RD ed. *Anaesthesia.* 1996 New York, Raven Press, pp 1851-1895.
7 Samain E, Coriat P, Baron JF. L'anesthésie du coronarien en chirurgie générale. In: *Encycl Med Chir. Anesthésie-Réanimation* 36352 F20, 1993, 17 p.
8 Mangano DT, Layug EL, Wallace A, Tateo I. Effects of atenolol on mortality and cardio-vascular morbidity after non cardiac surgery. Multicenter study of perioperative ischemia research group. *N Engl J Med* 1996; 85: 1713-1720.
9 Etienne J. Prophylaxie de l'endocardite infectieuse. *Ann Fr Anesth Reanim* 1994; 13: 567-572.
10 Anonymous. ACC/AHA task force report. Special report: guidelines for perioperative cardiovascular evaluation for noncardiac surgery. Report of the American College of Cardiology/American Heart Association Task Force on practice guidelines (Committee on perioperative cardiovascular evaluation for noncardiac surgery). *J Cardiothorac Vasc Anesth* 1996; 10: 540-552.
11 Wong DH, Weber EC, Schell MJ et al. Factors associated with postoperative pulmonary complications in patients with severe clinic obstructive disease. *Anesth Analg* 1995; 80: 276-284.
12 Diamantopoulos E, Grigoriadou M. Epidemiological and demographic characteristics of chronic peripheral arterial disease of the lower limbs. *Crit Ischaemia* 1996; 7: 5-13.
13 Raucoules-Aimé M, Lugrin D, Grimaud D. Neuropathie dysautonomique du diabétique et anesthésie. In: SFAR (ed.) *Conférences d'actualisation. 36ème Congrès National d'Anesthésie et de Réanimation.* Paris, Masson; 1994, pp 289-301.
14 Christopherson R, Beattie C, Frank SM, et al. Perioperative morbidity in patients randomized to epidural or general anesthesia for lower extremity vascular surgery. Perioperative ischemia randomized anesthesia Trial Study Group. *Anesthesiology* 1993; 79: 422-434.

15 Longrois D. Problèmes péri-opératoires posés par les patients atteints de pathologie vasculaire. Le point de vue de l'anesthésiste-réanimateur. *J Mal Vasc* 1998; 23: 49-53.
16 Rao TK, El-Etr AA. Anticoagulation following placement of epidural and subarachnoid catheters: an evaluation of neurologic sequelae. *Anesthesiology* 1981; 55: 618-620.
17 Frank SM, Fleisher LA, Breslow MJ et al. Perioperative maintenance of normothermia reduces the incidence of morbid cardiac events. A randomized clinical trial. *JAMA* 1997; 277: 1127-1134.

6

ASSESSMENT OF THE MICROCIRCULATION IN CRITICAL LIMB ISCHEMIA

BENGT FAGRELL

The final symptoms in patients with peripheral vascular disorders are not caused by the obstructions of the main vessels, but by a disturbed microcirculation of the region. Not until the blood flow in the nutritional capillaries decreases below the minimal demand of the tissue, will ischemic symptoms occur [1-3]. In patients with chronic, critical limb ischemia (CLI) the final cause for the symptoms of rest pain and ulcers is a totally abolished circulation in the nutritional skin capillaries. In order to be able to evaluate the nutritional status of these patients microcirculatory methods must consequently be used, and during the past two decades, new non-invasive techniques for performing such studies in clinical practice have been developed. The techniques most often used today are transcutaneous oxygen pressure measurements (tcpO2), laser doppler fluxmetry (LDf), and vital capillaroscopy (VC).

■ TRANSCUTANEOUS pO_2 ($tcpO_2$) MEASUREMENTS

The oxygen tension in unheated human skin is normally less than 5 mmHg. In order to be able to measure with some accuracy in clinical practice it is therefore necessary to heat the transducer and skin to an elevated and controlled temperature (usually between 37°-44°C). All available equipment nowadays has such a device incorporated into the transducer.

The principle of the tcpO_2 method is that it measures the amount of oxygen molecules that are transformed out through the skin. Most probably these molecules are delivered not only by the skin capillaries, but also by the vessels of the subpapillary vascular plexuses, and therefore the technique is only an indirect measure of the amount of oxygen that is delivered out to the skin surface from the oxygenated blood in the whole skin microvascular bed of the area.

TcPO_2 in Patients with Peripheral Arterial Obliterative Disease (PAOD)

The tcpO_2 method has during recent decades been used rather extensively in clinical practice for evaluating the skin microcirculation in patients with PAOD [4,5], and it has been demonstrated that it can be used successfully for evaluating the skin blood flow in patients with various degrees of PAOD. In 1980 Matsen et al. [6] showed that tcpO_2 was significantly lower in the foot of patients with PAOD than in normal subjects. They also found a correlation between the degree of arterial insufficiency as evaluated by the ankle/arm pressure index and tcpO_2 measurement, but a large scatter was seen in the patients. Others have also shown that there is a significant reduction of tcpO_2 already in the calf region in claudicants, but more so in patients with rest pain and/or depending gangrene, where the tcpO_2 of the foot was always less than 20 mmHg [7]. TcpO_2 was also found to be useful in discriminating between all degrees of ischemia, both in diabetic and non-diabetic patients [8,9].

The separation between groups may be improved by measuring tcpO_2 after a period of arterial occlusion. By using different parameters during the postocclusive reactive hyperemia response, total discrimination between different degrees of severity of disease have been seen in some studies [10]. This post-ischemic response test has also been shown to correlate nicely with the degree of arterial insufficiency [11]. Bongard and Krähenbühl [4] could not find any relationship between skin blood flow as evaluated by the 133Xenon clearance method and tcpO_2 in PAOD patients. They concluded that the 133Xenon clearance method considerably overestimates local blood flow in these patients because there is a significant reduction of fat in the subcutaneous tissue in areas with chronic ischemia.

In order to improve the applicability of tcpO_2 measurements in patients with PAOD, further dynamic interventions have been tried. It has been found that the tcpO_2 recovery half-time after ischemia discriminates well between patients with PAOD and normal subjects [10]. Harward et al. have also shown that the use of tcpO_2 determination at rest, coupled with the response to 100% oxygen inhalation, was an excellent predictor of the outcome of lower extremity amputation [12]. Similar results have also been presented by Bongard and Krähenbühl [13], and they concluded that the reaction of tcpO_2 to oxygen inhalation appears to be a good predictor for the prognosis of severely ischemic limbs.

The tcpO_2 technique has also been tested for its ability to predict the clinical outcome of vascular surgery or amputation. It has been found that the method itself, or in combination with laser doppler fluxmetry, is significantly better than, for example, ankle pressure measurements, especially in diabetics [14-17]. Caspary et al. [18] have shown that the method can also be used for studying the effect of vasoactive drugs on skin microvascular circulation in patients with PAOD, but then the electrode should not be heated to more than 37°C. At 44°C the effect could not be seen because of the local vasodilatation caused by the induced heat. In diabetic patients with PAOD and chronic foot ulcers tcpO_2 has also been found to be a better predictor for ulcer healing than e.g. systolic toe blood pressure. The probability of ulcer healing is very low (15%) when the tcpO_2 is below 25 mm Hg [19].

In summary, transcutaneous pO_2 measurements have been shown to be useful for predicting the degree of arterial insufficiency in patients with PAOD. However, as the measurements can virtually only be performed as a spotlight measurement on, for example, the dorsum of the foot in these patients, it is not a very sensitive method for predicting the risk of skin necrosis to develop in an ischemic region of the foot. As has been shown previously, most PAOD patients with rest pain or skin ulcers have ischemia in certain limited areas, most often the acral parts of the foot [3]. Consequently, it is unlikely that a spotlight measurement on the dorsum of an ischemic foot will be able to predict the nutritional status in the most ischemic part of the region, which is most often a toe. However, when the technique is used in combination with oxygen inhalation it has been shown to be helpful, for example in determining the best level of amputation in patients with PAOD [14].

■ LASER DOPPLER FLUXMETRY (LDf)

During the last decade laser doppler (LD) fluxmetry has been extensively used in clinical practice for evaluating skin microcirculation. However, it is still uncertain what actually generates the signal recorded from the skin with this method. The relative distribution of blood between the non-nutritional, subpapillary, thermoregulatory vascular bed and nutritional papillary capillaries in the skin differs markedly from one area to another. For example, in the skin of hands and feet, more than 90% of the blood flows only through the subpapillary vascular bed, and less than 10% goes through the nutritional capillaries (Fig. 1). There are now substantial data indicating that the dominant part of the LDf signal is generated by the movement of blood cells in the subpapillary vascular bed of the skin, and only a small portion is due to the motion of blood cells in the nutritional capillaries [20]. This is of extreme importance to have in mind when the data achieved are used in connection with clinical symptoms, especially in patients with disturbances of the peripheral circulation in extremities.

Problems with LDf in Evaluating Skin Microcirculation

The measuring depth of the LD technique when used on the skin is unknown. It has been postulated that it may measure only 1-2 mm deep. However, many factors, like the amount of blood and the composition of the skin in the region investigated, will influence the penetration and measuring depth [20]. Also the measuring volume may be changed by these factors. Consequently, actual flow in mm x minutes-1 x 100 g-1 of tissue cannot be obtained except during certain specific conditions.

Another confusing fact is that during complete arterial occlusion a significant flux signal is often recorded, indicating that a certain portion of the LD signal is not generated from tissue blood flow [1,21]. This value, which has been called the biological zero of the tissue, is especially large in patients with peripheral vascular disorders. It has been found that in patients with PAOD it can vary from 15% up to 80% of the total resting flux value! [21].

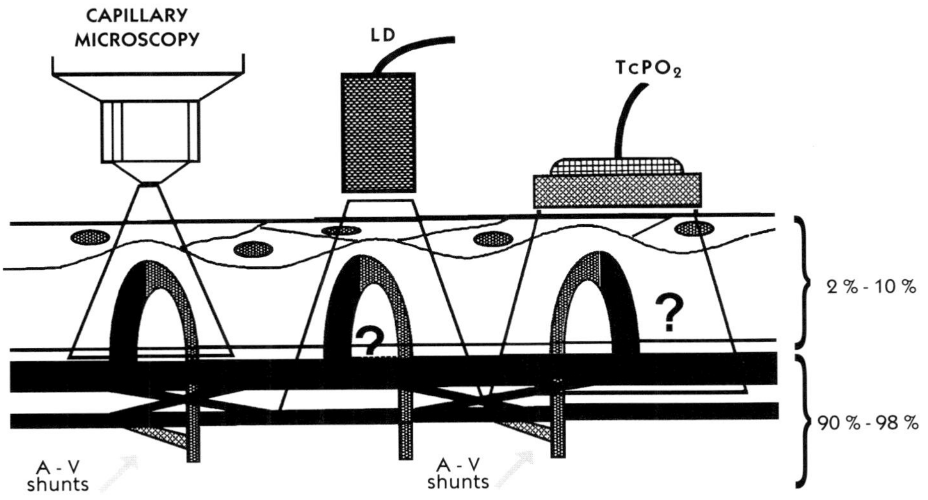

FIG. 1 Schematic drawing of the skin microvascular bed in humans and the measuring potential of three different methods: capillary microscopy, laser doppler fluxmetry (LD), and transcutaneous oxygen tension measurement ($TcpO_2$). The shaded areas indicate how much of the skin microcirculation that is covered by the three methods. Only 2 - 10% of the blood flow of the skin microcirculation goes through the nutritional skin capillaries, while 90-98% bypasses the nutritional capillaries and only enters the subpapillary vascular plexuses mostly through arterio-venous anastomoses.

■ THE USE OF LDf IN PATIENTS WITH PAOD

Resting LDf

In spite of the drawbacks mentioned, the LD technique, which is extremely simple to use clinically, has lately been utilized quite frequently in patients with PAOD. The resting LDf (rLDf) shows marked spatial and temporal variations, and it is therefore not surprising that no significant differences have been found between healthy subjects and patients with PAOD [22,23]. The overlap between the groups has been almost complete. Patients with moderate PAOD may sometimes even have a significantly higher rLDf than healthy subjects, while a significantly lower value has been recorded only in patients with critical limb ischemia, and when the measurements have been performed on the toe pulp [23,24].

rLDf during Leg Dependency

The reaction to leg dependency has also been tested. Most healthy subjects show a decrease in rLDf on dependency which is due to the so-called veni-vasomotor reflex [25,26]. Patients with PAOD also react to dependency with a decreased rLDf except in marked ischemia where the rLDf increases [8]. This shows that in patients with critical limb ischemia the normal reactivity in the microvascular bed of the skin is hampered or sometimes even abolished. This explains why an ischemic foot almost always becomes red or bluish in the dependent position, indicating that the skin microvessels are dilated and passively fill up with blood in this position. However, it is extremely important to stress that the color of the skin is not generated by the amount of blood in the skin capillaries, but almost exclusively dependent on the amount of blood in the subpapillary, predominantly venous plexuses, of the area.

Postocclusive Reactive Hyperemia

One of the most well known tests in clinical practice for evaluating the functional aspects of arterial circulation in a limb is the reactive hyperemia response to a certain period of arterial occlusion. This test has also been applied to patients with PAOD by using the LD technique. The occluding cuff has been placed at different levels of the extremity: the thigh, above and below the knee, the ankle and the toe [23,26].

Several different parameters have been tested for evaluating the skin microcirculation during the postocclusive reactive hyperemia response in patients with PAOD. One of the most discriminating parameters seems to be the time to reach the highest peak during hyperemia when the occluding cuff is applied at the ankle level [23]. Östergren et al. [26] have shown that the time to peak, in both the supine and sitting positions, discriminates well between healthy subjects and patients with PAOD.

Also, the latency time or flux reappearance time, which is defined as the time after an occlusion until start of the hyperemic response, has been found to be useful [23]. In healthy subjects this time is always less than three seconds, while in patients with (for example) critical ischemia the time is longer than 45 seconds. Long values seem to indicate that the distal vascular bed is affected by significant arterial obliterations, while patients with only proximal obstructions have a normal reappearance time [23].

Flow-motion Activity in LDf

As the technique of LDf can easily be applied to a certain skin area and used for measuring flux values for long periods, it is also very convenient for looking at variations of blood flow over time. The influence of myogenic activity on the human skin microcirculation (cf. vasomotion: see paragraph on vital capillaroscopy) can be studied over long periods of time. Oscillations in the skin blood flow have been found to be very sensitive to dynamic interventions and to be composed of waves with several different frequencies. Creutzig et al. [25] have shown that oscillations in the forefoot of healthy subjects was around 4 cycles/min while in patients with claudication this frequency decreased to 2.6 cycles/min. This low frequency increased in the patients if the skin was heated to 37 C, while no changes of frequency was seen in the healthy subjects.

The most extensive work on flow-motion activity in the LD signal has been reported by Hoffman et al. [27]. They found two different types of waves: large waves and small waves. The large waves were the same as described by Creutzig et al. [25] with a frequency of 3-4 cycles/min. These waves were present in both healthy subjects and different groups of PAOD patients. The prevalence of these waves was significantly decreased in the sitting position as compared to the supine position. The lowest values recorded were in patients with critical limb ischemia. The small waves had a frequency of about 20 cycles/min and were seen in 30 - 90% of patients, but in only 8% of healthy subjects. The prevalence of these waves increases successively with the severity of the disease. What generates these small waves is still under debate. Most probably they are present also in normal subjects, but overshadowed by the activity of other vascular beds [24]. When the large waves disappear they become more prominent. Consequently, the appearance of small waves is an indication of a severely impaired microcirculation in patients with PAOD.

The Effect of Therapeutic Procedures

The technique of LDf can also be used for studying the effect of drugs in PAOD patients. Caspary et al. [18] used a combination of $tcpO_2$ and LDf to study the effect of prostaglandin E_1 (PGE_1) and nucleotide phosphates on skin microcirculation in patients with PAOD. Both patients with claudication and ischemic skin ulcers were investigated. The results showed that the PGE_1 did not reduce the LDf signal, while the nucleotide phosphates reduced it by almost 30%. As both substances have been shown to induce an augmentation of the total flow to ischemic limbs when infused intra-arterially, the results indicate a steal phenomenon by the nucleotide phosphates while PGE_1 seems to improve the skin microcirculation in these patients. The results of the study very nicely demonstrate the importance of studying the local skin microcirculation when evaluating the effect of therapeutic procedures in these patients. However, it should also be stressed that an increase in LDf or $tcpO_2$ may only indicate an increase of the total microcirculation of the ischemic area. As will be shown later in this chapter, such improvement may not necessarily improve the nutritional circulation of the area in question. Consequently, other measures that study directly the blood flow in the nutritional capillaries of the skin have to be used in order to verify an effect also on the nutritional blood flow of the area [28].

■ VITAL CAPILLAROSCOPY

Two of the main symptoms in patients with PAOD are rest pain and skin necrosis or gangrene. These symptoms are due to a reduction of blood flow in the nutritional capillaries below the demand of the tissue [3,29]. The morphology of the nutritional skin capillaries can be easily studied by the technique of vital capillaroscopy. In the late sixties preliminary observations of the capillaries in the dorsal skin of the foot and toes showed significant changes of nutritional skin capillaries in patients with PAOD. Extensive investigations have later proved that capillaroscopy is one of the most sensitive methods for estimating the nutritional status of a certain skin area in patients with PAOD [3,28]. It has been shown that specific morphological changes of the nutritional capillaries appear with a successive decrease of the arterial circulation of the extremity. These changes can be classified and used for determining disturbances of the nutritional circulation in a particular skin area, and the risk of skin necrosis to develop can be accurately estimated [3,28,29].

The skin capillaries can be studied directly with an ordinary light, preferably stereo, microscope [28]. For morphological studies a microscopic magnification of about 5-100x is appropriate. A combination of a microscope and a television camera can be used, and the magnification then varies with the size of the monitor. Recently a *pen* microscope (Microwatcher ® Model VS-10, Mitsubishi Kasei Corp., Tokyo, Japan) working with fiber optics and a CCD chip has been found to be extremely handy for clinical investigations of the human skin capillaries. In order to make the skin transparent, and to further minimize the reflections from the skin surface, some kind of oil (e.g. paraffin oil) has to be applied onto the skin. When the feet are investigated, the patient should be in a sitting position to ensure a good blood filling of the region; also in patients with a markedly reduced arterial circulation of the leg.

Morphological Changes in Patients with PAOD

In patients with mild PAOD, the capillary bed of the forefoot does not seem to differ much from the normal pattern [3]. Only patients with moderately or severely reduced arterial circulation to the area show any marked capillary alterations. In patients with only a minor decrease of the arterial circulation the blood flow in the nutritional vascular bed is not reduced, and consequently no changes of the capillaries will be present. The morphological changes of the capillaries can be classified according to a simple three stage system [29], which has been shown to be very efficient for evaluating the viability of the skin in patients with ischemia. The different stages are shown in Fig. 2.

Stage A is the normal stage where the skin capillaries are seen as small dot- or comma-shaped structures in the skin papillae. All of the capillaries are filled with blood, and sometimes in patients with mild PAOD a moderate or marked dilatation *(micropool)* of the nutritional capillaries can be observed.

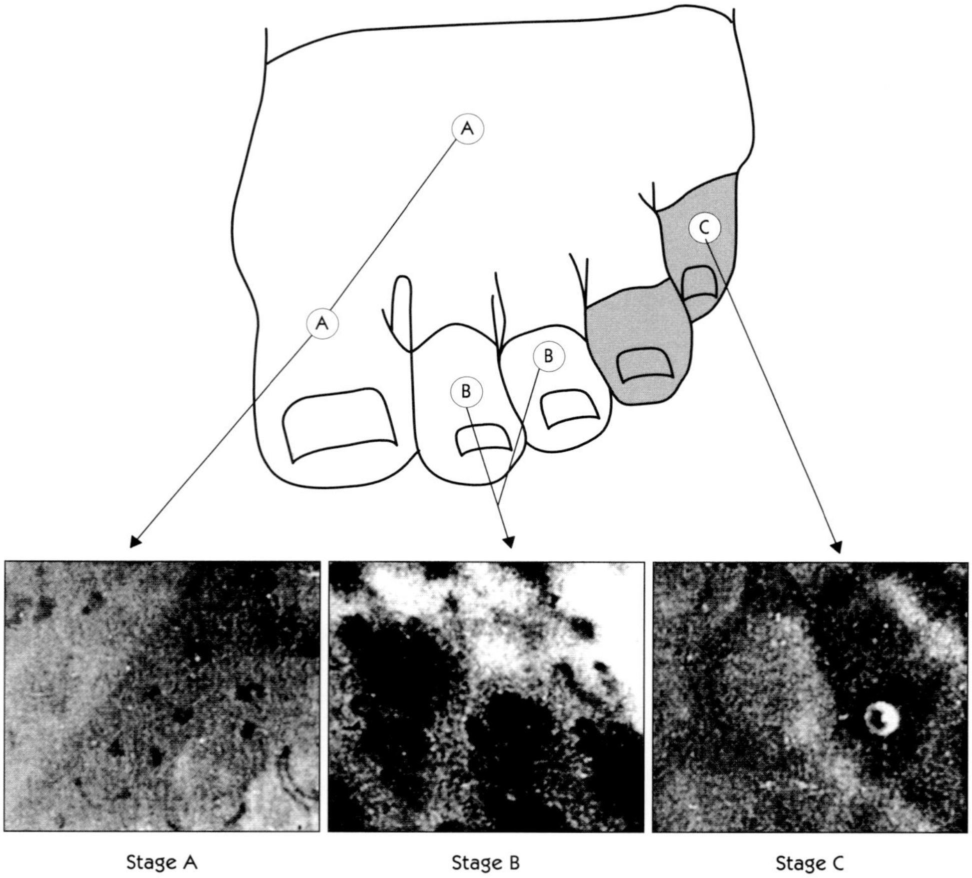

FIG. 2 The different capillary patterns in a patient with arterial insufficiency. The *mapping* of the capillary morphology reveals marked differences in the pattern from one limited area to another. Stage A (normally filled, but dilated capillaries) is seen in most of the foot skin, while toe 2 and 3 show edema formation and capillary hemorrhages. (Stage B). In toe 4 and 5 no blood filled capillaries are seen indicating a totally abolished nutritional circulation (Stage C). See further Figure 3.

Stage B represents indistinct capillaries where the capillaries cannot be sharply focused with the microscope. This is most often due to edema formation or structural changes in the skin papillae themselves, and is the earliest sign of an ischemic disturbance of the microvascular bed in the region. Capillary hemorrhages indicate a vast damage to the capillary wall, and these changes are seen in patients with more severe reduction of the arterial circulation to the region.

Stage C represents a marked reduction or complete absence of the number of blood filled capillaries, although the patients are in the sitting position during the investigation. These changes are seen in skin areas of patients with PAOD where the risk of tissue necrosis is impending.

Relation between Arterial Circulation and Capillary Stage

The correlation between the total arterial circulation of a leg, as evaluated by the systolic blood pressure of the digit, and the capillary stages of the same foot, has been studied in patients with PAOD [3,28]. In toes with a pressure less than or equal to 30 mmHg the capillaries are usually normal (Stage A). Edema or capillary hemorrhages (Stage B) are seen in only 4-5%. In toes with a pressure less than or equal to 30 mmHg the whole spectrum of capillary morphology can be seen. Severe changes (Stage C) are almost exclusively seen in toes with a high risk of, or already present, skin ulcers.

The described classification system was used in 69 patients with severe PAOD to test its value for predicting the risk of skin necrosis to develop in the area [29]. All patients had a toe pressure of less than or equal to 30 mmHg, and they were divided into three groups: 0 - 10, 11 - 20, and 21 - 30 mmHg. The morphological changes of the capillaries were classified as Stage A, B, or C. The patients were followed clinically for at least three months, to record if they developed skin necrosis or gangrene in any part of the ischemic foot. The highest predictive value for the toe pressure ranges was only 58%, while the corresponding values for Stage A, B, and C were 5, 16, and 93% respectively (Fig. 3). By using the combination of capillaroscopy and $tcpO_2$, Slagsvold et al. [30] found that there was a good agreement between the blood filling of the capillaries and the $tcpO_2$ value recorded. This also shows that the blood filling of the nutritional capillaries is a reliable indicator of the severity of skin ischemia.

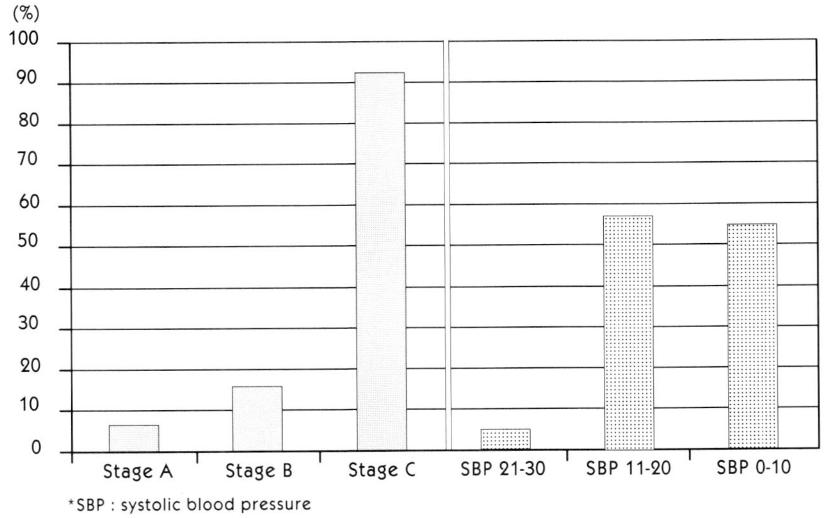

FIG. 3 The predictive value for the development of skin necrosis of capillary stages (Stage A, B, and C; see also Figure 2) and systolic blood pressure of the great toe (SBP: 0-10, 11-20, 21-30 mmHg). It can be seen that stage C has a very high predictive value (93%) for the risk of developing skin necrosis in an area over the observation period of three months. The highest value for SBP was 58% for SBP 11-20 mmHg.

It must be emphasized that ischemic skin necrosis most often appears in limited and localized areas, like one toe or over a bony prominence. This is an important reason why methods that can only measure the total circulation in a whole region, or that give only spot-like information of the microcirculation (like LDf and $tcpO_2$), have a low predictive value for the development of necrosis in a certain area of an ischemic foot. Blood pressure measurements for example can only be performed in one or two toes because of technical reasons. An evaluation of the circulation of other digits is consequently not possible. With capillaroscopy the *target organ* of ischemia, i.e. the nutritional skin capillaries, can be directly visualized in the whole area susceptible to necrosis, and a disturbed nutritional circulation of a limited region will be easily detected in the microscope. An example of how this *capillary mapping* can be performed is shown in Fig. 2.

Changes of Capillary Blood Cell Velocity (CBV)

By a more sophisticated and computerized technique of capillaroscopy (CapiFlow ®, Sweden) also the blood cell velocity can be studied in skin capillaries of humans [28,31]. Alternatively, a flying spot device can be used for this purpose, particularly in the lower velocity range [32]. It has been clearly shown that not only the morphology, but also the capillary blood flow is affected in patients with a reduced arterial circulation [28,33]. In patients with PAOD of the legs there is a discrepancy between the total and the nutritional skin microcirculation of the feet, which suggests a maldistribution of the local blood flow. In a study by Bongard et al. the total skin microcirculation was evaluated by laser doppler fluxmetry (LDf), and the blood flow in the nutritional skin capillaries (CBV) by dynamic capillaroscopy [24]. This combination was simultaneously used to study the nail fold microcirculation in the big toe of 12 legs with various degrees of PAOD, using 10 legs of healthy subjects as control. CBV was evaluated by the computerized, cross-correlation technique. The resting CBV was similar in patients and controls, whereas the total skin blood cell flux (LDf) in the same area was significantly increased in the patients in comparison to the normal subjects. The skin temperature was also significantly higher in the patients. However, the post-occlusive reactive hyperemia response was impaired for both CBV and LDf in the PAOD group. Also, the ratio between rCBV and rLDf was calculated to evaluate the fraction of total skin blood flow that reached the nutritional capillaries in the ischemic area. It was found that there was a marked decrease of this ratio in patients with PAOD as compared to controls, indicating a maldistribution of blood between non-nutritional and nutritional skin vessels of the ischemic area. The explanation for this is most probably that the ischemia triggers a dilatation of the arterioles and arterio-venous (AV) anastomoses of the non-nutritional, subpapillary vascular bed of the skin. The opening of the AV shunts will give rise to an increase of the pressure also in the venous system, keeping the AV pressure difference rather constant in the parallel coupled nutritional capillaries, which explains how the CBV can be unchanged or even decreased while the total skin blood flow is increased.

Consequently, in spite of a significantly higher skin temperature and resting total skin blood flow in PAOD patients, the capillary blood flow is not increased, indicating a maldistribution of blood between nutritive and non-nutritive skin vessels. This increase of the total skin microcirculation in toes of PAOD patients is in full agreement with earlier studies that have shown high resting skin blood flow in clinically ischemic feet as compared to normal feet [34].

In more recent studies in patients in different stages of PAOD using dynamic capillaroscopy the abovementioned changes in morphology as well as changes in dynamics were found: a significant decrease in resting CBV and reactive hyperemic response were observed with decreasing ankle blood pressure, especially in critically ischemic patients [8]. In addition, an increasing disturbance was seen of the physiological vasoconstriction response upon dependency of the leg, which response is meant to counteract the increase in orthostatic pressure upon dependency and is an important factor in (postoperative) edema prevention [35]. In critically ischemic patients with an ankle/brachial pressure index less than 0.2 this vasoconstriction disturbance was such that CBV was slightly higher in the sitting than in the supine position. Hence, nutritive perfusion improved upon dependency. This could be an explanation of the relief of pain these patients experience when they lower their ischemic leg [33].

■ FLOW MOTION ACTIVITY

Vasomotion is a rhythmic variation in the activity of the smooth muscle cells of the vascular walls and seems to be most prominent at the level of the arterioles. Its activity is centrally mediated through hormones and the nervous system, and locally controlled by myogenic mechanisms. It is present in the skin microcirculation to optimize the regulation of blood flow in different vascular compartments of the skin microcirculation, and to regulate tissue nutrition and fluid exchange. Its activity is seen as variations in the blood flow and can be studied by both capillaroscopy and the laser doppler technique (flowmotion; see paragraph on laser doppler fluxmetry) [36]. The observed fluctuations seen in the tracings of the two methods are generated by the sum of all vasomotion activity in the vascular bed proximal to the area of observation. The activity is markedly changed in pathological conditions, and it may be of importance for the treatment of the microvascular disturbances if vasomotion activity could be restored [23,37].

The flowmotion has been studied by capillaroscopy both in toes and fingers of healthy subjects and patients with PAOD [23]. There is a significant decrease of the flowmotion activity in the digits with a decreased arterial blood flow, and this is most probably due to the vasodilatation that is induced in the ischemic regions. The prevalence of flowmotion in toes of patients with PAOD was found to be significantly lower (25%) than in healthy subjects (80%), but the frequency is similar (4.5 - 5.5 cycles/min). With the laser doppler technique flowmotion waves with a frequency of more than 20 cycles/min (fast waves) have been found in patients with PAOD. The prevalence increased (33 - 92%) with the severity of ischemia [27,36]. Such fast flowmotion waves have never been seen in the CBV tracings, which could mean that they do not exist in these vessels, or that the technique of capillaroscopy is not sensitive enough to record these fast flowmotion components.

REFERENCES

1 Burton AC. Special features of the circulation of the skin. In Montagna W. & Ellis R A, eds. *Blood vessels and circulation. Vol. II. Advances in biology of the skin.* Pergamon Press, New York, 1961, pp 117-122.

2 Conrad MC. Abnormalities of the digital vasculature as related to ulceration and gangrene. *Circulation* 1968; 88: 568-581

3 Fagrell B. Vital capillary microscopy - A clinical method for studying changes of the nutritional skin capillaries in legs with arteriosclerosis obliterans. *Scand J Clin Lab Invest* 1973: 31; Suppl 133: 2-50.

4 Bongard O, Krähenbühl B. Pedal blood flow and transcutaneous pO_2 in normal subjects and in patients suffering from severe arterial occlusive disease. *Clin Physiol* 1984; 4: 393-401

5 Bongard O, Bounameaux H, Fagrell B. Effects of oxygen inhalation on skin microcirculation in patients suffering from peripheral arterial occlusive disease. *Circulation* 1992; 86: 878-886.

6 Matsen FA 3d, Wyss CR, Pedegana LR, et al. Transcutaneous oxygen tension measurement in peripheral vascular disease. *Surg Gynecol Obstet* 1980; 150: 525-528.

7 Cina C, Katsamouris A, Megerman J et al. Utility of transcutaneous oxygen tension measurements in peripheral arterial occlusive disease. *J Vasc Surg* 1984; 1: 362-371.

8 Ubbink DTh, Jacobs MJHM, Tangelder GJ et al. The usefulness of capillary microscopy, transcutaneous oximetry and laser doppler fluxmetry in the assessment of the severity of lower limb ischaemia. *Int J Microcirc: Clin Exp* 1994; 14: 34-44.

9 Ubbink DTh, Kitslaar PJEHM, Tordoir JHM et al. Skin microcirculation in diabetic and non-diabetic patients at different stages of lower limb ischaemia. *Eur J Vasc Surg* 1993; 7: 659-666.

10 Kram HB, White RA, Tabrisky J et al. Transcutaneous oxygen recovery and toe pulse reappearance time in the assessment of peripheral vascular disease. *Circulation* 1985; 72: 1022-1027.

11 Slagsvold CE, Kvernebo K, Stranden E, Kroese A. Postischemic transcutaneous oxygen tension response in assessment of peripheral atherosclerosis. *Vasc Surg* 1988; 22: 102-109.

12 Harward TRS, Volny J, Golbranson F et al. Oxygen inhalation-induced transcutaneous pO_2 changes as a predictor of amputation level. *J Vasc Surg* 1985; 2: 220-227.

13 Bongard O, Krähenbühl B. Predicting amputation in severe ischaemia. The value of transcutaneous pO_2 measurement. *J Bone Joint Surg (Br)* 1988; 70: 465-467.

14 Burgess EM, Matsen FA 3d, Wyss CR, Simmons CW. Segmental transcutaneous measurements of pO_2 in patients requiring below-the-knee amputation for peripheral vascular insufficiency. *J Bone Joint Surg (Am)* 1982; 64: 378-382.

15 Karanfilian RG, Lynch TG, Zirul VT et al. The value of laser doppler velocimetry and transcutaneous oxygen tension determination in predicting healing of ischemic forefoot ulcerations and amputations in diabetic and nondiabetic patients. *J Vasc Surg* 1986; 4: 511-516.

16 Wyss CR, Robertson C, Love SJ et al. Relationship between transcutaneous oxygen tension, ankle blood pressure, and clinical outcome of vascular surgery in diabetic and nondiabetic patients. *Surgery* 1987; 101: 56-62.

17 Ubbink DTh, Tulevski II, Den Hartog D et al. The value of non-invasive techniques for the assessment of critical limb ischaemia. *Eur J Vasc Endovasc Surg* 1997; 13: 296-300.

18 Caspary L, Creutzig A, Alexander K. Reaction of transcutaneous pO_2 and laser doppler flow signal on intraarterial infusion of PGE1 and nucleotide phosphates in patients with peripheral arterial occlusive disease. *VASA* 1987; Suppl 17: 17-22.

19 Kalani M, Brismar K, Fagrell B et al. Transcutaneous oxygen tension and toe blood pressure as predictors for outcome of diabetic foot ulcers. *Diabetes Care*, in press.

20 Fagrell B, Nilsson, G. Advantages and limitations with one-point laser doppler perfusion monitoring in clinical practice. *Vasc Med Rev*, 1995; 6: 97-101

21 Caspary L, Creutzig LA, Alexander K. Biological zero in laser doppler fluxmetry. *Int J Microcirc: Clin Exp* 1988; 7: 367-371.

22 Del Guercio R, Leonardo G, Arpaia MR. Evaluation of postischemic hyperemia on the skin using laser doppler velocimetry: Study on patients with claudicatio intermittens. *Microvasc Res* 1986; 32: 289-299.

23 Kvernebo K, Slagsvold CE, Gjoldberg T. Laser doppler flux reappearance time (FRT) in patients with lower limb atherosclerosis and healthy controls. *Eur J Vasc Surg* 1988; 2: 171-176.

24 Bongard O, Fagrell B. Discrepancies between total and nutritional skin microcirculation in patients with peripheral arterial occlusive disease (PAOD). *VASA* 1990; 19: 105-111.

25 Creutzig A, Caspary L, Hertel RF et al. Temperature-dependent laser doppler fluxmetry in healthy and patients with peripheral arterial occlusive disease. *Int J Microcirc: Clin Exp* 1987; 6: 381-390.

26 Östergren J, Schöps P, Fagrell B. Evaluation of a laser doppler multiprobe for detecting skin microcirculatory disturbances in patients with obliterative arteriosclerosis. *Int Angio* 1988; 7: 37-41.

27 Hoffmann U, Schneider E, Bollinger A. Flow motion waves with high and low frequency in severe ischemia before and after percutaneous transluminal angioplasty. *Cardiovascular Research* 1990; 24: 711-718.

28 Bollinger A, Fagrell B. *Clinical capillaroscopy - A guide to its use in clinical research and practice.* Stuttgart, Hogrefe & Huber, 1990, pp. 63-77.

29 Fagrell B, Lundberg G. A simplified evaluation of vital capillary microscopy for predicting skin viability in patients with severe arterial insufficiency. *Clin Physiol* 1984; 4: 403-411.

30 Slagsvold C-E, Rosèn L, Stranden E. The relation between changes in capillary morphology induced by ischemia and the postischemic transcutaneous pO_2 response. *Int J Microcirc Clin Exp* 1991; 10: 117-125.

31 Fagrell B, Fronek A, Intaglietta M. A microscope-television system for studying flow velocity in human skin capillaries. *Am J Physiol* 1977; 233: H318-321.

32 Boss Ch, Schneuwly P, Mahler F. Evaluation and clinical application of the flying spot in clinical nailfold capillary TV-microscopy. *Int J Microcirc: Clin Exp* 1988; 6: 15-23.

33 Ubbink DTh, Jacobs MJHM, Slaaf DW et al. Capillary recruitment and pain relief on dependency in patients with severe lower limb ischemia. *Circulation* 1992; 85: 223-229.

34 Morris-Jones W, Preston FE, Greaney M, Chatterjee DK. Gangrene of the toes with palpable peripheral pulses. *Ann Surg* 1981; 193: 462-466.

35 Ubbink DTh, Kitslaar PJEHM, Tordoir JHM et al. The relevance of posturally induced microvascular constriction after revascularisation in patients with chronic leg ischaemia. *Eur J Vasc Surg* 1992; 6: 525-532.

36 Seifert H, Jäger K, Bollinger A. Analysis of flowmotion by the laser doppler technique in patients with peripheral arterial occlusive disease. *Int J Microcirc: Clin Exp* 1988; 7: 223-236.

37 Östergren J, Fagrell B. Skin capillary blood cell velocity in patients with arterial obliterative disease and polycythemia - a disturbed reactive hyperemia response. *Clin Physiol* 1985; 5: 35-43.

7

ARTERIAL MAPPING BY DUPLEX SCANNING IN PATIENTS WITH CRITICAL LIMB ISCHEMIA

PETER BELL

Critical limb ischemia (CLI) is an increasing problem as society ages, affecting between 40-80 people per 100 000 population [1]. CLI usually refers to the chronic form of ischemia, which progresses fairly slowly and usually gives adequate time for investigation. Certain questions need to be answered before a rational program of treatment can be instituted. The most important of these is whether or not there is an acceptable runoff vessel below an occlusion which a reopening procedure or graft can resupply. Historically it has been the case that angiography was used to decide whether or not this was possible, but the use of other tests such as dependent doppler and pulse generated runoff (PGR) [2] have shown that angiography is not always required and can also be wrong. Many legs have been amputated unnecessarily for this reason. The advent of newer techniques such as digital subtraction angiography (DSA) and more recently magnetic resonance angiography have meant that if peripheral vessels are present, they can usually be seen [3]. However, these techniques are invasive and expensive. For these reasons it has increasingly been the case that other techniques of assessing whether or not such patients can be treated have evolved. As mentioned earlier, dependent doppler and PGR have been used for this purpose and dependent doppler can still be useful.

■ DUPLEX SCANNING

More recently, color-coded duplex scanning is being increasingly used instead of angiography prior to treatment, but not everyone agrees that duplex scanning is accurate enough to replace angiography for the treatment of CLI.

A number of publications now indicate that duplex scanning can be used not only to successfully image proximal arteries (such as the iliac and femoral arteries) without too much difficulty, but also to examine the crural vessels [4,5]. These vessels remain a problem in some centers, which is important as critical ischemia is caused most frequently by disease of the crural vessels (Fig. 1). Less frequently, proximal disease can cause critical ischemia usually by affecting multiple levels, in which case the proximal lesion can be seen and dealt with successfully, often alleviating the problem. However, the major problem is usually located in the crural vessels and therefore the importance of duplex in this area becomes a vital issue. Duplex scanning, as everyone agrees, is very much dependent upon the operator. In centers where the operators are experienced and do many such scans, the sensitivity and specificity of the examinations are excellent and provide a working alternative to angiography [6,7]. However, in centers where the scanning techniques are not as good, then regular checks and quality assurance programs are necessary to maintain those standards. In the right hands excellent pictures can be obtained (Fig. 2). In our institution we have used this technique for the last 5 years to the total exclusion of angiography which is only used now for therapeutic reasons.

When a patient is referred with critical ischemia, duplex scanning alone provides an additional advantage in that the patient can be diagnosed and dealt with after a single visit to the clinic without having to come back for an angiogram or suffering the invasion caused by that procedure which has the added advantage of reducing costs. Examination of the patient by the clinician can direct the technologist to the appropriate examination site, although it is wise for the entire arterial tree from the iliac vessels down to the foot to be examined properly. If the iliac arteries cannot be seen easily, bowel preparation will make this possible, but we find this is usually unnecessary. After the scan the technologist completes a report form (Fig. 3) which allows us to proceed directly to treatment without any other diagnostic method being used.

In our practice, angioplasty (usually subintimal angioplasty) is the treatment of choice [8] (Fig.4), and if suitable the patient has an angioplasty without further investigations. This has the advantage that the radiologist spends his entire time in therapeutic activity. An additional advantage is that treatments can be planned and localized to particular areas without the need for extensive angiography. If angioplasty is not possible, the surgeon can approach the patent vessel indicated by the technologist in the knowledge that the inflow is adequate. This has usually been assessed by duplex prior to the patient going to theatre. In addition, the duplex scan has been used to mark out the vein and indicate whether it is suitable prior to surgery allowing adequate preoperative planning. This policy started after we had assessed 1658 arterial segments in 148 limbs [9]. We studied these patients using color duplex and digital angiography. In that study individual arterial segments were classified on the basis of peak systolic velocity. Ratios of < 2 were classified as 0–49% stenosis and > 2 as 59–99% stenosis. The third category was an absent signal indicating an occlusion. The same arterial segments were classified

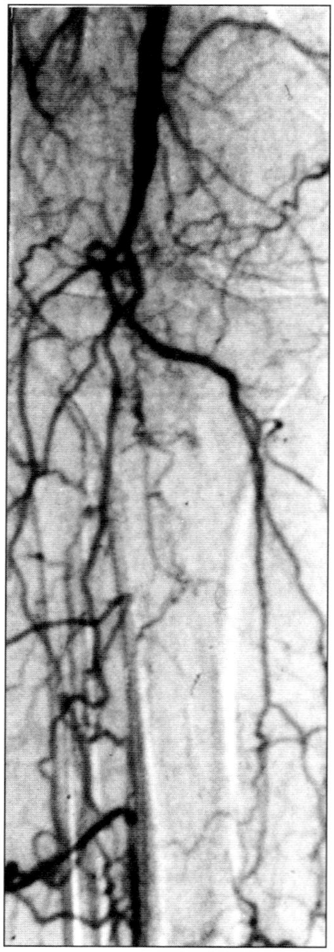

FIG. 1 An angiogram showing crural disease as a cause of critical ischemia.

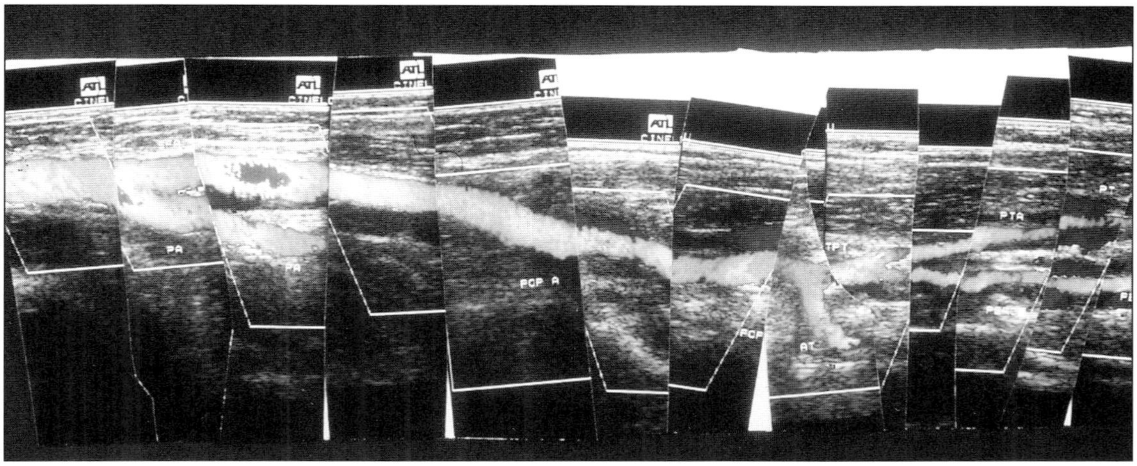

FIG. 2 Duplex scan showing the vessels of the lower limb.

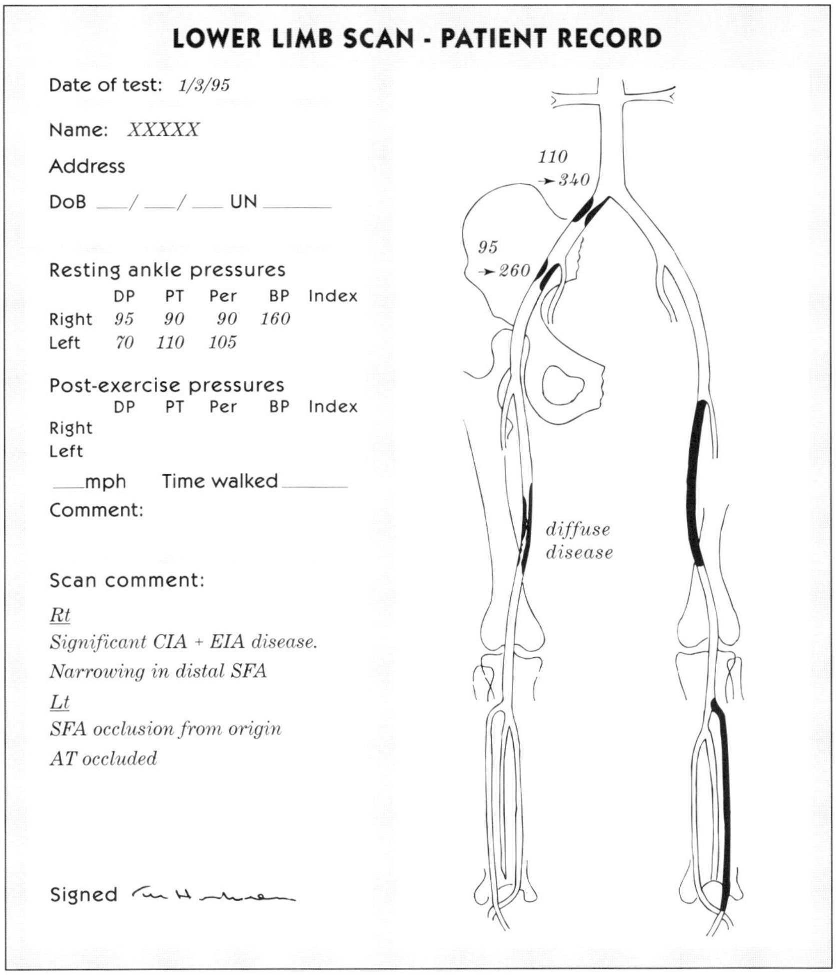

LOWER LIMB SCAN - PATIENT RECORD

Date of test: *1/3/95*

Name: *XXXXX*

Address

DoB ___/___/___ UN ______

Resting ankle pressures

	DP	PT	Per	BP	Index
Right	*95*	*90*	*90*	*160*	
Left	*70*	*110*	*105*		

Post-exercise pressures

	DP	PT	Per	BP	Index
Right					
Left					

___mph Time walked ______

Comment:

Scan comment:

Rt
Significant CIA + EIA disease.
Narrowing in distal SFA
Lt
SFA occlusion from origin
AT occluded

Signed

FIG. 3 The technologist completes a form which illustrates the area of occlusion or stenosis and outlines the distal patent segments.

using arteriography on a blind basis by two vascular radiologists. The overall agreement in this study was excellent with a k-value of 0.74, which allowed us to proceed and use this technique routinely in our clinical practice.

In addition we retrospectively examined the infrapopliteal vasculature in 51 patients using color duplex and digital angiography. A total of 204 arterial segments were available for analysis and each was graded as before. Two radiologists separately assessed the extent of the disease without knowing what the duplex findings were. At the conclusion of the study we found that agreement between duplex scanning and angiography never fell significantly below levels achieved between the two radiologists looking at the angiographic pictures alone. We concluded that duplex scanning could be used as effectively as angiography in below-knee situations [9].

■ DUPLEX SCANNING PRIOR TO ANGIOPLASTY

Once the patient has had a duplex scan performed and the segment that needs treatment identified, angioplasty (usually subintimal) can then take place. Over a 12-month period, 467 consecutive lower

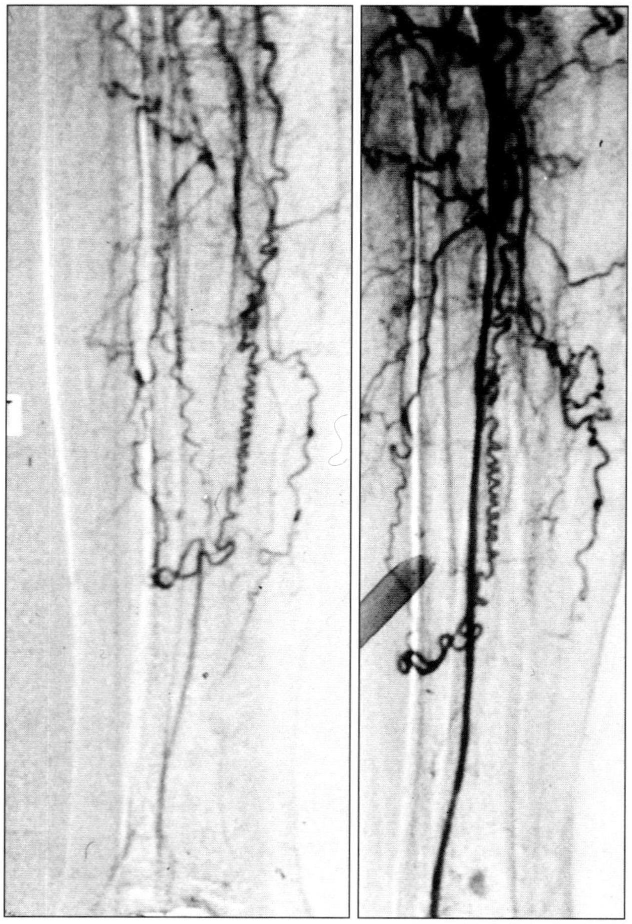

FIG. 4 Subintimal angioplasty of a crural obstruction diagnosed by duplex.

limb duplex scans were performed of which 437 were technically adequate. Of these patients, 39% were managed conservatively because they had claudication and 49% (230) were referred for percutaneous transluminal angioplasty, with the remainder being treated conservatively. In those patients referred for PTA in only 22 [10%] were there unexpected findings. There was agreement about the length of superficial femoral artery occlusion in 89%, and the presence or absence of a superficial femoral artery stump in 85%. Overall the accuracy of the procedure in therapeutic terms was found to be acceptable in 97% of cases [5].

■ PATIENTS TREATED BY SURGERY

If angioplasty is not thought to be possible either because of calcification or a recent thrombus, to give two examples, operation is the treatment of choice. A distal graft is usually performed and prior to that the long saphenous vein is assessed by duplex scanning to make sure it is acceptable and available. If it is not, the opportunity is taken to scan other veins such as the short saphenous or the arm veins so that the patient's vein situation is known before the operation starts. With this knowledge the patient is taken to the operating theatre and the artery shown to be patent on the duplex scan is exposed, usually one of the crural vessels. When this has been done, a green butterfly needle or a cannula is inserted into the artery and an on-table angiogram performed. This is best done using a C-arm with DSA capability, but if this machine is not available an X-ray can be performed using an x-ray plate wrapped in a sterile towel and an ordinary X-ray machine. If the picture shows continuity of the vessel to the foot nothing else needs to be done except for the operation to proceed. If there is some doubt about continuity, it might be reasonable to perform on-table pressure measurements to assess runoff before making a final decision about operability [10]. Using this technique we retrospectively reviewed all lower limb vascular reconstructions over a 2-year period. Eighty-five limbs underwent vascular reconstruction based on color duplex alone. Seventy percent of these patients were suffering from infrapopliteal disease and 29 of them had crural disease alone. In all infrapopliteal cases the duplex scan compared favorably to the on table angiogram, there being no false predictions based on duplex. The conclusion from this study was that duplex could be used with confidence prior to surgery [11].

Using this technique we now treat over 60% of our cases by primary angioplasty and relatively few with distal surgery. If, however, the stenosis cannot be dealt with in this way then surgery is undertaken without an angiogram. The sequence of events following admission for critical ischemia would be as follows: after duplex scanning a decision is made as to whether or not angioplasty should be undertaken. In our institution more than 65% of patients with critical ischemia can be dealt with in this way. If surgery is decided upon, the patient then has a long saphenous vein scan; if this is not suitable the short saphenous and arm veins are scanned and marked to provide the surgeon with an easy access to suitable venous material. The patent artery beyond the occlusion is next explored and an on-table DSA angiogram performed to ensure that there is continuity between this vessel and the arch. If there is no continuity it is reasonable to do a peripheral resistance measurement in the first instance. Postoperatively, the patient's vein graft is scanned using duplex to ensure that a stenosis does not occur. If it does then the patient's vein can be dilated by angioplasty. We have never found in any of our operated cases that the duplex has been wrong.

■ CONCLUSION

Duplex scanning is therefore an excellent technique for the investigation of patients with critical ischemia. Angioplasty can follow immediately after the duplex scanning and is suitable for over 65% of cases in our own series. If not suitable because fresh thrombus is present or calcification is a problem, then surgery can be undertaken on the basis of duplex scanning alone. This approach reduces the costs of treatment of these cases as well as the complications. The additional benefit is that veins can be identified by the scan at the same time and early postoperative surveillance will tell the surgeon whether or not the graft is patent. Angiography is no longer required prior to the treatment of patients with critical ischemia; duplex scanning suffices in most instances.

REFERENCES

1 Stranden E, Myhre O. Intraoperative diagnostics in critical limb ischaemia. *Critical Ischaemia* 1994; 4: 45-52.
2 Currie IC, Wilson YG, Davies AH et al. Pulse generated run off versus dependent doppler ultrasonography for assessment of calf vessel patency. *Br J Surg* 1994; 81: 1448-1450.
3 Mulligan SA, Matsuda T, Lanzer P et al. Peripheral arterial occlusive disease: prospective comparison of MR angiography and color duplex US with conventional angiography. *Radiology* 1991; 178: 695-700.
4 Wilson YG, George JK, Wilkins DC, Ashley S. Duplex assessment of runoff before femoro-crural reconstructions. *Br J Surg* 1997; 84: 1360-1363.
5 Pemberton M, Nydahl S, Hartshorne T et al. Colour-coded duplex scanning can safely replace diagnostic arteriography in patients with lower-limb arterial disease. *Br J Surg* 1996; 83: 1725-1728.
6 Koelemay MJ, Legemate DA, de Vos H et al. Can cruropedal colour duplex scanning and pulse generated runoff replace angiography in candidates for distal bypass surgery. *Eur J Vasc Endovasc Surg* 1998; 16: 13-18.
7 Karacagil S, Lofberg AM, Granbo A et al. Value of duplex scanning in evaluation of crural and foot arteries in limbs with severe lower limb ischaemia. A prospective comparison with angiography. *Eur J Vasc Endovasc Surg* 1996; 12: 300-303.
8 Sensier Y, Hartshorne T, Thrush A et al. A prospective comparison of lower limb colour-coded duplex scanning with arteriography. *Eur J Vasc Endovasc Surg* 1996; 11: 170-175.
9 Sensior Y, Fishwick G, Owen R et al. A comparison between colour duplex ultrasonography and arteriography for imaging infrapopliteal arterial lesions. *Eur J Vasc Endovasc Surg* 1998; 15: 44-50.
10 Parvin SD, Evans DH, Bell PRF. Peripheral resistance measurement in the assessment of severe peripheral vascular disease. *Br J Surg* 1985; 72: 751-753.
11 Pemberton M, Nydahl S, Hartshorne T et al. Can lower limb vascular reconstruction be based on colour duplex imaging alone? *Eur J Vasc Endovasc Surg* 1996; 12: 452-454.

8

DUPLEX SCANNING AND ANGIOGRAPHY OF THE LOWER LEG AND FOOT ARTERIES

MARK KOELEMAY, DINK LEGEMATE, JIM REEKERS, MICHAEL JACOBS

Patients presenting with critical limb ischemia tend to have extensive disease of the lower leg arteries. The long-term results of endovascular interventions or surgery for limb salvage justify an aggressive approach in patients with severe lower extremity ischemia. Before planning an intervention, accurate information must be obtained on the anatomy and function of the cruropedal outflow tract. Angiography is the traditional gold standard *for evaluation of the outflow vessels to the ankle and foot. Over the past 10 years the role of non-invasive techniques for assessment of the arteries of the lower extremity has become more important, mainly due to the developments in duplex ultrasonography and magnetic resonance angiography (MRA). This chapter provides a brief overview of angiography for assessment of the outflow vessels to the ankle and foot, and focuses on the role of duplex scanning in this respect.*

■ ANGIOGRAPHY

Conventional angiography is generally accepted as the routine examination before angioplasty or peripheral bypass surgery. The success of endovascular and surgical revascularization procedures below the level of the trifurcation requires adequate visualization of the very distal arteries. This poses a challenge for the radiologist because the technical success of angiography may be limited by a number of factors. Some of these are intrinsic; visualization of very distal vessels may be impaired by poor cardiac function or the administration of limited amounts of contrast agent in patients with renal insufficiency. Darcy [1] has written a comprehensive review on techniques to improve the accuracy of diagnostic angiography. These were divided in ways to deliver more contrast agent to the peripheral arteries and methods to improve the detection of contrast agents. Several techniques have been proposed to improve contrast delivery. Adjunctive procedures such as reactive hyperemia [2,3], pharmacological peripheral vasodilatation with tolazoline, nitroglycerin or prostaglandin E_1 [3] and angiography with balloon occlusion of the contralateral leg have been successfully applied. Antegrade selective catheterization of the superficial femoral artery also results in better depiction of runoff vessels compared to unselective digital subtraction angiography (DSA), as was demonstrated by Jacob et al. [4] in a randomized trial. The application of a stepping table and delayed filming has been proposed to improve contrast detection with screen film angiography. The development of intraarterial digital subtraction techniques (iaDSA) has provided the opportunity to prolong acquisition time. This is helpful in case of slow or retrograde flow. Sniderman et al. [5] found that iaDSA visualized more patent distal vessels than conventional angiography and reduced the contrast load. Disadvantages of iaDSA are small field size, susceptibility to movement artifacts and a lower spatial resolution than screen film studies [1].

Taking into account these factors, adequate arteriograms can be obtained and treatment can subsequently be planned. While some question the accuracy of iaDSA and advocate intraoperative angiography before bypass surgery, there is no convincing evidence from the literature that this may improve results. Three recent studies compared iaDSA with intraoperative or post-PTA angiography. In a multicenter trial by Baum et al. [6] iaDSA was evaluated in 117 patients with predominantly rest pain or tissue loss (82%) and compared with intraoperative angiography [6]. Sensitivity was 69% and specificity 94% for identification of normal tibial and pedal arterial segments. They found significant variation over the different participating centers. Oser et al. reported on a selected group of 86 patients with predominantly rest pain or ischemic tissue loss. iaDSA had a sensitivity for detection of patent named vessels of 95% and a 92% specificity compared to post-PTA angiography [7]. In 28/29 (97%) patients with a surgical exploration, iaDSA correctly identified the optimal recipient artery. Sayers et al. compared iaDSA and intraoperative angiography in 50 critically ischemic limbs in 45 patients before distal bypass surgery [8]. The positive predictive value of iaDSA for a feasible reconstruction to a tibial artery was 100% with only 74% negative predictive value. In the 39 patients correctly predicted to be candidates for operation by iaDSA there was one single discrepancy in selecting the optimal recipient artery. They concluded that iaDSA was suitable to predict a reconstruction but not to exclude the possibility of revascularization. In summary, adequate information can be obtained by a technically successful diagnostic iaDSA.

■ DUPLEX SCANNING

Duplex scanning (DS) is an established non-invasive technique for evaluation of the aortoiliac and femoropopliteal arteries. The B-mode provides anatomical information, whereas the doppler mode gives a functional assessment of the vascular tree. The Peak Systolic Velocity (PSV) ratio, calculated as the PSV in the stenosis divided by the PSV in the pre- or poststenotic region, has proven to be a simple and accurate parameter to quantify the hemodynamic significance of vascular lesions. This parameter is directly related to changes in luminal cross-sectional area and is independent of flow velocity. In a metaanalysis DS proved to have a sensitivity of 80-86% and over 95% specificity for detection of hemodynamically significant stenoses or occlusions in the aortoiliac and femoropopliteal arteries [9]. The addition of color doppler has significantly improved the diagnostic accuracy and has probably reduced the examination time [10]. From the same metaanalysis it became clear that the infragenicular arteries had not been studied extensively enough to appreciate the actual diagnostic accuracy of DS within these vessels [9]. Recently, however, more studies have reported on experience with DS of the infragenicular and pedal arteries. Because DS can detect very low flow velocities, it carries the potential to detect more patent arteries than iaDSA in patients with multilevel disease.

Scanning Technique

Duplex scanning of the infrapopliteal vessels requires ample experience of the vascular technologist. The diameter of the crural and pedal arteries may not be more than 2-3 mm in patients with critical ischemia. Moreover, significant superficial femoral artery disease will be present in the majority of these patients with subsequent low flow velocities. Finally, at some places the artery is located directly underneath the skin. Pressing too hard with the probe may mimic an occlusion in an actually patent artery. There is no standard position in which to examine the patient. The anterior tibial artery is nearly always examined with the patient supine. Examination of the tibial-peroneal trunk, posterior tibial and peroneal arteries is preferred in the prone or lateral decubitus position, whereas others keep the knee slightly flexed with the patient in the supine position. In case of very slow flow, examination of the tibial arteries with the patient in the upright position can be used for flow augmentation [11].

Typically, the infragenicular arteries are insonated at a 30-60 degree angle with 5.0-7.5 MHz linear array transducers and identified by the presence of a color signal or the vessel wall in case of an occlusion. The accompanying veins can be used as a road map for each of the arteries if their identification is difficult. The technologist must be careful to control the depth of the sample in order not to confuse the posterior tibial and peroneal arteries. We prefer to start the examination at the level of the ankle and follow each tibial artery into its origin from the tibial-peroneal trunk or popliteal artery. Downward from the ankle, the common plantar, dorsal pedal, and deep plantar artery can be examined at the level of the foot. We do not routinely study the entire lateral and medial plantar arteries.

Feasibility

Table I provides a summary of all studies on crural and pedal artery duplex scanning published until September 1998. The study populations comprise predominantly patients with rest pain and ischemic tissue loss. This is the typical category of patients for a complete examination of the lower leg. In patients referred for claudication examination of the crural and pedal arteries is generally not necessary because here DS will often reveal significant inflow disease.

A technically adequate DS examination is feasible in the majority of patients. Caster et al. reported a success rate of 94%, 96% and 83% for the anterior tibial, posterior tibial and peroneal artery, respectively [13]. This difference was statistically significant. The more laborious examination of the peroneal artery has been confirmed in other studies [11,15-17] and can be explained by the fascial borders surrounding the peroneal artery and its deeper anatomical plane. Here it is interesting to note that the peroneal artery shows the highest interobserver variability on angiography readings too [11,17] The presence of severely calcified arteries in diabetic patients may also obstruct adequate assessment. Not all studies recorded technical success rates of DS examination. From available data it can be noted that technical success rates range from 90 to 100% of all examinations.

DS of the crural arteries is a time consuming investigation. Table I lists the time needed for a complete crural examination. The 10-20 minutes mentioned by Koennecke et al. [12] and Langholz et al. [15] seems somewhat optimistic compared to our own experience. Koennecke et al. [12] used waveform analysis at the ankle level to infer disease and did not scan the crural arteries over their entire length. This is likely to explain the short examination time. We think that one should allow 45 minutes for a complete lower leg examination.

Criteria for Quantification of Disease

Table I lists the criteria applied by different investigators for quantification of disease. With the exception of Hatsukami et al. [14] all authors defined occlusion of a vessel if no flow could be detected with color doppler imaging within the lumen. Hatsukami et al. [14] added the presence of collateral vessels as a prerequisite to decide that the vessel was actually occluded. In case no collaterals could be detected DS was considered non-diagnostic. Six authors applied the standard PSV ratio > 2.0 to identify a 50% stenosis. Wilson et al. [19] reported the use of standard PSV ratio criteria but were not specific

Table I SUMMARY OF ALL STUDIES ON LOWER LEG DUPLEX SCANNING PUBLISHED UNTIL SEPTEMBER 1998

1st author [ref.]	*Year*	*N*	*Population CI/CLI/O %*	*Criteria*	*Non-diag %*	*Time (min) (Range)*
Sensier [11]	1998	51	49/47/4	PSV ratio > 2.0	0	-
Koennecke [12]	1989	53	NA	Spectral broadening	-	10-20
Caster [13]	1992	100	17/68/15	Presence of flow	9	-
Hatsukami [14]	1992	29	NA	Algorithm*	6	150
Langholz [15]	1993	31	0/100/0	PSV ratio > 2.0	-	10
Karacagil [16]	1996	40	12/88/0	PSV ratio > 2.0	3	45 (30-60)
Koelemay [17]	1997	23	9/81/0	Semiquantitative**	2	-
Larch [18]	1997	50	54/46/0	PSV ratio > 2.0	-	-
Wilson [19]	1997	43	0/100/0	NA	-	-
Koelemay [20]	1998	120	16/84/0	Semiquantitative**	2	45 (20-90)
Aly [21]	1998	90	90/10/0	PSV ratio > 2.0	-	-
Ligush [22]	1998	36	0/100/0	PSV ratio > 2.0***	-	30 (20-55)

N: number of patients CI: claudication CLI: critical limb ischemia O: other NA: not available
Criteria: criteria for significant lesions Non-diag: non-diagnostic Time (min): examination time in minutes
* See text
** 2: no or minor vessel wall irregularities seen with the B-mode and fully patent lumen on color flow
1: severe vessel wall irregularities (B-mode), diffuse intraluminal narrowing or isolated subtotal stenosis (doppler)
0: occlusion (no flow)
*** With the addition of color flow parameters, grading into: patent without stenosis, patent with stenosis, occluded

about its cut-off point. Koennecke et al. [12] used spectral broadening of the waveform recorded at the level of the ankle to indirectly quantify the degree of stenosis in the entire artery. The algorithm used by Hatsukami et al. [14] defined a < 50% stenosis if the waveform showed triphasic flow. In case of no triphasic signal and the presence of poststenotic turbulence the lesion was deemed to represent a 50-99% stenosis. Ligush et al. [22] added disturbances in color flow to a PSV ratio > 2.0 to grade arteries as patent without stenosis, patent with stenosis and occluded. In our vascular laboratory we apply a comparable semiquantitative scoring system combining B-mode and doppler parameters. Patency of the crural and pedal arteries is graded as 2 for no or minor vessel wall irregularities seen with the B-mode and fully patent lumen determined with color flow imaging; 1 for severe vessel wall irregularities (B-mode), diffuse intraluminal narrowing or isolated subtotal stenosis (Color and pulsed doppler, PSV ratio 3.0 to 4.0); and 0 for an occlusion.

DIAGNOSTIC ACCURACY

Table II lists the results of studies evaluating the diagnostic accuracy of DS compared to iaDSA as *gold standard.* The aims of the different studies showed somewhat variation. Caster et al. [13] aimed at DS for the detection of patent vessels. Of all arteries, 84% of angiographically patent arteries could be identified with DS. The low specificity of 72% in this case reflects the number of angiographically occluded arteries judged to be patent with DS. This phenomenon is recognized in nearly all studies and is reflected by the relatively low sensitivities for detection of occlusions in the five studies listed in table II. The sensitivities ranged from 64% to 94% and indicate that in a substantial number of cases DS judged an artery or arterial segment to be patent that did not show filling on the angiogram. The reverse occurs

Table II DIAGNOSTIC ACCURACY OF LOWER LEG DUPLEX SCANNING COMPARED TO ANGIOGRAPHY

1st author [ref.]	*Stenosis > 50%*		*Occlusion*		*Stenosis / occlusion*		*Patent*	
	Sensitivity %	*Specificity %*	*Sensitivity %*	*Specificity %*	*Sensitivity %*	*Specificity %*	*Sensitivity %*	*Specificity %*
Sensier [11]	20	98	70	90	-	-	-	-
Koennecke [12]	73	95	94	91	-	-	-	-
Caster [13]	-	-	-	-	-	-	84	72
Hatsukami [14]	-	-	87	100	78	79	-	-
Langholz [15]	-	-	-	-	-	-	-	-
Karacagil [16]	-	-	-	-	83	76	-	-
Koelemay [17]	-	-	73	83	-	-	-	-
Larch [18]	-	-	-	-	88	46	-	-
Wilson [19]	-	-	-	-	-	-	-	-
Koelemay [20]	-	-	64	88	-	-	-	-
Aly [20]	-	-	-	-	82	99	-	-
Ligush [22]	-	-	-	-	-	-	-	-

too, but less frequently. An explanation might be that the sensitive DS detects flow in an artery that cannot be filled with contrast agent due to reasons mentioned in the section on angiography. From our personal experience we must comment that the sensitive DS sometimes detects flow in what appears to be a strongly developed collateral on the angiogram (Fig. 1A-B-C).

Although not shown in detail here, the diagnostic accuracy for assessment of the peroneal artery is consistently lower than for the anterior and posterior tibial arteries [11,15,17,18,20]. However, as experience improves the peroneal artery may be examined with more confidence and accuracy.

From the literature exact criteria for quantification of hemodynamically significant stenoses cannot be derived. Only three studies aimed specifically at detection of stenoses >50% in the crural arteries. Koennecke et al. [12] used waveform analysis with reasonable results but did not scan the whole artery, Karacagil et al. [16] had too few patients with stenoses to appreciate the accuracy of PSV ratios and Sensier et al. [11] applied PSV ratios with disappointing sensitivity (20%). Our semiquantitative system cannot provide exact cut-off points to express the hemodynamic significance of crural artery lesions. However, the significance of a quantitative approach for assessment of vessels with 2-3 mm in diameter is debatable, both for assessment with DS and iaDSA. In practice, the vascular surgeon will use only fully patent vessels for bypass grafting.

MANAGEMENT

Considering the discrepancies between DS and iaDSA and their respective potentials and inadequacies, it can be decided that both diagnostic modalities can be used for clinical decision making. Experience with decision making based on DS as the sole diagnostic modality is limited and exists mainly of retrospective analysis. In the study by Wilson et al. [19] dependent doppler, angiography and DS were used to select the best crural artery for bypass surgery in 43 patients with critical ischemia. Selection with DS included conventional velocity criteria not otherwise specified and color flow images reflecting anatomical criteria such as caliber, calcification and continuity with the pedal arch. While angiography selected the best suitable runoff vessel in 32 cases, DS proved to be correct in all patients at operation. Ligush et al. [22] reported equivalent results for DS and angiography to guide infrainguinal

Links							
Arterie	CM	PSV	Ratio	EDV	Wand	PGR Voor	PGR Na
A. popl inf	- 6	37			1		
Truncus		30			1		
ATA prox		0			0		
mid							
dist		0			0	0	0
ATP prox							
	-25	30			1		
mid					2		
dist	-39	27			2	1	2
A. per prox		29			1		
mid		30			1	1	2
dist		27			1		
ADP		0			0		
RPP		0			0		
APC		25			2		

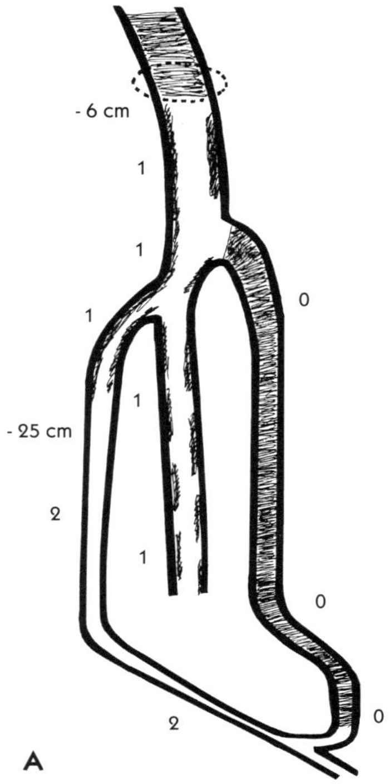

A

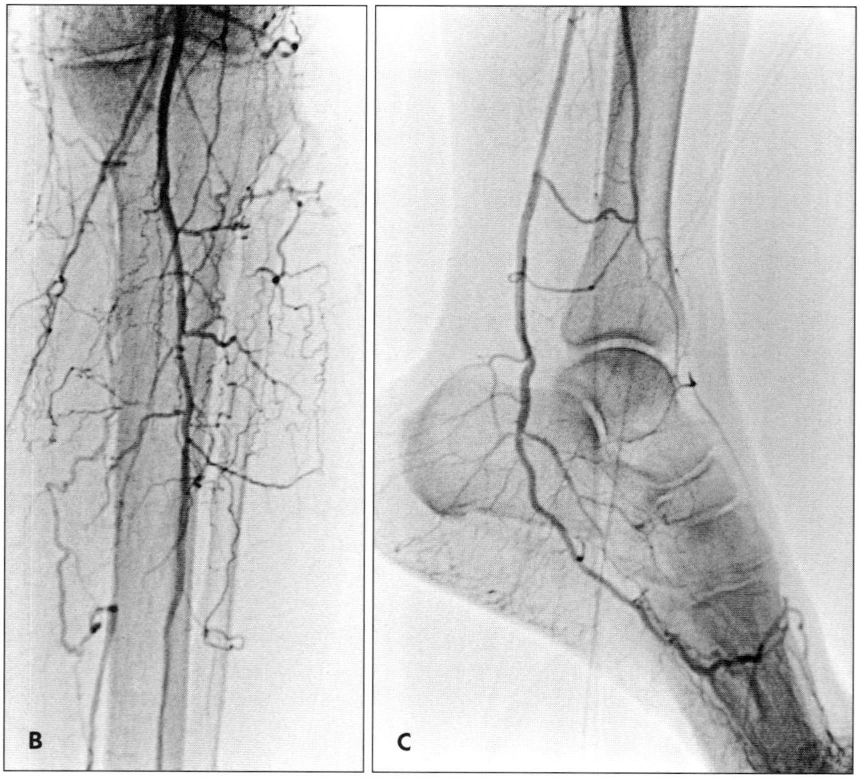

FIG. 1 Shows the duplex (A) and antero-posterior (B) and lateral (C) iaDSA of a patient with ischemic rest pain. The posterior tibial and common plantar arteries are fully patent on DS and iaDSA. DS detected a continuous posterior tibial artery, whereas iaDSA shows filling by collaterals half way the lower leg.

reconstructions in 36 patients with critical ischemia. DS was correct in 83% and angiography in 90% of revascularizations. In a retrospective analysis of our own data on 120 patients, we concluded that management decisions (conservative, PTA, surgery, or amputation) in a population with predominantly critical ischemia could be based on a work-up with DS and pulse generated runoff (PGR) [23] in 59% of the patients [20]. PGR was used to provide a functional assessment of the outflow to the foot. These decisions included 79% of the femoropopliteal bypasses and 41% of femorocrural reconstructions. Pemberton et al. [24] went one step further and, among other surgical reconstructions, reviewed the outcome of 44 infragenicular bypasses based on duplex scanning and dependent doppler sonography as the sole preoperative imaging modality. These patients represented a selected group of 34% of all operative revascularizations. Significant lesions were detected by a PSV ratio >2.0 and occlusions by the absence of flow. It must be noted that on-table prereconstruction angiography was performed in all patients to confirm the results of DS. However, during the operation there were no unexpected findings but in one patient.

In our hospital we prospectively studied 117 consecutive patients between April 1997 and September 1998 referred for severe claudication (24%) and ischemic rest pain or tissue loss (76%). Eleven patients had bilateral examinations. As standard policy, patients were evaluated with DS and PGR and management was based a priori on the non-invasive evaluation by the staff vascular surgeons and interventional radiologists. Diagnostic angiography was performed on indication only. With the results of our comparative study in mind, this meant that when a femorocrural bypass was considered diagnostic angiography was performed liberally. Definitive management comprised conservative treatment in 33 patients, all without diagnostic angiography. PTA was performed in 25 patients (4 after diagnostic angiography), surgery in 66 (27 after angiography), and primary amputation in 4 (all without angiography). Femoropopliteal bypass surgery was performed in 24/30 (75%) patients without diagnostic angiography. The popliteal segment in these patients was fully patent and there was at least one fully patent tibial artery to the level of the foot (Fig. 2). Femorocrural bypass surgery was performed in 11/28 (39%) patients with a fully patent single crural artery without diagnostic angiography. In conclusion, a 75% reduction in

LINKS

Arterie	CM	PSV	Ratio	EDV	Wand
A. popl inf		44			2
Truncus		70			2
ATA prox		+			Retro
mid	24	15			2
dist	32	15			2
ATP prox	16	30			2
mid	30	20			2
dist	42	25			2
A. per prox	16	60			2
mid	24	75			1
dist	29	0			0
ADP		10			2
RPP		10			2
APC		15			2

PGR	
Voor	Na

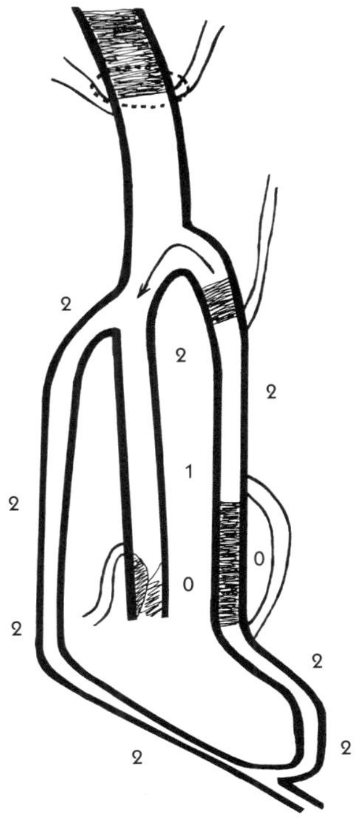

FIG. 2 Shows the duplex map of a patient with rest pain and an occluded SFA. The infragenicular popliteal artery is fully patent with a fully patent posterior tibial artery into the foot. Successful infragenicular femoropopliteal bypass surgery was performed without preoperative iaDSA.

diagnostic angiographies could be obtained with this strategy. Compared to the patients evaluated in the previous comparative study, there was no difference in in-hospital clinical outcome.

CONCLUSION

In the diagnostic work-up of patients with critical limb ischemia preinterventional information is required on the anatomy and function of the lower extremity outflow tract. Reliable information can be obtained both with iaDSA and non-invasive techniques such as duplex scanning. As the experience with duplex scanning is accumulating, the number of patients referred for diagnostic angiography can be reduced and patients can be directly scheduled for an endovascular or surgical intervention, provided that duplex expertise is available. This does not mean that angiography is to be supplanted by duplex scanning but that in patients with equivocal examinations both modalities can be used to supplement each other. Decisions regarding treatment should therefore be taken in cooperation by the vascular surgeon, interventional radiologist, and vascular technologist.

REFERENCES

1 Darcy MD. Lower extremity arteriography: current approach and techniques. *Radiology* 1991; 178: 615-621.
2 Feins RH, Roedersheimer LR, Baumstark AE, Green RM. Predicted hyperemic angiography: a technique of distal arteriography in the severely ischemic leg. *Surgery* 1981; 89: 202-205
3 Cohen M, Vogelzang RL. A comparison of techniques for improved visualization of the arteries of the distal lower extremity. *AJR* 1986; 147: 1021-1023.
4 Jacob AL, Stock KW, Proske M, Steinbrich W. Lower extremity angiography: improved image quality and outflow vessel detection with bilaterally antegrade selective digital subtraction angiography. A blinded prospective comparison with aortic flush digital subtraction angiography. *Invest Radiol* 1996; 31: 184-193.
5 Sniderman K, Morse SS, Strauss EB. Comparison of intra-arterial digital subtraction angiography and conventional filming in peripheral vascular disease. *J Can Assoc Radiol* 1986; 37: 76-82.
6 Baum RA, Rutter CM, Sunshine JH et al. Multicenter trial to evaluate vascular magnetic resonance angiography of the lower extremity. *JAMA* 1995; 274: 875-880.
7 Oser RF, Picus D, Hicks ME et al. Accuracy of DSA in the evaluation of patency of the infrapopliteal vessels. *J Vasc Interv Radiol* 1995; 6: 589-594.
8 Sayers RD, Naylor AR, London NJM et al. The additional value of intraoperative angiography in infragenicular reconstruction. *Eur J Vasc Endovasc Surg* 1995; 9: 211-217.
9 Koelemay MJW, den Hartog D, Prins MH et al. Diagnosis of arterial disease of the lower extremities with duplex ultrasonography. *Br J Surg* 1996; 83: 404-409.
10 Vries SO, Hunink MGM, Polak JF. Summary receiver operating characteristic curves as a technique for meta-analysis of the diagnostic performance of duplex sonography in peripheral arterial disease. *Acad Radiol* 1996; 3: 361-369.
11 Sensier Y, Fishwick G, Owen R et al. A comparison between colour duplex sonography and arteriography for imaging infrapopliteal arterial lesions. *Eur J Vasc Endovasc Surg* 1998; 15: 44-50.
12 Koennecke HC, Fobbe G, Hamed MM, Wolf KJ. Diagnostik Arterieller Gefasserkrankungen der unteren extremitäten mit der farbkodierten Duplexsonographie. *Rofo Fortschr Geb Rontgenstr Neuen Bildgeb Verfahr* 1989; 151: 42-46.
13 Caster JD, Cummings CA, Moneta GL et al. Accuracy of tibial artery duplex mapping. *J Vasc Technol* 1992; 16: 63-68.
14 Hatsukami TS, Primozich JF, Zierler RE et al. Color doppler imaging of infrainguinal arterial occlusive disease. *J Vasc Surg* 1992; 16: 527-531.
15 Langholz J, Stolke O, Behrendt Ch et al. Farbkodierte Duplexsonographie von Unterschenkelarterien - Darstelbarkeit in Zuordnung zu Fontainestadien. *Ultrschall Med* 1993; 14: 279-284.
16 Karacagil S, Lofberg AM, Granbo A et al. Value of duplex scanning in evaluation of crural and foot arteries in limbs with severe lower limb ischaemia - a prospective comparison with angiography. *Eur J Vasc Endovasc Surg* 1996; 12: 300-303.
17 Koelemay MJW, Legemate DA, van Gurp J et al. Colour duplex scanning and pulse-generated runoff for assessment of popliteal and cruropedal arteries before peripheral bypass surgery. *Br J Surg* 1997; 84: 1115-1119.
18 Larch E, Minar E, Ahmadi R et al. Value of color duplex sonography for evaluation of the tibioperoneal arteries in patients with femoropopliteal obstruction. A prospective comparison with anterograde intraarterial digital subtraction angiography. *J Vasc Surg* 1997; 25: 629-636.
19 Wilson YG, George KD, Wilkins DC, Ashley S. Duplex assessment of runoff before femorocrural reconstruction. *Br J Surg* 1997; 84: 1360-1363.
20 Koelemay MJW, Legemate DA, de Vos H et al. Can cruropedal duplex scanning and pulse-generated runoff replace angiography in candidates for distal bypass surgery ? *Eur J Vasc Endovasc Surg* 1998; 16: 13-18.
21 Aly S, Sommerville K, Adiseshiah M et al. Comparison of duplex imaging and arteriography in the evaluation of lower limb arteries. *Br J Surg* 1998; 85: 1099-1102.
22 Ligush J, Reavis SW, Preisser JS, Hansen KJ. Duplex ultrasound scanning defines operative strategies for patients with limb-threatening ischemia. *J Vasc Surg* 1998; 28: 482-491.
23 Beard JD, Scott DJA, Evans JM et al. Pulse generated runoff: a new method of determining calf vessel patency. *Br J Surg* 1988; 75: 361-363.
24 Pemberton M, Nydahl S, Hartshorne T et al. Can lower limb vascular reconstruction be based on colour duplex scanning alone ? *Eur J Vasc Endovasc Surg* 1996; 12: 452-454.

9

MEDICAL TREATMENT WITH PROSTANOIDS IN PATIENTS WITH CRITICAL ISCHEMIA

JEAN-LOUIS GUILMOT, CHRISTIAN BOISSIER

The classification of Leriche and Fontaine distinguish four stages of severity of lower limb arterial occlusive disease. One of the main drawbacks of this descriptive classification is the incomplete reflection of the severity of the arterial lesions and its pathophysiology. It also suggests a chronological sequence of stages, which in fact rarely exists. The term critical limb ischemia (CLI) was introduced some 10 years ago following the European Consensus meetings [1]. The recommended definition of CLI aimed for a more precise description of limb threatening ischemia also stating that inadequate treatment would inevitably lead to a major amputation. The definitions in the Consensus document did not solve all the problems, but they allowed to improve our comprehension of the pathophysiological mechanisms leading to tissue damage in critical ischemia.

PATHOPHYSIOLOGY OF CLI

In contrast to acute ischemia, the clinical manifestation of critical ischemia is more progressive and treacherous. The perfusion deficit caused by macrocirculatory obstructions is responsible for adaptive mechanisms at the level of the microcirculation in order to maintain adequate tissue perfusion. Whenever these adaptive reactions fail, jeopardized microcirculatory endothelium initiates an inappropriate defense mechanism with deliverance of vascoconstrictive substances and activation of blood cells. In the absence of microcirculatory recovery, the damage is irreversible and tissue ischemia will be definite. However, the rapidity of the development of irreversible damage is quite variable. It depends of the extend of macrocirculatory lesions and the degree of preexisting microcirculatory disease, for example caused by diabetes.

■ MEDICAL TREATMENT

The treatment of CLI consists of surgical revascularization or endovascular reopening procedures. Pharmalogical treatment only applies for those patients who face contraindications for revascularization or when an amputation is unavoidable after a failed procedure. Medical treatment is also justified in a certain number of patients in whom distal revascularization is not possible, either due to absence of recipient crural or pedal arteries or because no autologous vein is available. In the latter case synthetic grafts are not often used because of the poor results in distal bypass surgery. The potential number of patients for medical treatment is decreasing during the last decade due to the improved results of surgery and angioplasty [2]. Some 10 years ago it was estimated at 40% of patients with CLI, at present it is probably less than 20%. The European Consensus on CLI recommended to attempt a reconstructive procedure if there is a 25% chance of saving a useful limb for at least 1 year, based on local audited past experience.

The criteria of this reflection are not really objective and it has to be emphasized that the different therapeutic options for a patient with CLI can only be considered in medical centers experienced in surgical and endovascular techniques. This is important because the patient confronts an unacceptable risk if the indications for medical treatment are overestimated.

Acute limb ischemia is most often caused by a fresh thrombus in the crural arteries which can be adequately treated by local or general fibrinolysis. Vascular lesions in the major arteries are more complex, especially when associated with ulcerated plaques, atherosclerotic lesions and old thrombus. Consequently, a medical treatment which intends to ameliorate arterial lesions of the major arteries in patients with CLI is doomed to fail because the delay of the working mechanism to restore vascular perfusion will be too long. A medical treatment interfering at the macrocirculation, still to be invented, should be focussed on the prevention or aggravation of atherosclerotic lesions. In case of CLI, only a curative treatment carries the hope to have a chance of success. Microcirculatory disturbances secondary to major artery disease, however, are a target for medical treatment. Therefore it should be considered as a complimentary treatment to revascularization rather than an alternative.

■ ROLE OF PROSTANOIDS

Among all medication applied for CLI, the prostanoids and specifically prostacycline (PGI_2) are by far the most extensively studied pharmalogical substances. PGI_2 possesses the majority of necessary components to counteract the detrimental microcirculatory alterations associated with critical ischemia. Apart from a strong local vasodilatory action, PGI_2 has an antiplatelet activity which prevents or corrects the activation of platelets and leukocytes as well as their reciprocal interaction. Furthermore, PGI_2 inhibits the release of platelet growth factor and thus the myointimal hyperplasia, and restrains leukocyte induced endothelial damage. The short half-life of PGI_2, approximately 2 to 3 minutes, does not allow its therapeutic use. A stable prostacyclin analogue ilosprost (Ilomedine®), which has a half-life of 30 minutes, has been clinically used for more than 10 years. At present, Ilomedine is only available for intravascular administration and the oral form, currently in phase 3 studies, will probably be available in the year 2000. Sodium beraprost is another oral form of PGI_2 which has been studied in patients with intermittent claudication. To our knowledge it has not been applied in CLI. PGE_1 (alprostadil) is not commercially available in France for the treatment of patients with peripheral vascular disease. In Germany, however, PGE_1 has been extensively studied and now commercialized for the indication of severe lower limb arterial diseases.

SELECTION OF PATIENTS

The selection of patients who might benefit from medical treatment with prostanoids is crucial. Patients with irreversible ischemia, who are physically and morally weak, should not be treated with an expensive and prolonged medical treatment which only partially relieves the problem and delays an unavoidable amputation. The selection criteria are still not well defined and no consensus has been reached so far.

Several functional diagnostic tests can be used to assess the microcirculatory perfusion: transcutaneous oxygen pressure ($tcpO_2$) measurement, digital plethysmography, laser doppler and capillary microscopy. These techniques are commonly used but are limited in differentiating between nutritional flow and thermoregulatory flow. In patients with CLI the distribution of these two levels of microcirculatory flow are continuously disturbed.

Among the microcirculatory techniques, $tcpO_2$ is the most commonly used. If the transcutaneous pressure at the skin of the foot is less than 10 mmHg in supine position and remains less than 20 mmHg in leg dependency, most authors consider the ischemia irreversible and therefore propose to perform immediate amputation. However, other parameters, psychological arguments included, interfere with the clinical judgment and the skin oxygen pressure only contributes partially to the final decision. It should be emphasized that there is only a moderate correlation between the clinical signs and the hemodynamic parameters.

APPLICATION OF PROSTANOIDS

The present indication of prostanoid treatment for critical ischemia (France, September 1997) is as follows: *treatment of chronic severe lower limb ischemia in patients at risk for amputation in whom revascularization by surgery or angioplasty has failed or is not indicated, as judged by surgical and radiological expertise.*

Prostanoid treatment is only performed by intravenous infusion and requires daily administration for six hours for a period of 21 to 28 consecutive days. It is a quite onerous treatment because it demands in-hospital observation, strict supine position during infusion and continuous blood pressure monitoring. The most common and frequently (approximately 70% of patients) observed side effects are facial flush, headache, nausea and cramp-like abdominal pain. More serious side effects like arterial hypotension and angina have been reported in less than 50% of the patients. The maximal dosage of 2 ng/kg/mn should be progressively administered within 3 days.

■ GENERAL RESULTS

ILOPROST, A STABLE PGI2 ANALOGUE

The results of the main randomized clinical studies, comparing iloprost and placebo, are evaluated in the meta analysis of Loosemore et al. [3]. Between 1986 and 1989, six studies in six European countries were performed, including a total of 705 patients. All patients suffered from stage III and/or IV ischemia according to the classification of Leriche and Fontaine. The therapeutic protocol was similar in all studies except for the duration of treatment, which varied between 2 and 4 weeks (Table I). At the end of the study, all patients were classified as responders or non-responders, both in the iloprost and the placebo groups. The responders were patients who showed ulcer healing of at least 30% of ulcer surface and/or complete relief of pain. The results are summarized in Table II, showing that the number of responders was significantly higher in the iloprost group as compared to the placebo group.

At 3 months follow-up (French study) and at 6 months follow-up (Swedish and English studies), amputation rates were compared between both groups (Table III). Two studies showed a significant difference in favor of iloprost. Meta-analysis of the three studies confirmed a highly significant reduction of amputations ($p<0.001$) at 6 months in patients treated with iloprost. In two studies [6,7] mortality and major amputation were analyzed as a combination, showing its prevalence of 35% in the iloprost group (45/130) as compared to 55% in the placebo group (68/124). Patients in the iloprost group had a significantly higher chance ($p<0.001$) to be alive with both legs at 6 months follow-up as compared to patients treated with placebo.

Table I DATA ON CLINICAL ILOPROST STUDIES FOR CRITICAL LIMB ISCHEMIA

1st author Country [ref]	*Number of patients*	*Clinical stage*	*Treatment*	*Follow-up*
Diehm 1989 [4] *Germany*	101	IV	4 weeks	None
Brock 1990 [5] *Germany*	109	IV	4 weeks	None
Norgren 1990 [6] *Sweden*	103	IV	2 weeks	6 months
Bliss 1991 [7] *U.K.*	151	III and IV	2 - 4 weeks	6 months
Guilmot 1991 [8] *France*	128	III and IV	3 weeks	3 months
Balzer 1991 [9] *Germany*	113	III	2 weeks	1 months

Table II DATA ON RESPONDERS IN BOTH GROUPS AT THE END OF THE STUDY

1st author [ref.]	*Responders Iloprost group (%)*		*Responders Placebo group (%)*		*Significance*
Diehm 1989 [4]	32/53	(60)	8/48	(17)	p < 0.05
Brock 1990 [5]	31/56	(55)	12/53	(23)	p < 0.05
Norgren 1990 [6]	15/50	(30)	7/53	(13)	NS
Bliss 1991 [7]	30/80	(38)	20/71	(28)	NS
Guilmot 1991 [8]	47/87	(54)	15/41	(37)	p < 0.05
Balzer 1991 [9]	30/55	(55)	23/58	(40)	p < 0.05
Total	**185/381**	**(49)**	**85/324**	**(26)**	**p < 0.05**

9

66

Table III AMPUTATION RATE IN THREE EUROPEAN STUDIES

1st author [ref.]	*Amputations Iloprost group (%)*		*Amputations Placebo group (%)*		*Significance*
Norgren 1990 [6]	13/50	(26)	23/53	(43)	p < 0.05
Bliss 1991 [7]	23/80	(29)	33/71	(46)	p < 0.01
Guilmot 1991 [8]	11/87	(13)	8/41	(20)	NS
Total	**49/217**	**(22.6)**	**64/165**	**(38.8)**	**p < 0.001**

NS: not specified

PGE_1

In the first clinical studies, PGE_1 has been administered intraarterially in order to avoid pulmonary metabolism. This rather invasive technique, in which atherosclerotic diseased arteries are punctured, has never been performed in France and has now also been abandoned by others. The clinical studies with PGE_1 by venous access have been performed before those with PGI_2. These studies only evaluated patients with stage III arterial disease or compared its clinical value with pentoxifylline, which cannot be considered as an active reference treatment in critical ischemia. We therefore do not analyze these results.

More recently a clinical study was performed comprising 267 patients with skin ulcers, with and without diabetes. Intravenously administered PGE_1 during 4 hours was compared with 6 hours intravenous iloprost per day for a period of 21 to 28 days [10]. Relief of pain and ulcer healing were the main end points. According to these criteria, 52.7% of the patients successfully treated (responders) belonged to the iloprost group and 43.1% to the PGE_1 group ($p=0.148$). The number of responders was similar among patients with and without diabetes in the iloprost group: 53.3% versus 51.4%. In the PGE_1 group the number of diabetic responders was less: 36.6% versus 53.3%. At 6 months follow-up 62.2% of patients were alive without amputation in both groups. Amputation rate was slightly higher in the iloprost group (32.1% versus 27.2%) but the number of deaths was 50% lower in the iloprost group (7.5% versus 14.6%, $p=0.10$). Side effects like headache, facial flush, nausea, vomiting and diarrhea occurred significantly more frequently in the iloprost (73.9%) than in the PGE_1 (31%) group ($p<0.05$).

The limited number of patients, however, does not allow to proof a significant difference between PGE_1 and iloprost but it shows a tendency in favor of iloprost with respect to the clinical efficacy in the treatment of stage IV arterial disease, especially in diabetic patients.

■ THROMBOANGIITIS OBLITERANS (BUERGER'S DISEASE)

Thromboangiitis obliterans or Buerger's disease is an inflammatory, non-atherosclerotic arterial disease, affecting young patients before the age of 40 years, which clinically manifests with severe ischemic symptoms: rest pain, digital ulcerations, and distal gangrene. The exact etiology still remains unknown but the role of tabacco in the cause of aggravation of the lesions is preponderant. The majority of these patients has a history of heavy cigarette smoking. Buerger's disease can be differentiated from the atherosclerotic arterial disease of a young subject by the more dismal functional prognosis but by a better vital prognosis, similar to non-atherosclerotic subjects with the same age.

The inflammatory aspect of the arterial lesions and the distal appearance of the disease explain why only a few patients might benefit from a surgical or endovascular revascularization. The medical treatment of thromboangiitis obliterans patients with iloprost has been evaluated in a European study [11], showing very beneficial clinical results. This double blind study comparing iloprost and aspirin comprised 133 patients with critical ischemia of whom 98 suffered from ulceration or distal gangrene. After 3 weeks of treatment, 85% of patients medicated with iloprost improved with partial or complete ulcer healing or relief of pain in contrast to only 17% in the aspirin group. Complete ulcer healing occurred in 35% of patients in the iloprost group as compared to 13% in the aspirin group. At 6 months follow-up the beneficial results maintained in 88% of the iloprost group against 21% in the aspirin group.

After completion of the study, iloprost was officially approved in France in 1991 for the indication of Buerger's disease. The treatment offers a considerable benefit to these patients, facilitating ulcer healing in the majority of cases and reducing the necessity for repetitive amputations. However, these beneficial results are only achieved or maintained if the patients completely refrain from smoking.

■ THE FUTURE OF THE PROSTANOIDS

The results of the clinical studies with prostanoids for the treatment of critical ischemia are disappointing. By all means it is obvious that a significant benefit is achieved in the patients who are still alive with both legs at 6 months follow-up. Although the results appear positive the benefit remains insufficient. Therefore, several questions merit asking.

Did the patients who were enrolled in the clinical studies already have too advanced a stage of arterial disease-in other words, did a considerable number of patients have irreversible ischemic lesions? It is difficult to answer this question without having addressed this particular issue in clinical studies. However, it can be stated that the results in the stage III patients [7,9] are not better than in stage IV patients. Should the medical treatment be complementary to revascularization techniques? An open multicenter study was performed in 713 patients who already underwent a below knee amputation. Intravenously administered iloprost during 3 weeks and conventional treatment was compared with conventional treatment only [12]. The results were disappointing since there were no differences at 6 months follow-up with respect to wound healing of the stump and the number of additional amputations. At present there are no relevant clinical studies reporting on the influence of prostanoids on distal bypass patency and its functional outcome.

Should the protocol of 3 to 4 weeks intravenous treatment be changed or extended in order to obtain better results? The answer should be provided shortly by the results of a multicenter study performed in Europe and USA with an oral form of iloprost. If this therapy is effective it will also limit the hospital stay considerably and therefore reduce the costs of the medical treatment of patients with critical limb ischemia.

CONCLUSION

The prostanoids constitute a modest but indisputable therapeutic improvement in patients with critical limb ischemia. Although the obtained results did not come up to the initial expectations, it is not acceptable to reject a treatment which allows to save a lower leg in approximately 15% of the patients [13]. The cost/benefit analysis merits regular evaluation in order to assess the demanding medical policy in patients with critical limb ischemia.

REFERENCES

1 Anonymous. Second European Consensus Document on Chronic Critical Leg Ischemia. *Circulation* 1991; 84 (suppl 4): 1-26.
2 Bull PG, Mendel H, Hold M et al. Distal popliteal and tibioperoneal transluminal angioplasty: long term follow-up. *J Vasc Interv Radiol* 1992; 3: 45-53.
3 Loosemore TM, Chalmers TC, Dormandy JA. A meta-analysis of randomized placebo control trials in Fontaine stages III and IV peripheral occlusive arterial disease. *Int Angiol* 1994; 13: 133-142.
4 Diehm C, Abri O, Baistsch G et al. Iloprost a stable prostacyclin derivative, in stage 4 arterial occlusive disease. A placebo-controlled multicenter study. *Dtsch Med Wochenschr* 1989; 114: 783-788.
5 Brock FE, Abri O, Baitsch G et al. Iloprost in the treatment of ischemic tissue lesions in diabetics. Results of a placebo-controlled multicenter study with a stable prostacyclin derivative. *Schweiz Med Wochenschr* 1990; 120: 1477-1482.
6 Norgren L, Alwmark A, Angqvist KA et al. A stable prostacyclin analogue (iloprost) in the treatment of ischaemic ulcers of the lower limb. A Scandinavian-Polish placebo controlled, randomised multicenter study. *Eur J Vasc Surg* 1990; 4: 463-467.
7 U.K. Severe Limb Ischaemia Study Group. Treatment of limb threatening ischaemia with intravenous iloprost: a randomised double blind placebo controlled study. *Eur J Vasc Surg* 1991; 5: 511-516.
8 Guilmot JL, Diot E. Treatment of lower limb ischaemia due to atherosclerosis in diabetic and nondiabetic patients with iloprost, a stable analogue of prostacyclin. Results of a french multicenter trial. *Drug invest* 1991; 3: 351-359.
9 Balzer K, Bechara G, Bisler H et al. Reduction of ischaemic rest pain in advanced peripheral arterial occlusive disease. A double blind placebo controlled trial with iloprost. *Int Angiol* 1991; 10: 229-232.
10 Altstaedt HO, Berzewski B, Breddin HK et al. Treatment of patients with peripheral arterial occlusive disease Fontaine stage IV with intravenous iloprost and PGE1: a randomized open controlled study. *Prostaglandins Leukot Essent Fatty Acids* 1993; 49: 573-578.
11 Fiessinger JN, Schafer M. Trial of iloprost versus aspirin treatment for critical limb ischaemia of thromboangiitis obliterans. The TAO study. *Lancet* 1990; 335: 555-557.
12 Dormandy J, Belcher G, Broos P et al. Prospective study of 713 below-knee amputations for ischaemia and the effect of a prostacyclin analogue on healing. Hawaii Study Group. *Br J Surg* 1994; 81: 33-37
13 Iloprost (extension d'indication). Pas assez bien évalué. *Rev Presc* 1998; 189: 739-741.

10

NON-PROSTANOID MEDICAL THERAPIES IN CRITICAL LIMB ISCHEMIA

LARS NORGREN

The true incidence of critical limb ischemia is not known in all details, but from amputation registries and other sources it seems reasonable to assume that this disease affects at least 500-1000 per million inhabitants per year. A general finding is that about 60% of those suffering from critical limb ischemia are suitable for an intervention: open surgery or an endovascular procedure. Some 15-20% may require an immediate amputation, while the remaining 20-25% do not show technical prerequisites for an intervention or have contraindications for such procedures. As they do not require an immediate amputation there should be a place for other, non-interventional therapies.

Some evidence exists for the beneficial use of prostanoids and with certain prerequisites there may be a place for spinal cord stimulation. These treatments are covered elsewhere in the book. The present chapter intends to discuss other medical therapies that include drug treatment but also other non-interventional treatments.

■ GENERAL TREATMENT

Among patients with critical limb ischemia, 30-40% also suffer from diabetes. Many of these patients do not have rest pain, but show their first sign of critical limb ischemia when peripheral gangrene or an ischemic ulcer appears. Foot care and control of the diabetic disease is therefore of greatest importance.

Also, hypertension has to be treated. It is sometimes questioned whether a high blood pressure is required to improve tissue perfusion. This is never the case; instead hypertension is an important part of the pathologic process and should therefore be treated. The role of smoking is evident and all attempts should be made to convince the patient to quit smoking. Rest pain, usually causes bad sleeping and pain relief shall always be attempted, although the ischemic pain may be very difficult to treat.

■ DRUG TREATMENT

Previously, vasodilatation was the goal of treatment and vasodilators were prescribed to patients with critical limb ischemia. Little effect was achieved and frequently patients even suffered more from to steal phenomena. This kind of treatment has no place whatsoever.

Presently, drugs to be used exert their mode of action on the microcirculation with the aim to reduce blood cell aggregation and adhesion, but possibly there is also an effect on microvessel dilatation. Local thrombolysis might be another approach, however, it is more frequently used in acute ischemia than in chronic critical limb ischemia.

Besides prostanoids (Chapter IX) few compounds have been tried in appropriately performed randomized controlled studies including a reasonably long-term follow up. In the evaluation of drug treatment, placebo controlled studies are the only way to select drugs which have a beneficial effect. Depending on the structure of the study even placebo may be effective depending on the endpoints chosen for the study [1]. Endpoints to select for all kinds of studies on patients with critical limb ischemia have to be carefully evaluated. Even though patients may find rest pain their main problem, it is more or less impossible to use this symptom as an objectively assessed parameter. Also ulcer healing is difficult to assess, which means that mortality and amputation usually are the main endpoints to be selected for this severely diseased group of patients.

Antiplatelet Drugs

The effect of both aspirin (ASA) and Ticlopidin as well as Clopidogrel is a prevention of platelet aggregation to damaged endothelial surfaces. In the meta-analysis of the final report of the second cycle of the *Antiplatelet Trialists Collaboration* [2] there is evidence that these antiplatelet agents produce a significant reduction of stroke, myocardial infarction and vascular death, and also that they improve patency of peripheral arteries and grafts. However, there has not been any specific inclusion of patients with critical limb ischemia and symptoms, such as rest pain and ulcers.

Ticlopidin has been tried in a placebo controlled Japanese study with the finding that ulcers healed significantly better after ticlopidin than after placebo [3]. On the other hand, most patients included suffered from thromboangiitis obliterans and not from atherosclerosis.

Anticoagulants

Despite the fact that unfractionated heparin has been used frequently as prophylaxis and in conjunction with vascular surgical procedures, it has not been tried in an efficient way for the symptoms of critical limb ischemia. Two studies using low molecular weight heparin (LMWH) have, however, shown beneficial results concerning healing of ischemic ulcers [4,5].

Studies on oral anticoagulation have usually included patients with intermittent claudication and not those with critical limb ischemia. In one bypass study from Austria it was, however, shown that survival was slightly better for patients on oral anticoagulation [6].

Defibrinating Agents

Ancrod has been tried in two placebo-controlled studies but failed to show any benefit with respect to healing of ischemic ulcers. Neither could they show any reduction of subsequent amputations [7,8].

Other Vasoactive Drugs

Naftidrofuryl has been shown to increase walking distance and also increase transcutaneous oxygen pressure ($TcpO_2$) in patients with rest pain [9,10]. However, in a small, placebo-controlled study in patients with critical limb ischemia, a subjective improvement was felt to the same extent in both groups [11].

Buflomedil has been evaluated in one placebo-controlled trial in patients with skin necrosis. Capillaroscopy showed an improvement of 72% versus 23% [12]. There is, however, no further study in patients with critical limb ischemia.

Pentoxifylline has been investigated in two studies in patients with critical limb ischemia, both double blind placebo controlled multicenter trials. In these trials 600 mg pentoxifylline was administered intravenously twice daily for up to 21 days. In the European study [13], rest pain was significantly lower in the treatment group while in the Norwegian study there was just a trend to amelioration of rest pain [14]. The follow up was, however, short in both studies.

The calcium channel blocker Nifedipine exerts an effect on the peripheral vascular system, although the clinical benefit is not yet proven [15-17].

Local Thrombolysis

Although most evidence exists for acute thrombolysis, the STILE study also treated patients with symptoms of critical limb ischemia lasting for more than two weeks [18]. Surgical revascularization was found to be more effective in this group. Intermittent low-dose urokinase therapy has been proposed, although not tried in any controlled trial [19].

New Compounds

L-arginine, the precursor of endogenous nitric oxide (NO) induces peripheral vasodilatation and inhibits platelet aggregation in humans. Although single infusions may increase blood flow in the femoral artery of patients with severe ischemia any long-term effect has not been proven as yet [20].

Quite another issue is the neuropeptide NPY, which is a sympathetic co-transmitter with noradrenalin. NPY-induced contraction is blocked by alpha-trinositol, and in a small clinical study in diabetics with critical limb ischemia it was found that microcirculatory parameters such as laser-doppler flux and $TcpO_2$ improved. However, any long-lasting effect remains to be proven [21].

■ GENE THERAPY

Since angiogenic growth factors can stimulate development of collaterals in animals that have been attributed to peripheral ischemia, various growth factors have been tried. A certain interest concerns the vascular endothelial growth factor (VEGF) [22]. In a recent clinical study gene transfer of VEGF was performed in nine patients with critical limb ischemia. Not only could new collaterals directly be documented by angiography, but also the ankle brachial index improved significantly and healing of ischemic ulcers was seen in four out of seven limbs [23].

■ OTHER TREATMENT MODALITIES

Techniques that act on the sympathetic nerve system, like a lumbar sympathetic block or a surgical sympathectomy, regularly increase blood flow, but mainly due to opening of arteriovenous shunts and not to an increase of nutritional blood flow. Despite these irrational reasons, patients may benefit in terms of reduced pain. In one study of 37 unselected patients with an ankle brachial index higher than 0.3, 78% of the patients had a long term relief of ischemic pain (24), while in a study from the late sixties only 6% had rest pain relief [25]. A review from 1985 concluded with a recommendation that sympathectomy could be beneficial in patients with rest pain and pregangrene [26]. A theoretical explanation for variations in results could be that both vasoconstriction and vasodilatatory fibers are to be found in the lumbar sympathetic ganglia [27]. It is, however, evident that there is no controlled study that can show any benefit of interruption of the sympathetic impulses.

Epidural spinal cord stimulation is discussed elsewhere in this book (Chapter XI). It should be mentioned that transcutaneous nerve stimulation has been tried in patients with severe ischemic disease [28]. Chelation therapy, implying the infusion of ethylene-diamine-tetra-acetate (EDTA) in patients with peripheral arterial occlusive disease, has been tried especially in patients with intermittent claudication. In a recent review of four randomized, double blind trials it was clearly established that the drug was not shown to be effective in treating the symptoms or in improving ankle brachial pressure index [29]. Transfusion of ultraviolet irradiated autologous blood has been described in both patients with critical limb ischemia and those with claudication. Uncontrolled studies have shown an improvement of rheology, but no controlled study exists as yet [30]. Hyperbaric oxygen has only been tried in uncontrolled studies although beneficial results have been presented in large series [31]. There is presently no reason to believe that local hyperbaric oxygen treatment is more effective than the generalized treatment [32]. Regional hyperthermic fibrinolytic perfusion utilizing recombinant tissue plasminogen activator has been used in a series of 19 patients, by means of a heart-lung-machine. From this uncontrolled study it is difficult to judge any benefit from this treatment compared to local thrombolytic treatment [33]. In 1993 the first experiences with heparin induced extracorporeal low density lipoprotein precipitation in critical limb ischemia were presented. The rationale is that elimination of excess fibrinogen and lipoproteins, especially LDL, should lead to an improvement of the microcirculation. The authors concluded that they were able to avoid 13 amputations [34]. The recent literature does not include further studies on this interesting topic.

■ CONCLUSION

Despite the fact that tremendous efforts have been made to find appropriate drugs to treat critical limb ischemia, little has been achieved. A major reason is most probably that only patients with end-stage ischemia are included in trials, and vascular morphology excludes interventions. Although it can be concluded that distal bypass procedures and endovascular interventions, as long as they are feasible, produce good outcome, it might be of value to try promising drugs before the end-stage of ischemia has appeared. From this overview it seems reasonable to conclude that the drugs which have been described are usually not useful, that sympathectomy may have a place, but without any possibility to predict who will respond.

Prostanoid treatment and spinal cord stimulation, discussed in other chapters, may be reasonable alternatives at present. Gene therapy is presently the most exciting area, and further studies are expected with great interest and some hope.

REFERENCES

1 Vayssairat M, Baudot N, Sainte-Beuve C. Why does placebo improve severe limb ischaemia? *The Lancet* 1988; 1: 356.

2 Anonymous. Collaborative overview of randomised trials of antiplatelet therapy - II: maintenance of vascular graft or arterial patency by antiplatelet therapy. Antiplatelet Trialists' Collaboration. *Br Med J* 1994; 308: 159-168.

3 Katsumura T. Therapeutic effet of ticlopidine for ischemic leg ulcers. In: Gordon JL, (ed.), *Ticlopidine: Quo Vadis?* Basel, Birkh user, 1984, pp. 167-172.

4 Gauthier O. Efficacy and safety of CY216 in the treatment of specific leg ulcers. In: Breddin K, Fareed J, Samarna M, (eds.) *Fraxiparine: analytical and structural data, pharmacology, clinical trials.* Stuttgart, Schattauer, 1987; 21.

5 Jörneskog G, Brismar K, Fagrell B. Low molecular weight heparin seems to improve local capillary circulation and healing of chronic foot ulcers in diabetic patients. *VASA* 1993; 22: 137-142.
6 Kretschmer G, Herbst F, Prager M et al. A decade of oral anticoagulant treatment to maintain autologous vein grafts for femoropopliteal atherosclerosis. *Arch Surg* 1992; 127: 1112-1115.
7 Martin M, Hirdes E, Auel H. Defibrinogenation treatment in patients suffering from severe intermittent claudication - a controlled study. *Thromb Res* 1976; 9: 47-57.
8 Tønnessen Kh, Sager P, Gormsen J. Treatment of severe foot ischemia by defibrination with ancrod: a randomised blind study. *Scand J Clin Lab Invest* 1978; 38: 431-435.
9 Abendroth D, Sunder-Plassman L. Einfluss einer intravenösen naftidrofurylbehandlung auf die Mikrozirkulation von Patienten im Stadium III and IV einer peripheren arteriellen Verschlusskrankheit. *VASA* 1988; Suppl 24: 33-37.
10 Horsch S. Uber den Einsatz von naftidrofuryl im Stadium III/IV einer peripher arteriellen Verschlusskrankheit. *VASA* 1988; Suppl 24: 38-43.
11 Greenhalgh RM. Naftidrofuryl for ischemic rest pain: A controlled trial. *Br J Surg* 1981; 68: 265-266.
12 Fagrell B, Hermansson IL. Effect of buflomedil on microcirculation of the skin in acral gangrene. *Fortschr Med* 1985; 103: 23-27.
13 Anonymous. Intravenous pentoxifylline for the treatment of chronic critical limb ischaemia. The European Study Group. *Eur J Vasc Endovasc Surg* 1995; 9: 426-436.
14 Anonymous. Efficacy and clinical tolerance of parenteral pentoxifylline in the treatment of critical lower limb ischemia. A placebo controlled multicenter study. The Norwegian Pentoxifylline Multicenter Trial Group. *Int Angiol* 1996; 15: 75-80.
15 Sjöberg T, Andersson KE, Norgren L, Steen S. Comparative effects of some calcium channel blockers on human peripheral arteries and veins. *Acta Physiol Scand* 1987; 130: 419-427.
16 Dormandy J. Clinical use of calcium antagonists in peripheral circulatory disease. *Ann NY Acad Sci* 1988; 522: 611-620.
17 Morgan RH, Psaila JV, Lewis P, Davies WT, Woodcock JP. The effects of nifedipine on blood flow in peripheral vascular disease of the lower limbs. *Eur Heart J* 1987; 8 (Suppl K): 87-91.
18 Anonymous. Results of a prospective randomised trial evaluating surgery versus thrombolysis for ischemia of the lower extremity. The STILE trial. *Ann Surg* 1994; 220: 251-268.
19 Leschke M, Schoebel FC, Strauer BE. Low-dose intermittent urokinase therapy in chronic symptomatic end-stage arterial disease - clinical relevance for patients with coronary artery disease or peripheral arterial occlusive disease. *Clin Hemorheol Microcirc* 1997; 17: 59-66.
20 Bode-Böger SM, Böger RH, Alfke H et al. L-arginine induces nitric oxide-dependent vasodilatation in patients with critical limb ischemia. A randomized, controlled study. *Circulation* 1996; 93: 85-90.
21 Nilsson L, Apelqvist J, Edvinsson L. Effects of α-Trinositol on peripheral circulation in diabetic patients with critical limb ischaemia. A pilot study using laser doppler fluxmetry, transcutaneous oxygen tension measurements and dynamic capillaroscopy. *Eur J Vasc Endovasc Surg* 1998; 15: 331-336.
22 Leung DW, Cachianes G, Kuang WJ et al. Vascular endothelial growth factor is a secreted angiogenic mitogen. *Science* 1989; 246: 1306-1309.
23 Baumgartner I, Pieczek A, Manor O et al. Constitutive expression of phVEGF165 after intramuscular gene transfer promotes collateral vessel development in patients with critical limb ischemia. *Circulation* 1998; 97: 1114-1123.
24 Persson AV, Anderson LA, Padberg FT Jr. Selection of patients for lumbar sympathectomy. *Surg Clin North Am* 1985; 65: 393-403.
25 Fulton RL, Blakeley WR. Lumbar sympathectomy. A procedure of questionable value in the treatment of arteriosclerosis obliterans of the legs. *Am J Surg* 1968; 116: 735-744.
26 Cotton LT, Cross FW. Lumbar sympathectomy for arterial disease. *Br J Surg* 1985; 72: 678-683.
27 Lundberg J, Norgren L, Ribbe E et al. Direct evidence of active sympathetic vasodilatation in the skin of the human foot. *J Physiol* 1989; 417: 437-446.
28 Kaada B. TNS in peripheral circulatory disorders. *L kartidningen (Swedish)* 1988; 85: 3370-3371.
29 Ernst E. Chelation therapy for peripheral arterial occlusive disease. A systematic review. *Circulation* 1997; 96: 1031-1033.
30 Karandashov VI, Petukhov EB, Karalkin AV. The regional hemodynamics in patients with chronic arterial insufficiency of the lower extremities after the transfusion of UV-irradiated autologous blood. *Vestn Khir Im II Grek* 1996; 155: 76-78.
31 Fredenucci P. Oxygénothérapie hyperbare et artéropathies. *J Mal Vasc* 1985; 10 (Suppl A): 166-172.
32 Diamond E, Forst MB, Hyman SA, Rand SA. The effect of hyperbaric oxygen on lower extremity ulcerations. *J Am Podiatry Assoc* 1982; 72: 180-185.
33 Mumme A, Maatz W, Walterbusch G. Regional hyperthermic fibrinolytic perfusion as ultima ratio in critical ischemia of the lower extremities. *Zentralbl-Chir* 1996; 121: 1069-1975.
34 Walzl M, Lechner P, Walzl B et al. First experiences with the heparin-induced extracorporeal low-density lipoprotein precipitation in the treatment of critical limb ischaemia: a new therapeutic approach? *Haemostasis* 1993; 23: 237-243.

11

SPINAL CORD STIMULATION IN CRITICAL LIMB ISCHEMIA

DIRK TH. UBBINK, MICHAEL J.H.M. JACOBS

The beneficial effect of electrical stimulation of the posterior parts of the spinal cord was initially discovered in the treatment of chronic, neurogenic pain syndromes [1,2]. Cook et al. in 1976 [3] published the first experience with spinal cord stimulation (SCS) in patients suffering from peripheral vascular insufficiency, showing positive effects on ischemic pain and healing of ulcers. Since then, two decades of research and application of neurostimulation have pullulated. The applications have widened to dorsal column, intrathecal or peripheral nerve stimulation in various other fields of interest, like angina pectoris, vasospastic disorders, sympathetic reflex dystrophy, epilepsy, spasticity, Parkinson's disease and incontinence.

This chapter will describe the technique of SCS itself, putative working mechanisms, and the clinical evidence presently available concerning the efficacy of this treatment in critical limb ischemia (CLI).

■ TECHNIQUE

Spinal cord or dorsal column stimulation utilizes low-voltage electrical pulses, which are generated by a pulse generator and conducted through a lead to an electrode positioned in the epidural space from where the spinal cord is stimulated and, thus, the desired effect is achieved.

Technical Specifications

The spinal cord stimulator nowadays consists of a pulse generator (Itrel III®, Medtronic, Minneapolis, USA), extension cable and a quadripolar lead (Fig. 1). The pulse generator provokes electrical pulses (nominal settings: amplitude 0.5 V, rate 85 Hz, width 210 s), which may be set continuous or cycling. The settings are fully programmable during and following implantation by means of a console programmer. The patient is supplied with a control magnet to toggle the stimulator on or off (Fig. 2). More recently, a remote control has become available by which the patient can (de-) activate the stimulator and adjust its amplitude in a certain range preset by the physician. The battery longevity largely depends on the settings, but is 3 years in general.

Effects, Precautions and Complications

The electrical pulses to the spinal cord induce paresthesiae in the area of the ischemic pain, after correct positioning of the electrode in the epidural space. A change in posture to the supine position usually increases the intensity of the paresthesiae, due to approximation of the spinal cord to the electrode. The device may affect the rate of cardiac demand pacemakers and, in turn, may be influenced by magnetic resonance imaging, surgical diathermy, power tools, and airport screening devices.

The most common complications are lead migration, requiring a change in amplitude, a change in activation of the four electrodes or, ultimately, repositioning of the lead. Lead infection, epidural hematoma, or spinal fluid leakage are exceptional.

Surgical Procedure

Implantation of the electrode is performed with the patient in the prone or lateral position. Routinely, prophylactic antibiotics are given intravenously for 24 hours perioperatively. Under local anesthesia a vertical incision of about five cm is made parallel to the spinal column at the level of L3-4. Through a Tuohy needle at the same level and under biplane fluoroscopic control the electrode is introduced and moved upwards to the midline of T10. Subsequently, the electrode is connected to an external stimulator, causing paresthesiae in the leg(s). The lead is then manipulated until the patient experiences pleasant paresthesiae extending down to the painful area of the foot. Surgical implantation under direct visual control is also an option, which allows a better prevention of lead migration by fixing it to the dura instead of the thoracolumbar fascia. The pulse generator is implanted in a subcostal, subcutaneous pouch and connected to the epidural electrode via the subcutaneous extension cable (Fig. 1). This can be performed in the same session or in two tempi after a test period of several days with an external stimulator to establish effectiveness. For this purpose a temporary epidural catheter with an electrode tip is also an option [4].

Cost of SCS

The price of the neurostimulation set varies among different countries between 6 000 and 9 000 dollars. This is about doubled when the cost of the implantation procedure is added.

■ WORKING MECHANISM

Hitherto, knowledge was scarce about how SCS exerts its allegedly anti-nociceptive and anti-ischemic effects. These effects suggest an influence on pain conducting nerve fibers and, possibly as a result of that, an improvement of local microcirculation.

Pain impulses are conducted by the larger, low-threshold A- and thinner, high-threshold C fibers to the cells of the substantia gelatinosa, the dorsal column fibers, and the first central transmission (T) cells in the dorsal horn. The dominating pain types in peripheral vascular disease (apart from possibly concomitant neuropathic pain due to nervous tissue damage) are deep ischemic pain and superficial pain from the area of ischemic ulcers and their penumbra. Both types respond to opiates. The deep pain appears to be alleviated by SCS, but the superficial pain seems to be merely masked by the paresthesiae and return as soon as the stimulation is discontinued. Frequently, a different response upon SCS is

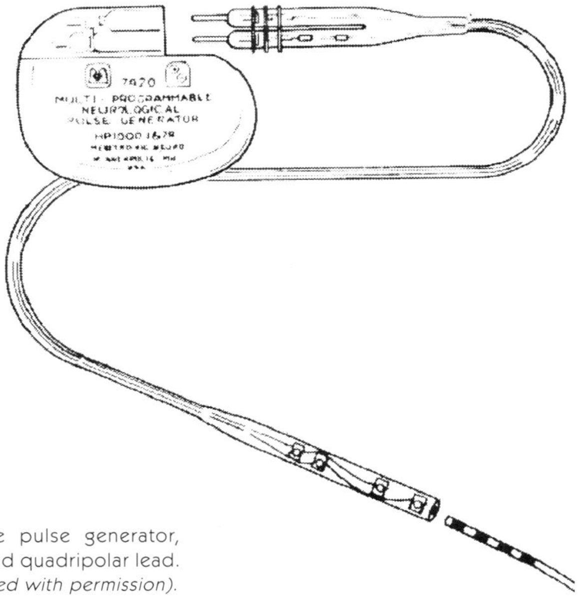

FIG. 1 Implantable pulse generator, extension cable, and quadripolar lead. *(Drawing reproduced with permission).*

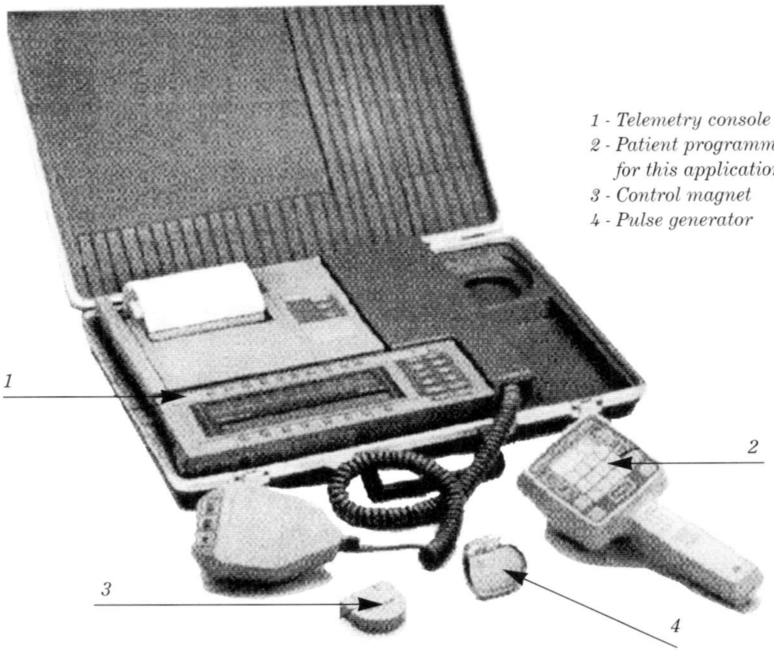

1 - Telemetry console programmer
2 - Patient programmer (not available for this application)
3 - Control magnet
4 - Pulse generator

FIG. 2 Remote programming equipment for pulse generator. *(Photograph reproduced with permission).*

observed between ischemic and neuropathic pain. Ischemic pain usually requires several days to diminish, whilst neuropathic pain responds in a few minutes.

SCS spreads electric pulses that activate fibers in the dorsal columns, giving rise to orthodromic and antidromic conduction of impulses (Fig. 3). The orthodromic transmission induces the paresthesiae as experienced by the patients. The antidromic impulses are transmitted to the dorsal horns, where they excite A fibers in the dorsal root, resulting in inhibition of the T cells, which in turn will no longer react upon pain stimuli from the smaller C fibers [5]. Autonomic pathways in the intermediate columns of the gray area of the spinal medulla regulate the peripheral vascular tone. Spread of the electrical stimulation into this area should release segmental spinal reflexes, thus inhibiting sympathetic impulses to the peripheral vessels.

Additionally, several hypotheses as to the alleviation of ischemic pain have been proposed, such as the following ones.

Direct inhibition of the pain signal going from the dorsal horn to the brain via the spinothalamic tracts, resulting in a reduced sympathetic tone and, subsequently, vasodilatation [6]. This explanation seems insufficient; firstly since only part of the pain components are relieved. Secondly, the analgesic effect of stimulation is usually not acute, as opposed to the response in neuropathic or vasospastic pain, indicating the peripheral vasodilatation and reperfusion of the ischemic area to counteract the pain-inducing mechanisms need some time. Thirdly, cutaneous vasodilatation may be observed also in the absence of detectable paresthesiae, illustrating that SCS not only blocks the pain signal.

Antidromic activation in sensory nerves, leading to release of vasoactive substances from thin nerve fibers and, hence, vasodilatation [7]. In animal studies the various conducting pathways were transected [8] (Fig. 3). Sectioning the dorsal roots appeared to preserve SCS-induced vasodilatation. Moreover, stimulating the peripheral nerve stump could not achieve this. This showed that antidromic activation is not mandatory. The observation in patients that the paresthesiae evoked by SCS are experienced as pleasant, pleads against recruitment of the high-threshold, pain-transmitting C fibers.

Release of vasoactive substances with effects on peripheral vascular beds [7,9,10], e.g. serotonin, substance P, vasoactive intestinal polypeptide (VIP), calcitonin gene-related peptide (CGRP), prostaglandins, or nitric oxide. However, at present insufficient evidence is at hand to support a role of any of these in SCS treatment.

Modulation of the autonomic (sympathetic) nervous system, inducing vasodilatation. The most plausible mechanism of SCS is a reduction of the sympathetic vasoconstrictor activity [11,12], although SCS seems effective also in patients who have undergone a previous chemical or surgical sympathectomy. In these cases, SCS may act on the remaining sympathetic activity, which usually is eliminated incompletely. Experimental evidence derived from studies transecting the ventral roots or the sciatic nerve [13], or performing lumbar sympathectomy [14] (see Fig. 4) confirmed an abolishment of the vasodilatory response otherwise induced by SCS.

Transmitters and Receptor Groups Influenced by SCS

Blocking the nicotine receptors in the ganglia appeared to abolish the SCS-induced vasodilatation, as opposed to the muscarine receptor. Blocking α-adrenergic, especially the $\alpha 1$-mediated transmission in the neuro-effector junction decreased the SCS effect. $\beta 2$-adrenergic blocking also caused an abolished response; $\beta 1$-blockers affected the skin, $\beta 2$-blockers the muscle perfusion [15].

Interpretation of Data

Much of the available evidence as to putative mechanisms of action is derived from animal studies. The validity of extrapolating these data to the clinical situation is murky. Most experiments studied acute ischemic situations, while clinically chronic ischemia with an uncertain neurogenic component needs to be treated, in which the positive effects may take some time to become manifest. Furthermore, the stimulation parameters used were not always comparable to those in the clinical setting.

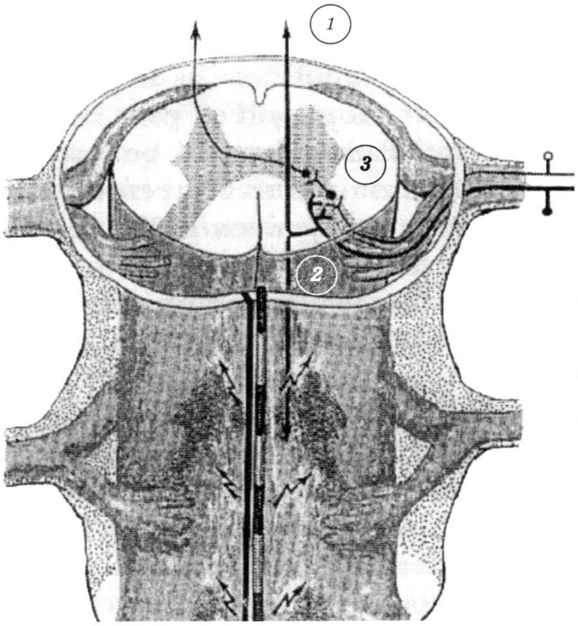

FIG. 3. Schematic drawing of the putative effects of SCS.

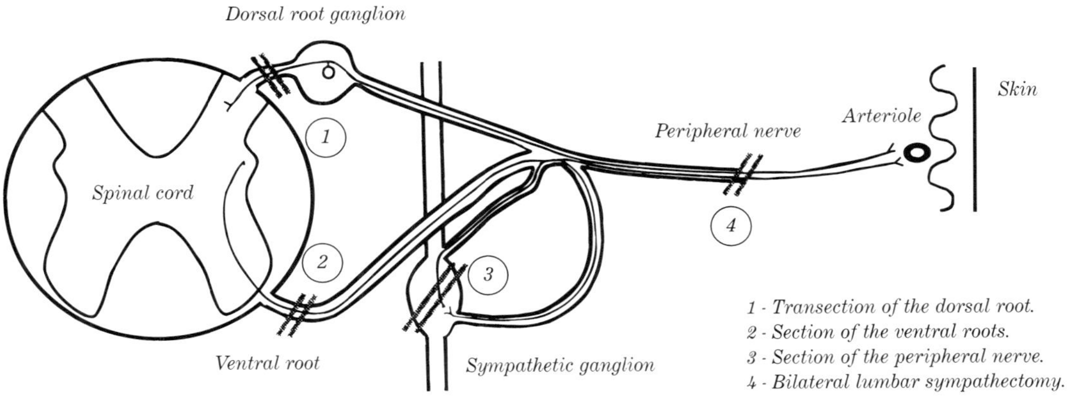

FIG. 4 Schematic diagram of (transections of) the spinal cord, ventral and dorsal roots, paravertebral sympathetic chain, and peripheral nerve supplying the skin. *(Drawing reproduced with permission).*

■ CLINICAL EXPERIENCE

Non-Controlled Trials

The alleged effect of SCS is largely based on cohort studies in patients with non-reconstructable CLI. An overview of these is presented in the Table. In general, these studies showed a clear reduction (60%-94%) of the ischemic pain during a follow-up period of up to 80 months. Healing of ischemic ulcers was seen in 39% to 67% of the cases, in particular of ulcers smaller than 3 cm^2. On the other hand, amputation frequency varied between 20% and 58%, which corresponded with a limb salvage rate after one year of 75%-80% and after four years of 42%-66%. No improvement of the ankle blood pressures was found except in the study of Fiume et al. [20]. They included claudicants as well, which may have benefited also from walking exercise. In contrast, an improved skin temperature and oxygen tension was observed in several studies in those who responded well to the therapy, suggesting the effect of SCS focuses mainly on the local microcirculation. However, an improvement of the $tcpO_2$ can be expected since it can be measured only in those patients who did not undergo an amputation. Some studies suggested $tcpO_2$ measurements could have a predictive value as to limb survival [31-33]. Objective assessment of the effect of SCS on skin microcirculation by means of capillary microscopy was performed in a study by Jacobs et al. [22]. Both capillary density and red blood cell velocity were found to increase in responders, which could explain relief of ischemic pain [34], healing of ischemic ulcers and an increased limb survival.

These studies, however, could not prove the efficacy of the SCS treatment due to the lack of a control group, while comparable pain relief might also have been achieved with analgesics and local wound care and limb salvage was not evidently better than the figures known from the natural history of patients with CLI. Furthermore, the inclusion of various causes and different stages of peripheral ischemia has obscured the judgment as to the best indication for this treatment.

Randomized Trials

Since the therapy of choice for CLI is vascular reconstruction and vascular surgeons nowadays tend to perform more aggressive distal bypass surgery, a relatively small fraction of patients remains that is non-reconstructable. Thus, a multicenter approach is indicated to recruit enough patients for a meaningful study. This is one of the main reasons why so far only a few prospective randomized controlled studies have been performed.

The first two-center randomized study was published by Jivegård et al. [35]. Fifty-one technically non-reconstructable patients with rest pain and/or ischemic ulcers (including ten diabetics) received analgesics with or without SCS. Pain relief, tissue loss, and limb salvage were investigated. Macrocirculatory parameters did not differ between both groups during a follow-up period of up to 18 months. Long-term pain relief occurred only in the group treated with SCS and tissue loss was less. After a mean follow-up of 108 months, 28 patients had undergone an amputation, of which 25 had died, and another 18 patients had died without amputation. After 5 years 35% of the patients were alive. Limb salvage after 18 and 60 months tended to be higher in the SCS than in the control group (62% vs. 45% and 51% vs. 35%, respectively). Subgroup analysis suggested a better limb salvage in normotensive patients.

In a study by Claeys et al. [33], 86 inoperable patients (including 13 diabetics) with an ischemic ulcer and an ankle systolic blood pressure less than 50 mmHg were treated with seven days intravenous prostaglandin E_1 (PGE_1) and were subsequently randomized into receiving 14 days PGE_1 with or without SCS. Ulcer healing occurred significantly more in the group treated with additional SCS (69%) than in the PGE_1 group (17%). Pain relief was also more frequently seen in the SCS group (40% vs. 10%). There was no difference in major amputation frequency (15% vs. 20%). After a follow-up duration of 12 months, ankle blood pressures did not improve, whereas $tcpO_2$ increased significantly in the SCS group. In retrospect, a $tcpO_2$ increase above 26 mmHg correlated with ulcer healing, whereas a $tcpO_2$ less than or equal to 10 mmHg predicted poor outcome.

Recently, the Dutch multi-center randomized controlled trial [36] was completed. In this study 120 patients with non-reconstructable CLI (i.e. Fontaine stages III or IV and ankle systolic blood pressure

Table SUMMARY OF NON-CONTROLLED STUDIES ON SCS IN CRITICAL LIMB ISCHEMIA

Author [ref.]	*Year*	*N*	*Fontaine*	*Follow-up*	*Clinical results*	*Perfusion parameters*
Tallis [16]	1983	10	2,3,4		Pain reduction in 60%	Improved ^{133}Xe skin perfusion in responders
Augustinsson [17]	1985	34	3,4*	Mean 16 mo.	Pain reduction in 94% Ulcer healing in 50% Amputation in 38%	Skin temperature increase in 70%
Broseta [18]	1986	41	2,3,4*	Up to 25 mo.	Pain reduction in 71% Ulcer healing when < 3 cm	Improved doppler signals, no improved doppler pressures. Increased skin temperature
Graber [19]	1987	9	3,4	Up to 16 mo.	Pain reduction in 78% Amputation in 33%	No improved doppler pressures
Fiume [20]	1989	45	2,3,4	Up to 16 mo.	Clinical improvement in 83% Ulcer healing when < 3 cm	Doppler pressure improvement in 42% $TcpO_2$ improved in responders
Sampere [21]	1989	19	3,4		Pain reduction in 71%	Improved cutaneous flow
Jacobs [22]	1990	20	3,4	Up to 36 mo.	Acute pain reduction in 90% Final pain reduction in 60% Ulcer healing in 67% when < 3 cm Amputation in 30% Limb salvage 80% at 1 year	No improved doppler pressures Skin nutritive micro-circulation improved in responders
Bunt [23]	1991	15	3,4		Pain reduction in 80% Amputation in 20% Limb salvage 75% at 1 year	Improved doppler pressures. $TcpO_2$ improved in responders
Steude [24]	1991	10	3,4		Acute pain relief	$TcpO_2$ and skin temperature increase
Galley [25]	1992	244	2,3,4*		Pain relief	Doppler pressures. $TcpO_2$
Mingoli [26]	1993	76	2,3,4	Up to 76 mo.	Pain relief in 75% Ulcer healing in 39% when < 3 cm Amputation in 58% Limb salvage 42%	
Horsch [27]	1994	177	3,4	Mean 36 mo.	Pain relief in 62% Amputation in 32% Limb salvage 66% at 4 years	No improved doppler pressures $TcpO_2$ improved in responders
Rickman [28]	1994	41	3,4*	Up to 24 mo.	Pain relief in 80% Ulcer healing when < 2 cm	Inclusion if ABI<0.40 TP<30 mmHg
Guarnera [29]	1996	35	2,3,4	Up to 80 mo.	Pain relief in 77% Limb salvage 80%	
Claeys [30]	1997	237	3,4	Mean 31 mo.	Pain relief in 73% Amputation in 27%	$TcpO_2$ improved in responders
Kumar [31]	1997	39	2,3,4	Up to 36 mo.	Pain relief in 77% Amputation when $tcpO_2$ < 10 mmHg	Improved doppler signals $TcpO_2$ improved in responders
Gersbach [32]	1997	20	3,4	Mean 14 mo.	Limb salvage 63% Limb salvage 88% when supine/sitting $tcpO_2$ gradient > 15 mmHg	No improved $tcpO_2$ or skin nutritive microcirculation

* Not only atherosclerotic patients included: e.g. also Buerger's disease, vasospastic disorders.

less than or equal to 50 mmHg or an ankle/brachial pressure index less than or equal to 35% or no palpable ankle arteries in diabetics) were allotted standard treatment (local wound care, optimal medical treatment) with or without SCS, according to the intention-to-treat principle. Diabetes was present in 37% of the patients. Mortality and risk factors were comparable in both groups. No serious side effects of the SCS treatment were observed. In eight patients no effective SCS could be applied due to implantation problems. Limb salvage in the SCS and standard treatment groups was 60% vs. 46% after one year and 52% vs. 46% after two years, respectively, without significant differences. In the SCS group significantly less analgesics were necessary. Ulcer healing did not differ between both groups. Quality of life analysis in these patients during the follow-up period of 18 months showed a general improvement, but no significant differences between both groups. No substantially contributing prognostic factors for limb salvage were found. A cost-effectiveness study showed that SCS treatment during the follow-up period (including hospital stay, amputations, revalidation, extramural care) was significantly more expensive (26%), mainly due to the cost of the stimulator and implantation per se. The total cost of treatment of these patients was approximately $ 30 000 during 18 months of follow-up. Hence, the overall impression was that SCS was not more effective than standard conservative treatment.

Simultaneously, the effect of both treatment modalities on skin microcirculation was investigated using capillary microscopy (CM), $tcpO_2$ and laser doppler (LD), which was assessed before and during the follow-up period. Before the start of the treatment, patients were divided in three microcirculatory groups on the basis of pre-defined cut-off values, derived from previous studies [34,37]: *poor* (capillary density $<20/mm^2$, absent post-occlusive reactive hyperemia in CM and LD and a $tcpO_2$ <10 mmHg), *good* (capillary density $>20/mm^2$, present reactive hyperemia in CM and LD and $tcpO_2$ >30 mmHg), and *intermediate*, when not meeting the above mentioned criteria (e.g. having a $tcpO_2$ between 10 mmHg and 30 mmHg). During follow-up, skin microcirculation showed a significant improvement of the $tcpO_2$ and reactive hyperemic response in CM, irrespective of the treatment given. Limb survival at 12 months was 17% in the *poor* group, 63% in the *intermediate* and 88% in the *good* group, indicating a high predictive value of microcirculatory classification as to an imminent amputation during follow-up (hazard ratio 0.28; $p<0.0001$, log rank) [38]. In the *intermediate* subgroup, comprising half of the patients in the trial, amputation frequency after 6 and 12 months was twice as high in the standard treatment group, but this did not reach statistical significance (48% vs. 24%, $p=0.10$). A treatment effect was apparent in the *intermediate* category for the parameters concerning the hyperemia response in CM and LD, which deteriorated in the standard treatment group, but not in the SCS group. These data are to be published this year.

GENERAL CONSIDERATIONS

Spinal cord stimulation has been launched as a promising treatment for patients with intractable critical limb ischemia. So far, evidence from the literature shows a positive effect of SCS on ischemic pain and healing of small ulcers. Despite uncertainties as to its working mechanism, SCS appears to mobilize the reserve capacity of the local skin microcirculation. SCS tends to prolong limb survival, but convincing evidence is scarce. In addition, SCS therapy for this purpose is expensive and not reimbursed in every country. Therefore, patients who are most likely to benefit from SCS appear to be those with a borderline skin microcirculation. Patients with a poor local microcirculation have a high amputation risk. The yield of SCS in these will be low. Patients with a good microcirculation appear to benefit from any therapy. In these patients SCS is also not indicated.

The therapy of choice in patients with chronic CLI is and remains vascular reconstructive surgery and/or a reopening procedure [39]. This holds also when very distal (pedal) bypass surgery is indicated even in diabetic patients [40], because this treatment has an acceptable long-term patency and a lower morbidity and mortality than primary amputation. Hence, SCS therapy is only indicated in patients in whom the diagnosis is clear and a surgical or angioplastic intervention cannot be performed. This is the case if all crural and pedal arteries are unsuitable as receptor artery (as should be confirmed by means of selective angiography), no autologous vein is available for distal bypass surgery, or when the patient's condition or concomitant diseases form a serious contraindication for surgical intervention. Usually a considerably lowered ankle systolic blood pressure (less than 50 mmHg) or an ankle/brachial pressure index less than 0.35 should confirm the presence of critical ischemia. In patients with diabetes and incompressible ankle arteries the absence of arterial pulsations in the foot can be used as a selection criterion. One should bear in mind that the use of different cut-off values for rest pain and ulceration may be more discriminative, since the healing of an ischemic ulcer requires a higher perfusion pressure than the mere elimination of rest pain. Hence, amputation risk is different between patients with and without an ischemic ulcer, despite similar ankle pressures. Therefore the recently proposed North American reporting standards may be more suited for this purpose [41]. Microcirculatory investigation can be useful to further refine patient selection [38,42]. The presence of a large (greater than 3 cm in diameter) or rapidly progressing ulcer or gangrene is regarded as a contra-indication for SCS [18,20,22,26,28].

This fairly easy, last-resort therapeutic option deserves future studies, which should have the format of randomized controlled studies, involving a clearly defined target population and comprising investigation of quality of life, and a cost-benefit analysis. Since ischemic ulceration expresses the insufficiency of the local microcirculation, investigation of the skin microcirculatory perfusion is important so as to appreciate the status of the skin as to its viability, reserve capacity, and healing potential. For this purpose, the $tcpO_2$ measurement appears to be the best stand-alone technique to select eligible patients. A $tcpO_2$ between 10 and 30 mmHg indicates CLI that has the highest probability to respond favorably to SCS. As a practical and financially tenable policy, external stimulation can be used for a few days as a testing period. When this appears effective final implantation is to be performed. If not, other means of treatment should be considered.

REFERENCES

1 Long DM, Hagfors N. Electrical stimulation in the nervous system: the current status of electrical stimulation of the nervous system for relief of pain. *Pain* 1975; 1: 109-123.

2 Kumar K, Nath R, Wyant GM. Treatment of chronic pain by epidural spinal cord stimulation: a ten-year experience. *J Neurosurg* 1991; 75: 402.

3 Cook AW, Oygar A, Baggenstos P et al. Vascular disease of the extremities. Electrical stimulation of spinal cord and posterior roots. *NY State J Med* 1976; 76: 366-368.

4 Racz TA, McDaniel KE, Racz GB. Trial spinal cord stimulation via stimulating epidural catheter. *Curr Rev Pain* 1997; 1: 383-389.

5 Melzack R, Wall PD. Pain mechanism: a new theory. *Science* 1965; 150: 971-979.

6 Hosobuchi Y. Treatment of ischemic pain by neurostimulation. In: Lipton S, Tunks E, Zoppi M (eds.). *Advances in pain research and therapy.* Raven Press, New York, 1990, pp 223-226.

7 Hilton SM, Marshall JM. Dorsal root vasodilatation in cat skeletal muscle. *J Physiol* 1980; 299: 277-288.

8 Linderoth B. Spinal cord stimulation in ischemia and ischemic pain. Possible mechanisms of action. In: Horsch S, Claeys L (eds.). *Spinal Cord Stimulation II.* Steinkopff Verlag, Darmstadt, 1995, pp 19-35.

9 Linderoth B, Gazelius B, Frank J, Brodin E. Dorsal column stimulation induces release of serotonin and substance P in the cat dorsal horn. *Neurosurgery* 1992; 31: 289-296.

10 Kjartansson J, Lundeberg T. Effects of electrical nerve stimulation (ENS) in ischemic tissue. *Scand J Plastic Reconstr Surg* 1990; 24: 129-134.

11 Jänig W. Spinal cord stimulation in patients: basic anatomical and neurophysiological mechanisms. In: Horsch S, Claeys L (eds.). *Spinal cord stimulation. An innovative method in the treatment of PVD.* Steinkopff Verlag, Darmstadt, 1994, pp 37-58.

12 Naver H, Augustinsson LE, Elam M. The vasodilating effect of spinal dorsal column stimulation is mediated by sympathetic nerves. *Clin Autonom Res* 1992; 2: 41-45.

13 Linderoth B, Fedorcsak I, Meyerson BA. Peripheral vasodilatation after spinal cord stimulation: animal studies of putative effector mechanisms. *Neurosurgery* 1991; 28: 187-195.

14 Linderoth B, Gunasekera L, Meyerson BA. Effects of sympathectomy on skin and muscle microcirculation during dorsal column stimulation: animal studies. *Neurosurgery* 1991; 29: 874-878.

15 Linderoth B, Herregodts P, Meyerson BA. Sympathetic mediation of peripheral vasodilatation induced by spinal cord stimulation. Animal studies on the role of cholinergic and adrenergic receptor subtypes. *Neurosurgery* 1994; 35: 711-719.

16 Tallis RC, Illis LS, Sedgwick EM et al. Spinal cord stimulation in peripheral vascular disease. *J Neurol Neurosurg Psych* 1983; 46: 478-484.

17 Augustinsson LE, Carlsson CA, Holm J, Jivegard L. Epidural electrical stimulation in severe limb ischemia. Pain relief, increased blood flow, and a possible limb-saving effect. *Ann Surg* 1985; 202: 104-110.

18 Broseta J, Barbera J, de Vera JA et al. Spinal cord stimulation in peripheral arterial disease. A cooperative study. *J Neurosurg* 1986; 64: 71-80.

19 Graber JN, Lifson A. The use of spinal cord stimulation for severe limb-threatening ischemia: a preliminary report. *Ann Vasc Surg* 1987; 1: 578-582.
20 Fiume D, Palombi M, Sciacca V, Tamorri M. Spinal cord stimulation (SCS) in peripheral ischemic pain. *PACE* 1989; 12: 698-704.
21 Sampere CT, Guasch JA, Paladino CM et al. Spinal cord stimulation for severely ischemic limbs. *PACE* 1989; 12: 273-279.
22 Jacobs MJHM, Jörning PJ, Beckers RC et al. Foot salvage and improvement of microvascular blood flow as a results of epidural spinal cord electrical stimulation. *J Vasc Surg* 1990; 12: 354-360.
23 Bunt TJ, Holloway GA, Lawrence P et al. Experience with epidural spinal stimulation in the treatment of end-stage peripheral vascular disease. *Sem Vasc Surg* 1991; 4: 216-220.
24 Steude U, Abendroth D, Sunder-Plassmann L. Epidural spinal electrical stimulation in the treatment of severe arterial occlusive disease. *Acta Neurochir* 1991; 52 (suppl): 118-120.
25 Galley D, Rettori R, Boccalon et al. Electric stimulation of the spinal cord in arterial diseases of the legs. A multicenter study of 244 patients. *J Mal Vasc* 1992; 17: 208-213.
26 Mingoli A, Sciacca V, Tamorri M et al. Clinical results of epidural spinal cord electrical stimulation in patients affected with limb-threatening chronic arterial obstructive disease. *Angiology* 1993; 44: 21-25.
27 Horsch S, Claeys L. Epidural spinal cord stimulation in the treatment of severe peripheral arterial occlusive disease. *Ann Vasc Surg* 1994; 8: 468-474.
28 Rickman S, Wuebbels BH, Holloway GA Jr. Spinal cord stimulation for relief of ischemic pain in end-stage arterial occlusive disease. *J Vasc Nurs* 1994; 12: 14-20.
29 Guarnera G, Mascellari L, Bianchini G et al. The role of spinal cord electrical stimulation in critical ischemia of the extremity. *Minerva Cardioangiologica* 1996; 44: 663-667.
30 Claeys LG. Improvement of microcirculatory blood flow under epidural spinal cord stimulation in patients with nonreconstructable peripheral arterial occlusive disease. *Artif Organs* 1997; 21: 201-206.
31 Kumar K, Toth C, Nath RK et al. Improvement of limb circulation in peripheral vascular disease using epidural spinal cord stimulation: a prospective study. *J Neurosurg* 1997; 86: 662-669.
32 Gersbach P, Hasdemir MG, Stevens RD et al. Discriminative microcirculatory screening of patients with refractory limb ischaemia for dorsal column stimulation. *Eur J Vasc Endovasc Surg* 1997; 13: 464-471.
33 Claeys LG, Horsch S. Transcutaneous oxygen pressure as predictive parameter for ulcer healing in end-stage vascular patients treated with spinal cord stimulation. *Inter Angio* 1996; 15: 344-349.
34 Ubbink DT, Jacobs MJHM, Slaaf DW et al. Capillary recruitment as the cause of pain relief on dependency in patients with severe lower limb ischemia. *Circulation* 1992; 85: 223-229.
35 Jivegård LE, Augustinsson LE, Holm J et al. Effects of spinal cord stimulation (SCS) in patients with inoperable severe lower limb ischaemia: a prospective randomised controlled study. *Eur J Vasc Endovasc Surg* 1995; 9: 421-425.
36 Klomp HM, Spincemaille GH, Steyerberg EW et al. Design issues of a randomised controlled clinical trial on spinal cord stimulation in critical limb ischaemia. *Eur J Vasc Endovasc Surg* 1995; 10: 478-485.
37 Ubbink DTh, Jacobs MJHM, Tangelder GJ et al. The usefulness of capillary microscopy, transcutaneous oximetry and laser doppler fluxmetry in the assessment of the severity of lower limb ischaemia. *Int J Microcirc Clin Exp* 1994; 14: 34-44.
38 Ubbink DT, Spincemaille GH, Reneman RS et al. Prediction of imminent amputation in patients with non-reconstructable leg ischemia by means of microcirculatory investigations. *J Vasc Surg* 1999; in press.
39 LoGerfo FW. Epidural spinal cord electrical stimulation: an unproven methodology for management of lower extremity ischemia. *J Vasc Surg* 1991; 13: 518-519.
40 Pomposelli FB Jr. Marcaccio EJ. Gibbons GW et al. Dorsalis pedis arterial bypass: durable limb salvage for foot ischemia in patients with diabetes mellitus. *J Vasc Surg* 1995; 21: 375-384.
41 Rutherford RB, Baker JD, Ernst C et al. Recommended standards for reports dealing with lower extremity ischemia: revised version. *J Vasc Surg* 1997; 26: 517-539.
42 Bunt TJ, Holloway GA. $TcpO_2$ as an accurate predictor of therapy in limb salvage. *Ann Vasc Surg* 1996; 10: 224-227.

12

PERCUTANEOUS TRANSLUMINAL THERAPY FOR CRITICAL LIMB ISCHEMIA

AMMAN BOLIA

Chronic critical limb ischemia (CLI) endangers (part of) the leg. The European consensus document defines CLI as the presence of rest pain, ulceration or gangrene with maximum ankle or toe systolic pressures of 50 or 30 mm Hg, respectively [1]. In an aging population, CLI is now an increasingly common condition with an estimated incidence of 400 to 1000 per million per year [2,3]. The outlook for these patients reflects the multifocal nature of atherosclerosis with a high mortality rate from cardiovascular disease (46%) after 5 years [4] and a 25% amputation rate after 1 year. Despite attempts to revascularize [5], the modern management of these high-risk patients demands a team approach, combining the skills of non-invasive vascular imaging, interventional radiology and vascular surgery.

PERCUTANEOUS TRANSLUMINAL ANGIOPLASTY

Although percutaneous transluminal angioplasty (PTA) of the iliac, femoral and popliteal arteries is a well established therapeutic option in peripheral vascular disease [6], experience with infrapopliteal PTA has been limited and its role until recently has been less well defined. This is because multiple lesions are a common finding in the infrapopliteal vessels, and there has been concern that failed PTA at this level could jeopardize subsequent bypass surgery [7,8].

Although a recent study [9] has described a fivefold increase in the use of PTA for critical ischemia, the reported results are surprisingly variable. Some authors [10-12] describe poor results and caution against the use of PTA in critical ischemia; some report results significantly worse than those in claudicants [13-19]. Others, however, reported good results and support the use of PTA in critical ischemia [20-27].

The method of percutaneous treatment of atherosclerotic lesions in arteries below the knee initially suffered from technical disadvantages relating to the large size of catheters and conventional Teflon-coated guidewires. The advent of coronary artery PTA triggered a number of technological advances that facilitated infrapopliteal angioplasty. These advances included high-pressure balloon catheters on a small shaft (3.5F) together with a low profile of the balloons. The availability of steerable hydrophillic guidewires, road mapping facility and the use of vasodilators to overcome arterial spasm has extended the scope of treating crural arteries, from the calf into the foot. A large recent series in the form of an audit on the management of critical limb ischemia from this unit has produced extremely encouraging results in favor of PTA in the treatment of CLI [28-29].

■ CLINICAL INDICATIONS

All patients with CLI, including both diabetics and non-diabetics, who had ischemic rest pain requiring adequate analgesia for more than 2 weeks and an ankle systolic pressure of < 50 mmHg and toe systolic pressure of < 30 mmHg or ulceration or gangrene of the foot or toe are included for possible revascularization with angioplasty, so long as there are suitable lesions. Acute ischemia, either due to recent thrombosis or an embolic episode are not suitable for PTA, but may be suitable for lysis with possible embolectomy, either percutaneously or surgically.

■ ANATOMIC CONSIDERATIONS

Patients with limb-threathening ischemia commonly have multilevel atherosclerotic disease. Whilst in critical limb ischemia, the disease tends to be distal in the crural vessels, or in tandem with femoropopliteal disease, isolated disease in the femoropopliteal segment may be present. Disease in the iliac or profunda arteries is likely to be a contributory factor rather than the sole cause of critical ischemia. Lesions in any of these territories can be stenotic or occlusive.

■ MEDICATION

Prior to the procedure, most patients are on aspirin, 150-300 mg per day. For infrapopliteal angioplasty, 10-20 mg of nifedipine is used for the prevention of spasm, but nowadays nifedipine is used only for patients with malignant hypertension, in order to reduce puncture site complications following PTA.

During the procedure, intraarterial heparin is given as a matter of routine between 3000 and 5000 units and tolazoline up to 25mg is administered to prevent arterial spasm. It also helps to dilate the peripheral vessels so that when the revascularization procedure is completed, the distal vessels can accommodate the improved flow which is also enhanced by reduced peripheral resistance from the vasodilatation. Glyceryl tinitrate (GTN) is given in the doses of 100 g (up to 500 g) only when severe spasm occurs.

We apply GTN patches (5mg over 24hrs) to the treated leg if the patient has undergone infrainguinal angioplasty for occlusive disease in order to maintain a sustained vasodilatation peripherally, thus reducing peripheral resistance and subsequent improved flow through the angioplastied segment. Four doses of subcutaneous heparin over 24 hours are also given to prevent acute thrombosis of the angioplastied segment.

■ TECHNIQUE

Iliac and Profunda Femoral Artery Disease

Iliac and profunda femoral artery stenotic or occlusive disease is unlikely to be the only cause of critical ischemia, but may be contributory. Thus, treatment of these lesions might be helpful to improve the inflow before any treatment is carried out distally. Stenotic lesions are straightforward to deal with and virtually all lesions can be crossed using hydrophillic guidewires in order to achieve a successful dilatation. However, occlusive disease in the iliac vessels frequently results in a dissection from an ipsilateral approach and in such situations, a contralateral approach may be helpful in achieving recanalization through a dissection [30].

PTA of isolated profunda femoral artery stenoses can improve inflow and may be just sufficient to relieve the patient of critical ischemic symptoms. An ipsilateral puncture to treat profunda lesions is usually only possible when the disease is located more distally. If a stenotic lesion is present at the common femoral bifurcation or very near to the origin of the profunda femoral artery, then a crossover approach through a contralateral puncture is indicated. The preferred catheters for this approach may be a shaped catheter (sidewinder, cobra, renal) which can be hooked over the aortic bifurcation to allow the crossover approach.

Femoropopliteal Disease

Whilst isolated femoral or popliteal artery lesions may cause critical ischemia, more commonly it will be in tandem with tibial lesions. An ipsilateral puncture will allow access to most lesions in the infrainguinal region. Stenotic lesions and occlusions of less than 10cm in length may be dealt with intraluminally, but occlusions of longer segments or those that start from the origin of the superficial femoral artery without a stump require the technique of subintimal percutaneous transluminal angioplasty (S-PTA). For S-PTA, a 1.5mm J-wire (0.035, 180cm long, Boston Scientific, U.K.) allows the formation of a loop which is required to not only cross the lesion, but allow re-entry back into the lumen distally in the patent part of the artery (Fig. 1).

Popliteal Occlusion Extending into Tibial Arteries

This is often encountered in critical limb ischemia with the occlusion extending into one or more of the crural vessels. S-PTA allows recanalization to be carried out into one or more of the crural vessels depending on what is available distally.

The achievement of a successful recanalization into all three crural vessels provides the proximally recanalized artery the best chance of a good long-term patency, since it is common knowledge that a three vessel runoff is better than a single vessel outflow (Figs. 2 and 3).

Tibial Arterial Disease

The atherosclerotic process is most often localized in the crural arteries. Isolated stenotic lesions are easily dealt with by conventional 5F catheters, available in 2.5, 3 and 3.5mm diameters and the hydrophillic 0.035 guidewires. Multiple stenotic lesions, particularly in a single runoff vessel, may demand the use of fine balloon catheters (3.5F) and 0.014 guidewires.

Generous use of vasodilators is recommended to prevent or treat arterial spasm. Tolazoline up to a dosage of 25mg is helpful in the prevention of spasm, but in case spasm occurs GTN in 100 g boluses, up to a maximum of 500 g, should be considered.

Over the past six years, we have treated up to 30cm long total occlusions of tibial arteries with excellent results in the treatment of critical limb ischemia [31-33]. The commonly used system involves the use of 5F 3mm balloon and 120cm long catheter (Schneider U.K. Ltd.) and a curved tip 0.035, 180cm long hydrophillic guidewire (Terumo, Tokyo, Japan). As in the femoropopliteal segment, the aim is to loop the hydrophillic guidewire, which then allows progress through the entire length of the occlusion. This loop in the guidewire also allows re-entry back into the true lumen distally, as long as there is a reasonable distal patent vessel and the outflow arteries are not too heavily calcified or diffusely diseased (Fig. 4).

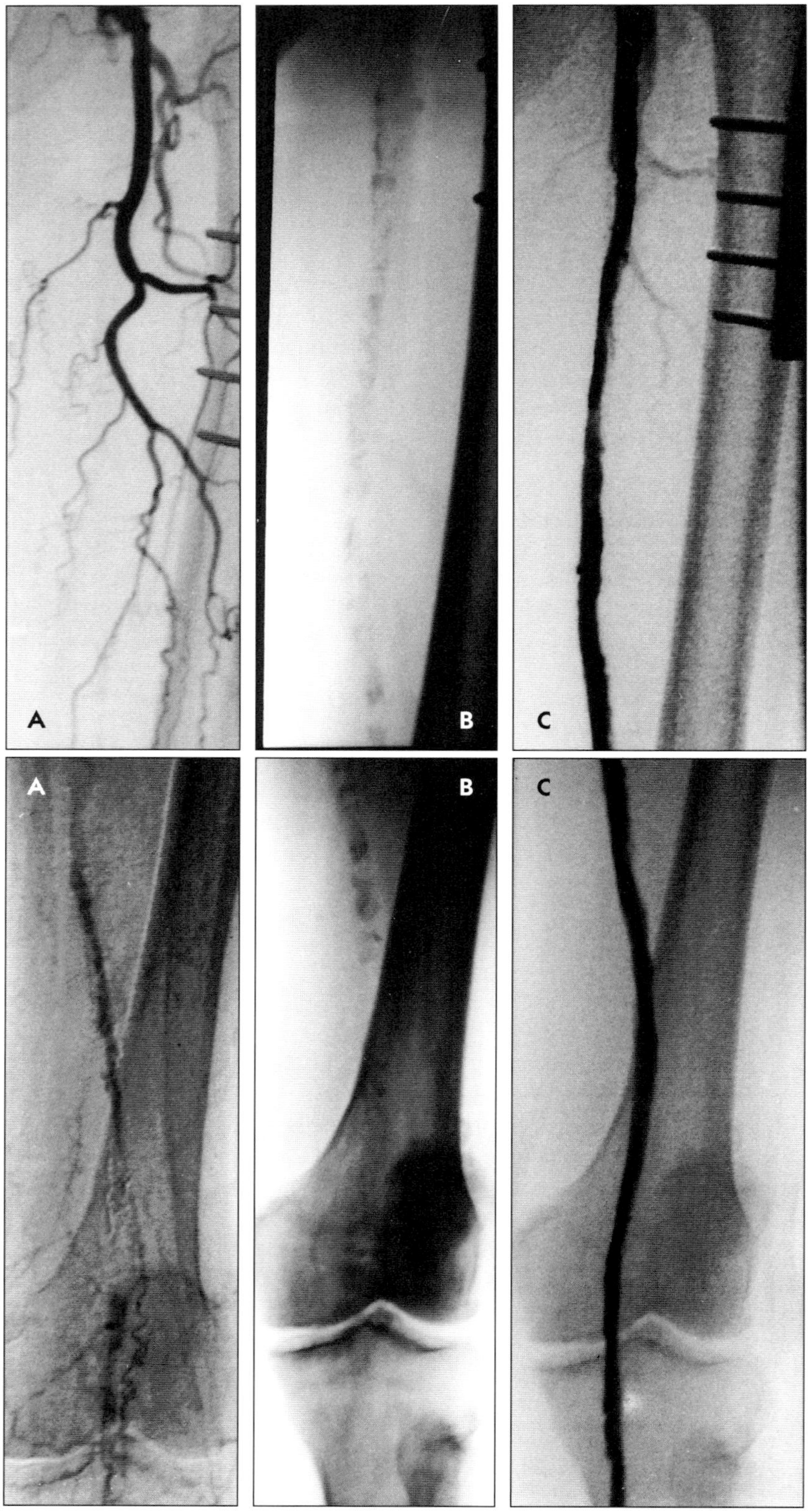

FIG. 1 (A and B) This 65-year-old patient with ischemic ulceration in his leg had a full length occlusion of the superficial femoral artery and a further occlusion of the popliteal artery demonstrating tandem disease. In addition, extensive calcification is noted throughout the length of the artery. The superficial femoral and popliteal artery occlusions were successfully recanalized by subintimal angioplasty with satisfactory resolution of his critical ischemia (C).

■ CLINICAL RESULTS

In 1993, a large national survey was carried out by the Vascular Surgical Society in the United Kingdom on critical limb ischemia and its management and outcome in 679 critically ischemic limbs. This survey indicated a prevalence of CLI of 1 in 2,500 annually. Thirty percent of the patients were diabetics and almost 70% of patients were offered some form of revascularization as the primary treatment option, with a 75% chance of successful limb salvage. Seventy-four percent of the patients had rest pain, 34% had gangrene and 31% had ulceration. Forty-two percent of the patients underwent bypass surgery, 21% received angioplasty, 16% had primary major amputation and 21% had conservative management. These results are summarized in the Table.

A similar survey was carried out locally at the Leicester Royal Infirmary. In a prospective study, 188 patients with 222 critically ischemic limbs were presented to a single vascular unit in a 12 month period. An attempt for revascularization was made in 73%. PTA was carried out in 42% of the patients, bypass surgery in 24%, a combination of surgery and PTA in 6%, primary major amputation in 10% and conservative management in 17%. The results are summarized in the Table.

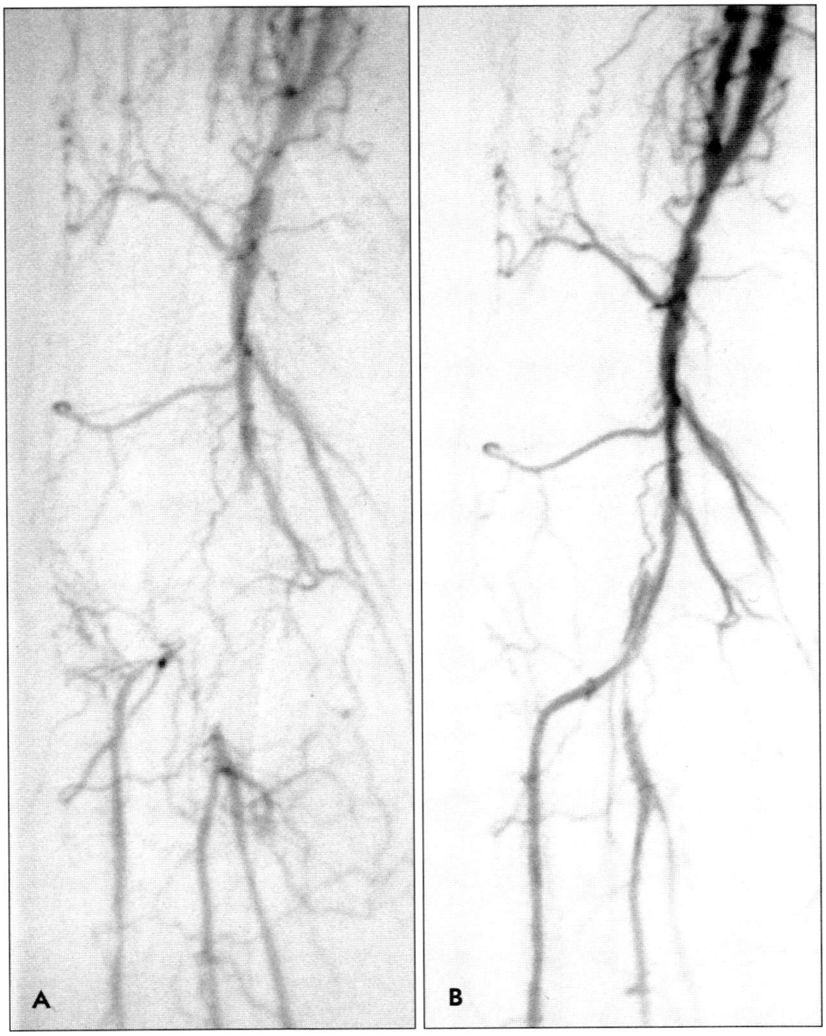

FIG. 2 This patient with a femoropopliteal above-knee graft has a distal popliteal occlusion that extends into all runoff vessels. Satisfactory recanalization of the popliteal artery into all the right off vessels was carried out.

The impact of S-PTA in critical limb ischemia is most noticeable in this study where, even in 1994, 42% of the patients with critical limb ischemia were managed by PTA alone, compared to the national average of 22%. This difference can only be explained by our extensive experience in the treatment of femoropopliteal and tibial occlusive disease by S-PTA. In fact at the present time, about 60% of CLI are treated with PTA, of which S-PTA forms a substantial proportion. This high proportion of patients with CLI treated with PTA is possible due to a deliberate policy to attempt PTA wherever possible as a first line treatment.

As a result of a large proportion of the patients being treated with minimally invasive techniques, the overall hospital stay was 10 days (national average: 24 days), with a 30-day mortality of 10% (national average: 14%), but a broadly similar overall limb salvage rate of 79% (national average: 68%). The mean hospital stay for angioplasty was in fact only 4.5 days compared to 16 days for bypass surgery and 18 days for primary amputation.

As a follow-up to the above study, a report was recently published [29]. In this study the results of surgery and angioplasty for the treatment of chronic severe lower limb ischemia were presented. Out of

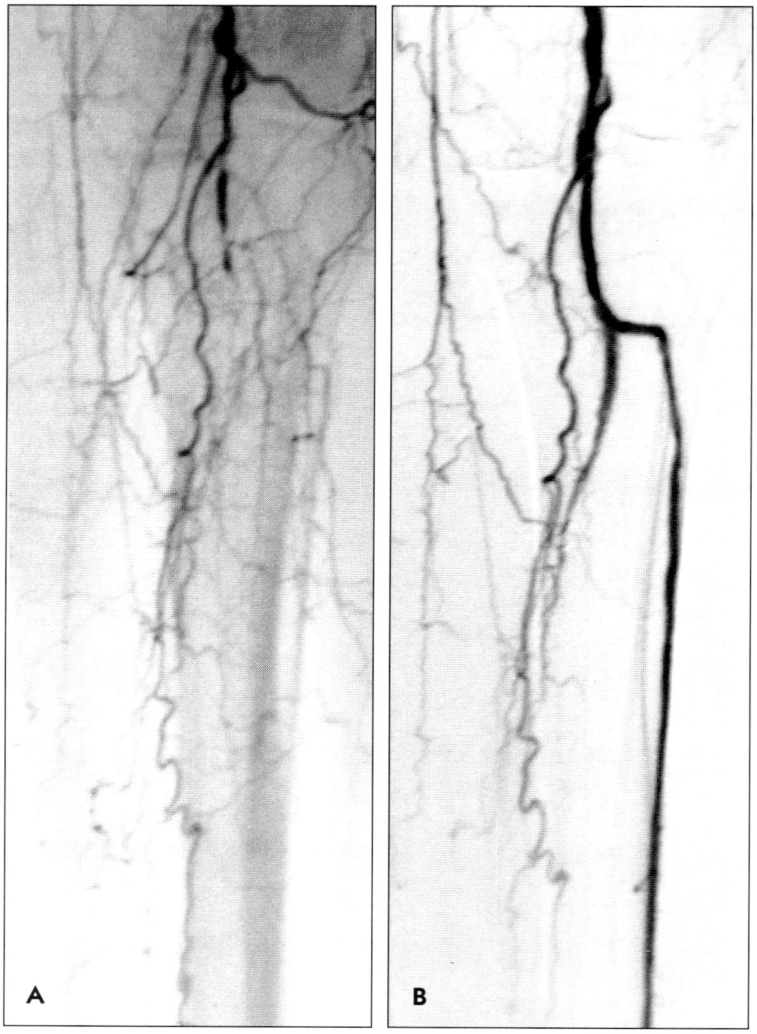

FIG. 3 There is severe disease of the distal popliteal artery, which terminates into an occlusion that extends into all runoff vessels. Successful recanalization into the anterior tibial artery was carried out with good results.

Table DATA FROM A NATIONAL AND LOCAL AUDIT ON DIFFERENT TREATMENT MODALITIES FOR CLI AND CLINICAL OUTCOME

	Vascular Surgical Society National Survey (1993)	*Leicester Royal Infirmary Audit (1994)*
	679 limbs	**222 limbs**
Bypass surgery	42%	24%
PTA	21%	42%
Surgery + PTA	6%	6%
Primary amputation	16%	10%
Conservative treatment	21%	17%
Hospital stay	24 days	10 days
Mortality (30d)	14%	10%
Overall limb salvage	68%	79%

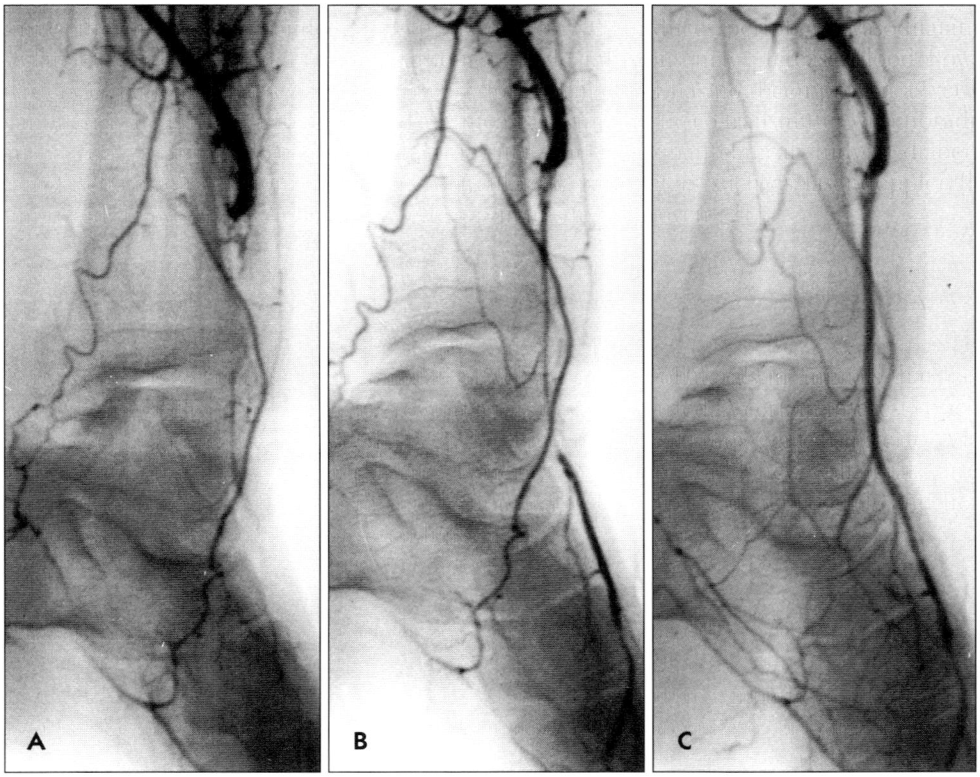

FIG. 4 There is a short occlusion of the anterior tibial artery beyond the femoral distal graft, thus threatening patency of the graft. Successful recanalization was carried but there was spasm in the proximal part of the artery. Further angioplasty and the use of antispasmodic agents has resulted in a satisfactory recanalization of the anterior tibial artery. Notice the dramatically improved foot circulation.

the 188 patients from the previous study, it was possible to obtain follow-up in 180 patients. The overall patient survival and limb salvage rate at 12 months were 75% and 71%, respectively. More significantly, the limb salvage rate for surgery and PTA was similar at 76% (Fig. 5).

There have been other, small series reporting treatment of tibial arteries with high success rates, though the majority of treatments involved stenotic rather than occlusive disease [7,21,34,35]. Most of these series indicate high technical success rates, and encouraging results in the management of CLI.

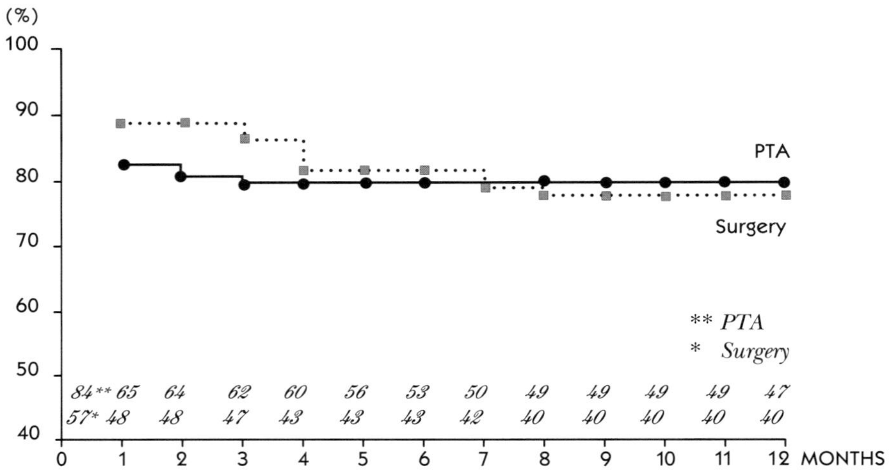

FIG. 5 Life table showing cumulative limb salvage for surgery and PTA, standard error less than 10% at all time points.

■ GENERAL CONSIDERATIONS

Several advances in the percutaneous management of peripheral vascular disease have allowed successful treatment of lesions in the distal arterial tree. Hydrophillic steerable guidewires, low profile balloon catheters, the availability of digital subtraction angiography equipment and the generous use of antispasmodic therapy has allowed extensive use of PTA as a means of treating crural arteries in critical ischemia.

The technique of S-PTA and its extensive use in occlusive disease of the femoropopliteal segment and the crural vessels has extended the scope of treating long occlusions and complicated lesions in the management of CLI. As a result of this, a large number of patients can now be treated by this method, thus allowing a large proportion of the patients with CLI to be treated with minimally invasive techniques.

While the treatment using S-PTA has been confined to occlusive disease in the femoropopliteal segment and the crural arteries, a number of patients have been treated with extensive long diffuse stenotic disease in the presence of CLI, in the hope of providing long-term benefits. Some patients with severe claudication rather than CLI and infrapopliteal arterial occlusive disease have also been treated by this method. Claudicants have been treated only if symptoms are severe enough to justify treatment.

There are only a limited number of reports on the treatment of crural artery disease, particularly occlusive disease. Most of the series include a small number of patients, but the recent series have shown encouraging results for the treatment of CLI.

Angioplasty of tibial artery occlusive disease and the treatment of tibial and femoropopliteal disease in the presence of CLI appear to be safe and successful with good initial technical results. In our hands, the experience of S-PTA has indicated that difficult and complicated femorodistal surgery can be avoided in a large number of cases.

Minimally invasive treatment of CLI by means of PTA offers many advantages. Avoidance of general anesthesia, reduced morbidity and mortality, shorter hospital stay, quick recovery, and minimal costs are some of the benefits of PTA. Finally, following failed PTA, surgical options are not jeopardized and can still be offered to the patient.

REFERENCES

1 Anonymous. Second European Consensus Document on Chronic critical leg ischaemia. European Working Group on Critical leg ischaemia. *Circulation* 1991; 84 (Suppl IV): 1-26.

2 Norgren L. Definition, incidence and epidemiology. In: Dormandy JA, Stock G (eds.). *Critical leg ischaemia: its pathophysiology and management.* Berlin, Springer Verlag, 1990, pp 7-13.

3 Anonymous. Critical limb ischaemia: management and outcome. Report of a national survery. The Vascular Surgical Society of Great Britain and Ireland. *Eur J Vasc Endovasc Surg* 1995; 10: 108-113.

4 Dormandy JA, Thomas PRS. What is the natural history of a critically ischaemic patient with and without his leg? In: Greenhalgh RM, Jamieson CW, Nicolaides AN (eds.). *Limb salvage and amputation for vascular disease.* Philadelphia, WB Saunders, 1988, pp 11-26.

5 Tyrrell MR. Wolfe JH. Critical leg ischaemia: an appraisal of clinical definitions. Joint Vascular Research Group. *Br J Surg* 1993; 80: 177-180.

6 Moore WS. Therapeutic options for femoro-popliteal occlusive disease. *Circulation* 1991; 83 (Suppl): I191-I193.

7 Schwarten DE. Clinical and anatomical considerations for nonoperative therapy in tibial disease and the results of angioplasty. *Circulation* 1991; 83 (Suppl): I186-I190.

8 Brown KT, Moore ED, Getrajdman GI, Saddekni S. Infrapopliteal angioplasty: long-term follow-up. *J Vasc Interv Radiol* 1993; 4: 139-144.

9 Pell JP, Whyman MR, Fowkes FG et al. Trends in vascular surgery since the introduction of percutaneous transluminal angioplasty. *Br J Surg* 1994; 81: 832-835.

10 Cooper JC, Welsh CL. The role of percutaneous transluminal angioplasty in the treatment of critical ischaemia. *Eur J Vasc Surg* 1991; 5: 261-264.

11 Currie IC, Wakely CJ, Cole SE et al. Femoropopliteal angioplasty for severe limb ischaemia. *Br J Surg* 1994; 81: 191-193.

12 Hasson JE, Acher CW, Turnipseed WD et al. Lower extremity percutaneous transluminal angioplasty: multifactorial analysis of morbidity and mortality. *Surgery* 1991; 110: 5A.

13 Adar R, Critchfield GC, Eddy DM. A confidence profile analysis of the results of femoropopliteal percutaneous transluminal angioplasty in the treatment of lower-extremity ischemia. *J Vasc Surg* 1989; 10: 57-67.

14 Beck AH, Muhe A, Ostheim W et al. Long-term results of percutaneous transluminal angioplasty: a study of 4750 dilatations and local lyses. *Eur J Vasc Surg* 1989; 3: 245-252.

15 Hunink MG, Donaldson MC, Meyerovitz MF et al. Risks and benefits of femoropopliteal percutaneous balloon angioplasty. *J Vasc Surg* 1993; 17: 183-194.

16 Capek P, McLean GK, Berkowitz HD. Femoro-popliteal angioplasty. Factors influencing long-term success. *Circulation* 1991; 83 (Suppl): 170-180.

17 Cambria RP, Faust G, Gusberg R et al. Percutaneous angioplasty for peripheral arterial occlusive disease. Correlates of clinical success. *Arch Surg* 1987; 122: 283-287.

18 Jeans WD, Armstrong S, Cole SE et al. Fate of patients undergoing transluminal angioplasty for lower-limb ischemia. *Radiology* 1990; 177: 559-564.

19 Johnston KW, Rae M, Hogg Johnston SA et al. Five-year results of a prospective study of percutaneous transluminal angioplasty. *Ann Surg* 1987; 206: 403-413.

20 Motarjeme A. PTA and thrombolysis in leg salvage. *J Endovasc Surg* 1994; 1: 81-87.

21 Buckenham TM, Loh A, Dormandy JA, Taylor RS. Infrapopliteal angioplasty for limb salvage. *Eur J Vasc Surg* 1993; 7: 21-25.

22 Rush DS, Gewertz BL, Lu CT et al. Limb salvage in poor-risk patients using transluminal angioplasty. *Arch Surg* 1983; 118: 1209-1212.

23 Glover JL, Bendick PJ, Dilley RS et al. Balloon catheter dilatation for limb salvage. *Arch Surg* 1983; 118: 557-560.

24 Lu CT, Zarins CK, Yang CF, Turcotte JK. Percutaneous transluminal angioplasty for limb salvage. *Radiology* 1982; 142: 337-341.

25 Saab MH, Smith DC, Aka PK et al. Percutaneous transluminal angioplasty of tibial arteries for limb salvage. *Cardiovasc Intervent Radiol* 1992; 15: 211-216.

26 Schneider E, Gruntzig A, Bollinger A. Percutaneous transluminal angioplasty for limb salvage - progress in therapy for severe leg ischaemia with rest pain and gangrene. In: Dotter CT, Gruntzig AR, Schoop W, Zeitler E (eds.). *Percutaneous transluminal angioplasty. Technique, early and late results.* Berlin, Springer Verlag, 1983, pp 222-226.

27 Jorgensen B, Henriksen LO, Karle A et al. Percutaneous transluminal angioplasty of iliac and femoral arteries in severe lower-limb ischaemia. *Acta Chir Scand* 1988; 154: 647-652.

28 Varty K, Nydahl S, Butterworth P et al. Changes in the management of critical limb ischaemia. *Br J Surg* 1996; 83: 953-956.

29 Varty K, Nydahl S, Nasim A et al. Results of surgery and angioplasty for the treatment of chronic severe lower limb ischaemia. *Eur J Vasc Endovasc Surg* 1998; 16: 159-163.

30 Bolia A, Fishwick G. Recanalization of iliac artery occlusion by subintimal dissection using the ipsilateral and the contralateral approach. *Clinical Radiology* 1997; 52: 684-687.

31 Bolia A, Sayers RD, Thompson MM, Bell PRF. Subintimal and intraluminal recanalization of occluded crural arteries by percutaneous balloon angioplasty. *Eur J Vasc Surg* 1994; 8: 214-219.

32 Nydahl S, London NJM, Bolia A. Technical Report: recanalisation of all three infrapopliteal arteries by subintimal angioplasty. *Clinical Radiology* 1996; 51: 366-367.
33 Nydahl S, Hartshorne T, Bell PRF et al. Subintimal angioplasty of infrapopliteal occlusions in critically ischaemic limbs. *Eur J Vasc Endovasc Surg.* 1997; 14: 212-216.
34 Flueckiger F, Lammer J, Klein GE, et al. Percutaneous transluminal angioplasty of crural arteries. *Acta Radiologica* 1992; 33: 152-155.
35 Bakal CW, Sprayegen S, Scheinbaum K et al. Percutaneous transluminal angioplasty of the infrapopliteal arteries: results in 53 patients. *AJR Am J Roentgenol* 1990; 154: 171-174.

13

CHANGING PATTERNS IN THE TREATMENT OF LOWER LIMB ISCHEMIA

SAMUEL S. AHN, KYUNG M. RO

Recent advances in endovascular technology have generated numerous alternative procedures for the treatment of arterial occlusive disease. In the United States, there are reportedly more than 400 000 hospitalizations for peripheral arterial occlusive disease per year, including approximately 50 000 angioplasties, 110 000 bypasses, and 69 000 foot or lower-limb amputations [1,2]. Currently, approximately 100 000 procedures are performed each year for chronic lower limb ischemia; moreover, it is estimated that greater than 10% of the population over the age of 70 and 1-2% of the population between the ages of 37 to 69 suffer from claudication [3-5].

While the number of patients with vascular disease has continued to rise over the years, increasing treatment options have paralleled this growth. The advent and expansion of endovascular techniques have brought changing patterns to the treatment of peripheral arterial occlusive disease, with similar trends in both the U.S. and in Europe. The proportion of aortobifemoral bypasses performed have reportedly decreased with a corresponding increase in the number of extra-anatomic grafts; in addition, the number of atherectomies, stents, endografts and particularly percutaneous angioplasties used to treat aortoiliac and femoropopliteal disease have markedly increased over the years [6]. A recent study reported by Whiteley et al. [6] showed a multifold increase in the number of percutaneous angioplasties and iliac stent procedures performed for iliac occlusive lesions from 1988-1994. The use of these less invasive procedures offers the advantage of shorter hospital stay, decreased morbidity and mortality, and minimal tissue trauma, however, these procedures have different indications, which are restricted further by their inherent limitations.

■ PERCUTANEOUS TRANSLUMINAL ANGIOPLASTY

The use of percutaneous transluminal angioplasty (PTA) for the treatment of iliac and femoropopliteal stenosis and short occlusions have been well established and long-term results for this treatment modality are now readily available. It appears that patients with short lesions in larger vessels have better patency rates that those with longer lesions localized to the superficial femoral artery [7,8]. PTA is a good option for those patients with mild disease who are not surgical candidates or patients with disseminated disease for whom surgery would come at a tremendous risk. In addition, PTA is often brought to bear as an alternative to or in conjunction with open reconstruction. However, in comparing the results of PTA with bypass surgery, the reports are conflicting [9-12]. Pell et al. [13] have shown that although there has been an increasing rate of arterial reconstruction over the years, this is attributable to the increase in the use of PTA for femoropopliteal disease. However, the extended use of PTA for critical limb ischemia did not decrease the frequency of arterial reconstruction. Furthermore, the authors reported that the rate of arterial reconstruction for critical limb ischemia did not increase, suggesting that the increase in reconstruction rates may be due to treatment of patients with claudication.

Wilson et al. [14], in their life table analysis comparing percutaneous transluminal angioplasty versus surgery for patients with iliac and femoral artery lesions, found that PTA had a higher initial technical failure rate, however, in those patients whose stage of disease allowed a choice between PTA or open revascularization for symptomatic peripheral arterial disease, both PTA and surgery had similar results at a mean follow-up of two years. The reported initial technical failure rate for PTA was 15%, which was comparable to the failure rates reported at that time. However, a long-term analysis performed by Johnston [15] in 1993 has indicated otherwise, with three-year patency rates falling below that anticipated for open revascularization.

In an analysis performed by Hunink et al. [16] in 1995 comparing PTA versus bypass for femoropopliteal disease, the authors concluded that angioplasty should be performed in patients who present with disabling claudication and femoropopliteal stenosis or occlusion as well as those patients with chronic critical ischemia and stenosis. However, they indicated that bypass is best suited for patients with chronic critical ischemia and a femoropopliteal occlusion. This recommendation has been corroborated further by Hartnell et al. [17], who have suggested that surgical revascularization or combined therapy be used in the presence of critical ischemia. In a recently published study, Zdanowski et al. [18] reported the shifting trend of the use of PTA for arterial occlusive disease in Sweden. This study, which evaluated the influence of revascularization on the mortality rate in patients with critical limb ischemia, concluded that the outcome following surgery versus PTA were comparable with respect to short-term survival and amputation. Furthermore, the authors noted that the number of percutaneous angioplasties performed for iliac, femoropopliteal and popliteotibial artery occlusive disease increased in Sweden from 25 to 75%, 5 to 50%, and 0 to 25%, respectively, between 1980 and 1995. In the analyses performed by Jeans et al. [11] and Tunis et al. [19] indicated an increase in the rate of arterial reconstruction associated with an increase in the number of PTAs performed; however, they could not delineate those procedures performed for claudication from chronic limb ischemia. They went further to conclude that PTA resulted in more patients being treated, indicating that the use of PTA for chronic limb ischemia could not effectively attenuate the relative number of reconstruction.

■ ATHERECTOMY

Mechanical atherectomy has been developed as an alternative to conventional percutaneous balloon angioplasty (PTA) because of the limitations of PTA. Theoretically, atherectomy has several advantages over percutaneous balloon angioplasty, namely greater immediate success rate with less dissection and acute occlusion, ability to treat complex lesions, and lower restenosis rates. Atherectomy physically removes occlusion by cutting or pulverizing by means of mechanical catheter-deliverable endarterectomy devices. Since its inception, several atherectomy devices, both extirpative and ablative, have been designed to meet the current challenges, however, critical examination of these devices indicate that atherectomy currently has limited application for the treatment of peripheral arterial occlusive disease.

According to recent reports, the use of atherectomy for the treatment of vein graft stenosis appears promising [20,21]. Porter et al. [20] evaluated their outcome with the use of atherectomy or the treatment of infrainguinal vein graft stenoses over a 6-year period. With a mean follow-up of 21 months in a group of 42 patients (52 procedures and 67 stenoses), the primary patency rates were 82%, 78%, and 78%, at 1, 2, and 3 years, respectively. In a study conducted by Dolmatch et al. [21] for atherectomy of anastomotic bypass graft stenosis, the authors reported a technical success rate of 92%, with an 88% graft patency rate at 14 months. In 74% of the areas of stenosis, there was less than 50% restenosis at those sites. The authors concluded that directional atherectomy is an effective treatment method with intermediate patency rates superior to those reported for PTA and comparable to those reported for conventional open revision. Furthermore, in their 4-year study of 17 patients, they concluded that the patency rates for those treated with directional atherectomy alone were the same as those who were treated with a combined therapy of atherectomy and PTA.

In contrast, Osborn et al. [22] performed a combined therapy of balloon angioplasty and atherectomy in a group of 88 highly selective patients (96 limbs) for the treatment of superficial femoral and popliteal artery stenoses. They reported that at a mean follow-up of 24 months, 71% of the limbs maintained their postoperative improvement according to the categorical reporting standards of the SVS/ISCVS. In addition, a recent randomized clinical trial comparing the use of balloon angioplasty versus atherectomy, Ahn et al. found that a combined therapy of balloon angioplasty and atherectomy yields higher intermediate patency rates than previously reported for either treatment modality alone.

■ STENTS

The use of expandable metallic stents for the treatment of occlusive disease was refined during the 1990s following the first published experience placing metallic stents in the peripheral circulation in 1987 by Sigwart et al. [23]. A common prototype for peripheral implantation is one that can be constrained on a small diameter delivery catheter and can be either balloon or self-expanding; this model allows flexibility in the length of the segment but the unsupported segments may be prone to compression. Before the development of intraluminal stents, patients in whom PTA proved unsuccessful were treated with open revascularization. Stents provided a less invasive alternative to treat the complications of PTA and improved long-term patency.

Intraluminal metallic stents are considered to be of particular use in the management of PTA-related complications such as dissection, recoil, and restenosis and are now a commonly used intervention against peripheral arterial occlusive disease. Expandable metallic stents placed into a stenotic or occluded iliac artery are technically feasible and reported outcomes have been generally favorable. Nonetheless, stents should be used with caution as the procedure enhances the development of myointimal hyperplasia [24]. Stents allow for both the treatment of residual stenosis following PTA as well as PTA-induced dissection without open surgery.

Ballard et al. [25] reported their series of 147 iliac artery stents (both *Wallstent* and *Palmaz*) placed in 98 limbs for severe claudication or limb-threatening ischemia and achieved cumulative patency rates of 87.6%, 61.9%, and 55.3% at 12, 18, and 24 months, respectively. Ideal candidates appeared to be those with common iliac artery occlusive disease or eccentric stenoses, patent superficial femoral artery, and three-vessel tibial runoff. Furthermore, long-term success of stent placement was limited by an occluded superficial femoral artery occlusion and poor infrainguinal runoff. As suggested in the report by Vorwerk et al. [26], late obstruction of iliac artery stents may be treated with percutaneous reintervention before open surgery is attempted.

Limitations of stenting include progression of arteriosclerosis in regions distal and proximal to stent placement as well as the development of intimal hyperplasia within the stented segment. Stenotic lesions distal to stent placement are reported to occur at a rate of 5% per year [27,28], as is the case with the patency of all vascular reconstructions, the progression of atherosclerotic disease will also undoubtedly be a limitation in stent patency. In addition, various investigators have reported the occurrence of endovascular stent infection. Although rare, stent infections can be potentially devastating and surgeons should consider use of prophylactic antibiotics [29,30].

■ ENDOVASCULAR GRAFTS

The development of metallic stents led to the expansion and evolution of numerous prototype endoluminal *stent-grafts* in an effort to extend this technology for the treatment of arterial occlusive disease. Because the endoluminal stent-graft acts as an endovascular bypass, patency rates are expected to surpass those obtained with conventional balloon angioplasty. Furthermore, an endoluminal bypass provides the potential advantage in that the graft can be tailored to the lesion length.

Peripheral endoluminal stent-grafts, both supported and unsupported, are distinguished by the type of graft fabric and framework used in their construction, however, most are composed of polytetrafluoroethylene (PTFE) or polyester (Dacron). PTFE, which has the ability to dilate over a wider range, has both the advantage of inserting a smaller diameter graft and enlarging it with a balloon and the disadvantage associated with recoil following balloon dilatation performed in-situ. Dacron, which has a known histocompatibility, is cumbersome and lacks the elasticity of PTFE. In a prospective randomized study conducted by Friedman et al. [31] comparing the PTFE and Dacron grafts, the authors were unable to determine any significant differences in long-term results between the two grafts.

In the iliac artery, intraluminal stent-graft placement have improved patency when compared to PTA, which is attributed to the greater technical success achieved by creating a wider lumen and thus dampening restenosis rates [32,33]. Pernes et al. [34], in their experience with the treatment of iliac occlusions using the *Cragg Endopro System 1* (Minimally Invasive Technologies, La Ciotat, France), reported a patency rate of 80% at a mean follow-up of seven months. Cynamon et al. [35] successfully placed 18 endoluminal stent-grafts for advanced aortoiliofemoral disease in 17 limb ischemic patients; at a mean follow-up of 21 months, they had a primary patency of 81% with a reported seven deaths resulting form myocardial infarct and two occlusions. These results have led to the heightened use of intraluminal stents in the iliac artery, however, results in the femoropopliteal arteries are less promising. Henry et al. [36] reported primary patency rates of 98% at the iliac level for both stenotic and occlusive lesions but a dismal 59% primary patency for lesions at the femoral level.

As stent-grafts become more readily available for use, clinical applications have increased. Numerous investigators have reported their experience with various endoluminal stent-grafts in patients with occlusive disease with the presumption that endoluminal bypass performed with a stent-graft may provide improved patency by minimizing the development of intimal hyperplasia [34-36]. Others, in their experience with the stent-graft system for the treatment of aneurysmal disease, found the stent-graft to be useful in isolating the aneurysmal portion of the artery with an endovascular lining and decreasing the risk of rupture [37-39]. In 1995, Marin et al. [40] reported their series of patients (n=11) treated with endoluminal stent-grafts for iliac aneurysms. They achieved a 100% patency rate at a mean follow-up of 11 months. Henry et al. [36] reported that of the seven aneurysms treated with the *Cragg Endopro System 1 stent-graft,* one patient expired from a myocardial infarct, another thrombosed requiring a bypass, and five remained patent.

Although considered beneficial in the treatment of aneurysmal disease, the benefits of the use of stent grafts for occlusive disease is uncertain. Its application at the femoropoliteal level is limited due to the small diameter of the vessel. Currently, primary patency rates for grafts placed in the infrainguinal region are similar to those obtained with conventional PTA. For the treatment of occlusive disease, stent-grafts serve to isolate the occlusive region from the flow lumen and minimizing the development of intimal hyperplasia following intervention with PTA. However, stent-grafts do not prevent injury occurring at the end regions of the grafts, which is actually the most common restenosis site. With modifications in stent design, these potential complications can be minimized. The endoluminal approach helps to avoid major abdominal aortic bypass or extraanatomic bypass in patients with advanced disease and those who are at considerable risks for conventional open surgery. Although short and mid-term results appear promising, long-term follow-up with larger patient series are required to truly assess the benefits of this less invasive technique.

GENERAL CONSIDERATIONS

The patterns in treatment for aortic occlusive disease have changed dramatically over the years. The past several decades have brought about immense changes in the management of peripheral arterial occlusive disease. No longer are we faced with the limited option of open reconstruction or no surgery. The diversity in treatment options have created interventions that can be utilized at various stages of the disease. Despite the fact that the overall mortality has not changed significantly, as complications from arterial rupture, bleeding and death still occur, endovascular interventions have the additional benefit of lower cost, minimal hospital stay, and decreased mortality and morbidity.

Coupled with the growing variety in treatment options, the number of procedures performed each year has increased as the indications have broadened. This may be attributed, in part, to the liberal application of criteria for indications. This progression towards minimally invasive techniques for the treatment of peripheral arterial disease has brought about additional changes in the preoperative investigation of arterial disease. In a recent study examining the sole use of non-invasive tests prior to lower extremity reconstruction and endovascular surgery, Sarkar et al. [41] found that arterial reconstruction without a preoperative angiogram is not only safe, but also a feasible alternative to a routine preoperative angiogram. The life table analysis demonstrated primary patency rates of 92% and 82% for iliac and infrainguinal reconstruction, respectively, signifying the accuracy of the combined non-invasive tests in delineating the arterial lesions and allowing to plan the revascularization without a separate preoperative diagnostic angiogram.

With the dramatic transition from the standard open reconstruction for chronic limb ischemia towards less invasive therapies, there is minimal change in the overall complication rates. Although open surgery is associated with longer hospitalization and increased risk of systemic complications such as myocardial infarcts and respiratory failure, endovascular surgery has the related risk of hematomas at the groin puncture site, pseudoaneurysms, and arteriovenous fistulas. However, technology improves and instruments become readily available, thus decreasing intra- and postoperative morbidity and mortality and improving the results of these endovascular treatment modalities. In addition, the combination of various endovascular therapies, such as atherectomy performed as an adjunct to PTA and endarterectomy performed along with stent-graft placement, may improve overall technical success and patency rates. Additional randomized clinical trials with long-term follow-up are clearly warranted.

REFERENCES

1 Gillum RF. Peripheral arterial occlusive disease of the extremities in the United States: hospitalization and mortality. *Am Heart J* 1990; 120: 1414-1418.

2 U.S. dept. of health and human services. Detailed diagnosis and procedures: national hospitalization discharge survey, 1990. *Vital Health Stat* 1992; 13.

3 Rutkow IM, Ernst CB. An analysis of vascular surgical manpower requirements and vascular surgical rates in the United States. *J Vasc Surg* 1986; 3: 74-83.
4 Droller H, Pemberton J. Cardiovascular disease in a random sample of elderly people. *Br Heart J* 1953; 15: 199-201.
5 Peabody CN, Kannel WB, McNamara PM. Intermittent claudication. Surgical significance. *Arch Surg* 1974; 109: 693-697.
6 Whiteley MS, Ray-Chaudhuri SB, Galland RB. Changing patterns in aortoiliac reconstruction: a 7-year audit. *Br J Surg* 1996; 83: 1367-1369.
7 Murray RR Jr, Hewes RC, White RI Jr et al. Long-segment femoropopliteal stenoses: is angioplasty a boon or a bust? *Radiology* 1987; 162: 473-476.
8 Capek P, McLean GK, Berkowitz HD. Femoropopliteal angioplasty: factors influencing long-term results. *Circulation* 1991; 83 (suppl 2): 170-180.
9 Wolf GL, Wilson SE, Cross AP et al. Surgery or balloon angioplasty for peripheral vascular disease: a randomized clinical trial. *J Vasc Intervent Radiol* 1993; 4: 639-648.
10 Holm J, Arfvidsson B, Jivegard L et al. Chronic lower limb ischemia: a prospective randomized study comparing the 1-year result of vascular surgery and percutaneous transluminal angioplasty (PTA). *Eur J Vasc Surg* 1991; 5: 517-522.
11 Jeans WD, Danton RM, Baird RN, Horrocks M. The effects of introducing balloon dilatation into vascular surgery practice. *Br J Radiol* 1986; 59: 457-459.
12 Anderson JB, Wolinski AD, Wells IP et al. The impact of percutaneous transluminal angioplasty in the management of peripheral vascular disease. *Br J Surg* 1986; 73: 17-19.
13 Pell JR, Whyman MR, Fowkes FG et al. Trends in vascular surgery since the introduction of percutaneous transluminal angioplasty. *Br J Surg* 1994; 81: 832-835.
14 Wilson SE, Wolf GL, Cross AP. Percutaneous transluminal angioplasty versus operation for peripheral arteriosclerosis: report of a prospective randomized trial in a selected group of patients. *J Vasc Surg* 1989; 9: 1-9.
15 Johnson KW. Iliac Arteries: reanalysis of results of balloon angioplasty. *Radiology* 1993; 186: 207-212.
16 Hunink MGM, Wong JB, Donaldson MC et al. Revascularization for femoropopliteal disease: a decision and cost-effectiveness analysis. *JAMA* 1995; 274: 165-171.
17 Hartnell G, Wakely C, Wyatt M, Horrocks M. Is iliac angioplasty indicated in the treatment of critical limb ischemia? *J Intervent Radiol* 1992; 7: 101-103.
18 Zdanowski Z, Troëng T, Norgren L. Outcome and influence of age after infrainguinal revascularisation in critical limb ischaemia: the Swedish vascular registry. *Eur J Vasc Endovasc Surgery* 1998; 16: 137-141.
19 Tunis SR, Bass EB, Steinberg EP. The use of angioplasty, bypass surgery, and amputation in the management of peripheral vascular disease. *N Eng J Med* 1991; 325: 556-562.
20 Porter DH, Rosen MP, Skillman RG et al. Mid-term and long-term results with directional atherectomy of vein graft stenoses. *J Vasc Surg* 1996; 23: 554-567.
21 Dolmatch BL, Gray RJ, Horton KM et al. Treatment of anastomotic bypass graft stenosis with direction atherectomy: short-term and intermediate-term results. *J Vasc Intervent Radiol* 1995; 6: 105-113.
22 Osborn JJ, Pfeiffer RB Jr, String ST. Directional atherectomy and balloon angioplasty for lower extremity arterial disease. *Ann Vasc Surg* 1997; 11: 278-283.
23 Sigwart U, Puel J, Mirkovitch J et al. Intravascular stents to prevent occlusion and restenosis after transluminal angioplasty. *N Engl J Med* 1987; 316: 701-70-6.
24 Cikrit DF, Gustafson PA, Dalsing MC et al. Long-term follow-up of the Palmaz stent for iliac occlusive disease. *Surgery* 1995; 118: 608-614.
25 Ballard Jl, Sparks SR, Taylor FC et al. Complications of iliac artery stent deployment. *J Vasc Surg* 1996; 24: 545-555.
26 Vorwerk D, Guenther RW, Schurmann K, Wendt G. Late reobstruction in iliac arterial stents: percutaneous treatment. *Radiology* 1995; 197: 479-483.
27 Vorwerk D, Guenther RW. Stent placement in the iliac arterial lesions: three years of clinical experience with the Wallstent. *Cardiovasc Intervent Radiol* 1992; 15: 285-290.
28 Palmaz JC, Laborde JC, Rivera FJ et al. Stenting of the iliac arteries with the Palmaz stent: experience from a multicenter trial. *Cardiovasc Intervent Radiol* 1992; 15: 291-297.
29 Deiparine MK, Ballard JL, Taylor FC, Chase DR. Endovascular stent infection. *J Vasc Surg* 1996; 23: 529-533.
30 Whelan DM, van Beusekom HM, van der Giesson WJ. Foreign body contamination during stent implantation. *Cath Cardiovasc Diag* 1997; 40: 328-332.
31 Friedman SG, Lazzaro RS, Spier LN et al. A prospective randomized comparison of Dacron and polytetrafluoroethylene aortic bifurcation grafts. *Surgery* 1995; 117: 7-10.
32 Raillat C, Rousseau H, Joffre R, Roux D. Treatment of iliac artery stenosis with the Wallstent endoprosthesis. *AJR Am J Roentgenol* 1990; 154: 613-616.
33 Martin EC, Katzen BT, Benenati JF et al. Multicenter trial of the Wallstent in the iliac and femoral arteries. *J Vasc Intervent Radiol* 1995; 6: 843-849.
34 Pernes JM, Auguste M, Hovasse D et al. Long iliac stenosis: initial clinical experience with the Cragg endoluminal graft. *Intervent Radiol* 1995; 196: 67-71.
35 Cynamon J, Marin ML, Veith FJ et al. Stent-graft repair of aorto-iliac occlusive disease co-existing with common femoral artery disease. *J Vasc Intervent Radiol* 1997; 8: 19-26.
36 Henry M, Amor M, Cragg A et al. Occlusive and aneurysmal peripheral arterial disease: assessment of a stent-graft system. *Radiology* 1996; 201: 717-724.
37 Cragg AH, Dake MD. Percutaneous femoropopliteal graft placement. *J Vasc Intervent Radiol* 1993; 4: 455-463.
38 Diethrich EB, Papazoglou CO. Endoluminal grafting for aneurysmal and occlusive disease for limb salvage. *Am J Surg* 1994; 168: 156-162.
39 Marin Ml, Veith FJ, Cynamon J et al. Transfemoral endovascular stented graft treatment for aorto-iliac and femoropopliteal occlusive disease for limb salvage. *Am J Surg* 1994; 168: 156-162.
40 Marin ML, Veith FJ, Lyon RT et al. Transfemoral endovascular repair of iliac artery aneurysms. *Am J Surg* 1995; 170: 179-182.
41 Sarkar R, Ro KM, Obrand DI, Ahn SS. Lower extremity vascular reconstruction and endovascular surgery without preoperative angiography. *Am J Surg* 1998; 176: 203-207.

14

ADJUNCTIVE TECHNIQUES TO PTA: THROMBOLYSIS, SUCTION EMBOLECTOMY AND ROTABLATOR

GIOACCHINO COPPI, ROBERTO SILINGARDI
ROBERTO MORATTO, STEFANO CAMPARINI, GIOVANNI RAGAZZI
STEFANO GENNAI, ROBERTO PACCHIONI

The initial introduction of fibrinolytic agents like streptokinase and urokinase in acute and subacute ischemia seemed to offer a wide range of medical prospectives, even though systemic use showed slow and frequently unsuccessful results, numerous complications and high costs. The advent of the balloon catheter created in 1963 by Fogarty [1] known simply as "the Fogarty" completely changed the therapeutic approach to acute limb ischemia. This catheter, considered to be the first therapeutic endovascular procedure, raised great enthusiasm for its simplicity of use and effectiveness and superseded the use of fibrinolytic therapy.

Towards the end of the 1970s and the following decade the limits of this method became evident and various proposals for a different medical approach were put forward. Even though the balloon catheter rapidly resolved embolisms in healthy arteries, the results it obtained when applied to arteriosclerotic arteries were very different: the number of amputations increased from 6% to 30% and the mortality rate rose from 18% to 25% [2]. Completion angiography during surgical interventions showed a frequent incompleteness in the attempt to unblock using the Fogarty catheter as well as the difficulty or impossibility to unblock small vessels.

In 1972 Greep et al. [3] had already shown that a Dormia catheter was capable of extracting more thrombi, even after the use of a Fogarty. In 1986 Zimmerman et al. [4] suggested that the embolectomy of the lower limb should be carried out under fluoroscopic guidance. In 1996 van Ommen et al. [5] officially confirmed the suspected intimal damage caused by the Fogarty catheter.

Growing evidence of the great effectiveness and fewer complications associated with the local application of fibrinolytic agents along with the evolution of the arterial catheters, which allow a more secure localized administration, have brought about a significant launch in thrombolytic therapy over the last decade, especially in combination with other surgical or endoluminal procedures.

■ THROMBOLYSIS

Fibrinolytic Agents

The thrombolytic agents presently used in clinical practice are streptokinase, urokinase, and rt-PA.

Streptokinase (SK), the leader in thrombolytic agents, is made up of a single protein chain produced by a number of strains of group A betahemolytic streptococcus. The latter may account for the allergic reaction encountered in 15%-20% of cases and even the anaphylactic shock in 4% of cases. Very often (in over 30% of cases) it can also cause fever [6-8]. Different from other thrombolytic agents, SK is not a direct activator of plasminogen, but activates the fibrinolytic system by a SK-plasminogen complex and shows a diphasic half-life of an initial 4 minutes and a final 30 minutes.

Urokinase (UK) is a purified protease extract obtained from human urine or from kidney cell cultures. It is a direct activator of plasminogen obtained with a DNA recombinant method, of which the half-life is 14-16 minutes. In comparison with SK it has some advantages, like a lower risk of bleeding and distal embolization, rare allergic symptoms, and it also permits the administration of lengthy treatments within shorter time limits. Its disadvantages are characterized by a modest stability at room temperature and relatively high costs as compared to SK [9,10].

Rt-PA (human recombinant plasminogen activator) is a second-generation thrombolytic agent obtained from the methodic DNA recombinant of tPA which is mainly responsible for the extrinsic physiologic activation of the plasminogen. It holds an elevated affinity for the plasminogen at the presence of fibrin, which renders it particularly active in the thrombus and ideal for local thrombolysis. Its half-life is about six minutes. Despite its high cost, it does present one notable advantage in comparison to other thrombolytics, and that is the fact that it achieves the lysis of the thrombus in less time: 1-6 hours compared to 48-72 hours with SK and 12-24 hours with UK. Rt-PA proves to be most favorable and to show the best results. There is, however, an on-going debate as to the actual risk of hemorrhage associated to rt-PA (11-36%) [7,11]. Actually there is no clinical proof to suggest a significant difference in the effectiveness and safety between UK and rt-PA.

Systemic Thrombolytic Therapy

The first clinical studies [12] using the systemic access of streptokinase showed modest results and revealed a high percentage of complications, like hemorrhages, stroke, thromboembolism and allergic reactions. The most recent case records underline just how the choice of patients, which undergo systemic thrombolysis, has reduced complications; nonetheless the results do not waver from a 30% success rate. The frequency and seriousness of these complications has changed the systemic administration of thrombolytic agents in the treatment of lower limb ischemia into localized applications.

REGIONAL FIBRINOLYTIC THERAPY

Regional thrombolysis has progressively found support on the grounds of clinical experience and pathophysiological knowledge. The knowledge about the half-life of the drug, the activator, the activated substrate and about the formation of the thrombus combined with a better knowledge of the interference on the thrombus, have allowed the development of appropriate drugs, like heparin and antiplatelet agents. Based on these principles, the concept of local administration has further developed, like selective infusion of the drug and activating the substrate inside the thrombus, thus allowing higher concentrations of the drug to act over a wider surface, reducing systemic effects. This has led to administering under pressure (Pulse Spray-AngioDynamics) which, with a vaporizing effect of the drug and destruction of the thrombus, results in a further acceleration of the process of thrombolysis [13].

Technique

The technique of pharmacological infusion of the fibrinolytic drug at a localized level foresees an arterial access by means of injection and the application of the introducers in the following ways: ipsilateral percutaneous femoral access (anterograde); contralateral femoral access (cross over) of the lesion to be treated; and brachial access. The puncture should be limited to the anterior wall of the vessel using small needles and small caliber introducers (4-5 Fr). A metallic guide wire is used to enter the site of the arterial thrombosis and if possible, also to pass through the thrombus in order to introduce the thrombolysis catheter to the inside of the thrombus. In the event that the guide wire does not penetrate the thrombus it is sufficient to use an infusion *end-hole* catheter to be able to infuse the trombolytic agent as close as possible to the thrombus. If the guide wire penetrates the thrombus, other types of catheters can be used such as *multiside hole* or *pulse spray* (with lateral microholes). The latter allows, by means of a particularly well devised pump, the injection of highly concentrated quantity of trombolytic material into the thrombus under high pressure. At difficult sites, coaxial catheters or infusion guide wires can also be of help. The infusion dosages are listed schematically in Table I.

TABLE I DOSAGES AND ESTIMATED EFFICACY TIMES

Low-dosage protocol	
Streptokinase : 5-10 000 U/hour Urokinase : 100 000 U/hour Rt-PA : 0.5-1 mg/hour	Lysis : 35 - 82% Time : 7-71 hours
High dosage protocol	
Urokinase : 4 000 U/min. first 2 hours 1 000-2 000 U/min. following hours	Lysis : 77 - 83% Time : 18-26 hours

The effectiveness and safety of one method as compared to another is still subject to further study even though McNamara et al. proposed that high dosage and reduced time treatment is providing better results [6,14].

In low-dosage protocols SK is infused at 5 000-10 000 U/h, UK at 100 000 U/h, and rt-PA at 0.5-1 mg/h. Looking at numerous case studies, a complete lysis of the thrombus is obtained in 35-82% of cases with infusion times between 7-71 hours and with significant hemorrhages in 8-24% of cases. In the high-dosage infusions that McNamara et al. proposed [6,14], the authors infuse UK in a dosage of 4 000 U/min. for 2 hours, followed by a dosage of 1 000-2 000 U/min. until the thrombus has resolved. A high-dosage regime requires an infusion time between 12 and 18 hours and brings an immediate success rate in 77-83% of cases without an increase in hemorrhage related complications.

In the most recent *pulse spray* technique, 20 mg of rt-PA up to 1 000 000 U of UK are injected under pressure through a special multiple-hole catheter system. A success rate of 90% can be observed with a lowered risk of hemorrhage [15-17]. The pulse spray (Fig. 1) system is particularly indicated in acute peripheral obstructions.

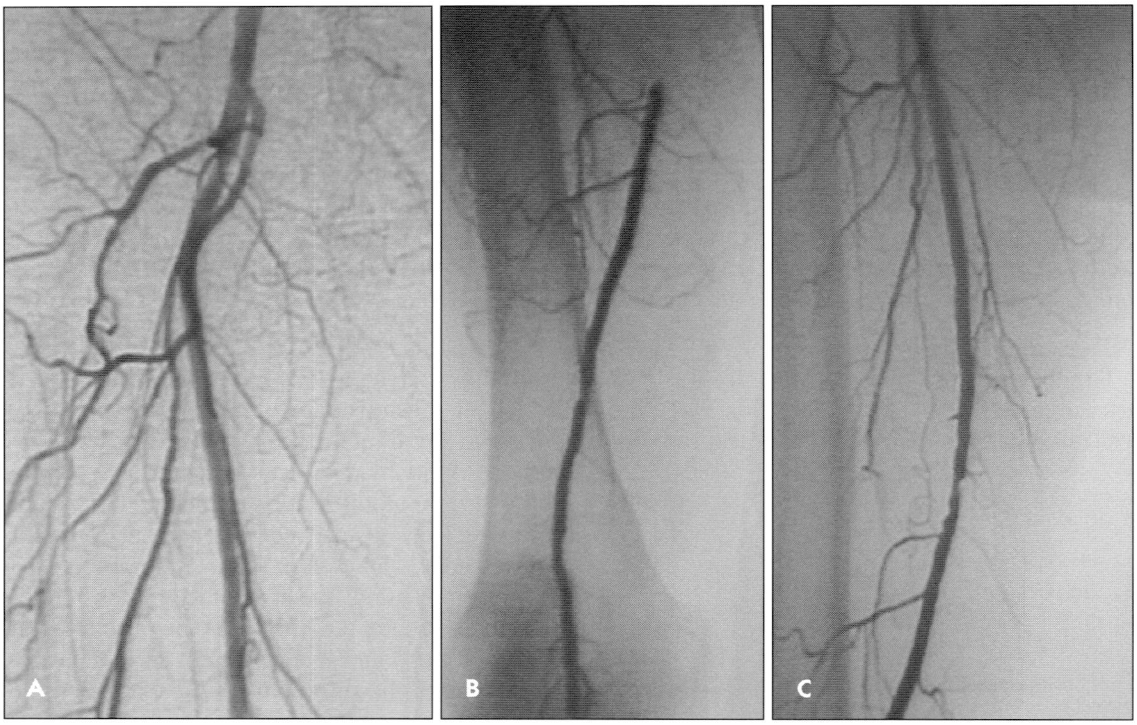

FIG. 1 Locoregional thrombolytic therapy of an acute occlusion of the superficial femoral artery. A, B - Angiography before treatment. C - Completion angiogram after thrombolysis: an artherosclerotic stenosis accessible for PTA.

In the field of thrombolysis there is still a controversy about the concomitant administration of heparin, as some would argue that it increases the risk of bleeding. This is usually used at a dosage of 500-1 000 U/h in order to avoid pericatheter thrombosis, to minimize potential peripheral embolizations or migration of the clot. In the absence of official guidelines our strategy is based on the following steps:

- maximum attention to access with small needles(19G) suitable for puncturing only the anterior wall of the vessel,
- use of thin introducers and catheters (3-5 Fr),
- preferably adopt contralateral arterial access (cross-over) whenever possible,
- avoid brachial access,
- use of the *pulse spray* (525 000 U of UK in 90 min.) whenever possible,
- include the use of heparin (10.000-20.000 U within 24 hours),
- discontinue the infusion if there are no signs of lysis after 2-4 hours.

Monitoring

The action of fibrinolytic agents on coagulation factors and subsequent possible depletion and risk of hemorrhage imposes a careful laboratory control. Attention should also be focused on the lytic effect on the target vessel in order to adapt the dosage of the drug and the position of the catheter. The monitoring of thrombolytic therapy is therefore performed by means of hematochemical tests and angiography. The levels of fibrinogen and FDP, APTT (Activated Platelet Thromboplatin Time) ratio, and platelet count during the procedure provide information on the state of the systemic lysis, but not on the effects on the thrombus; fibrinogen values to 100 mg% indicate a risk of bleeding even in the absence of actual hemorrhages, while values < 50 mg% require discontinuation of the infusion.

The progressive lysis of the thrombus is verified and visualized by selective angiography; the timing of the angiographic control is in function with the speed at which the thrombolysis proceeds, which itself is correlated to the dosages and to the method of infusion. Frequent controls are necessary during the pulse-spray infusion: every one to two hours with high dosage infusion and every four to six hours with a low dosage system. The times are doubled for simple localized infusion (extra thrombus), whilst a control every 6-12 hours is sufficient for systemic infusion.

Contraindications and complications

Based on the complications reported [9,18,19], the following contraindications for fibrinolytic therapy can be proposed:

- advanced ischemia (relative),
- hemorrhagic stroke within the first two months (absolute),
- gastrointestinal ulcer (relative),
- pregnancy (relative),
- serious hemorrhagic diathesis (absolute),
- recent vascular prosthetic surgery (absolute),
- uncontrolled hypertension (relative),
- hemorragic diabetic retinopathy (relative).

Complications are observed in 6-22% of cases [2,14,18,20] and include gastrointestinal hemorrhage and, less frequently, urological hemorrhage. However, when applied locally the incidence of complications is low. Other complications include severe bleeding at the arterial catheter site and retroperitoneal hemorrhages, often requiring surgical intervention.

CLINICAL SITUATIONS

Treatment of acute embolic occlusions

Being predominantly of cardiogenic origin, most embolic occlusions are treated surgically by means of the Fogarty catheter. The best results are obtained in patients that undergo embolectomy within eight hours of the acute event. With respect to limb salvage the results are favorable in 85% of the cases [6,21,22]. In recent years alternative and percutaneous techniques have been proposed like simple thrombosuction [7,23] or mechanical (*TEC, Kinsey, Amplaz*) or hydrolytic (*Hydrolyser Angiojet*) disintegration of the thrombus.

Treatment of acute thrombotic occlusions

The majority of thrombi occur on preexisting atherosclerotic lesions. In cases of acute thrombosis, urgent angiographic diagnostic approach is necessary, and today the loco-regional fibrinolytic treatment is the therapy of choice (Fig. 1). In the near future percutaneous thrombolysis will also be available to serve the same goal. Our own experience refers, in the last 3 years, to 105 cases of loco-regional fibrinolytic treatment with the following results: 80% of good results in the aortoiliac segment; 75% of good results in the femoropopliteal segment; and 60% of good results in the femorotibial segment.

Treatment of sub-acute or chronic arterial occlusions

Longer existing arterial thrombi can in some cases still be treated by means of thrombolysis. Several authors advocate thrombolysis to visualize the causing lesion with subsequent percutaneous transluminal angioplasty (PTA) [24-26]. Unfortunately, no diagnostic tests are available to differentiate between lesions, which can be treated with lysis and those not amenable for this approach. Localized arterial segments in the common iliac or superficial femoral artery can be pretreated with thrombolysis prior to PTA. Several authors proposed the administration of thrombolytic agents in case the guide wire cannot pass the occlusion [27]. There is, however, a potential risk of distal embolization, which on the other hand can be treated with suction or thrombolysis. We have encountered 5% of embolization in primary dilatation, but no embolization when thrombolysis was administered prior to PTA.

The success rate of endovenous systemic thrombolysis on arterial occlusions varies between 13% and 43% [12,21], whilst loco-regional administration of fibrinolytics inside the thrombus determines a recanalization in 60-90% of cases [28] in nonrecent occlusions with a patency of 80% after 2 years.

Our experience is limited to 21 cases treated over the last 2 years; they were patients aged from 39 to 62 years, who exhibited nonrecent (average 9 months) uni- or bilateral aortoiliac thromboses. After locoregional fibrinolytic therapy with urokinase (on average 12-14 hours) an arterial reopening was obtained in 85% of cases with successful endovascular correction of the lesion causing the thrombosis. In the extensive lesions it was necessary, however, to include complex endoluminal procedures with a large consumption of stents (Fig. 2).

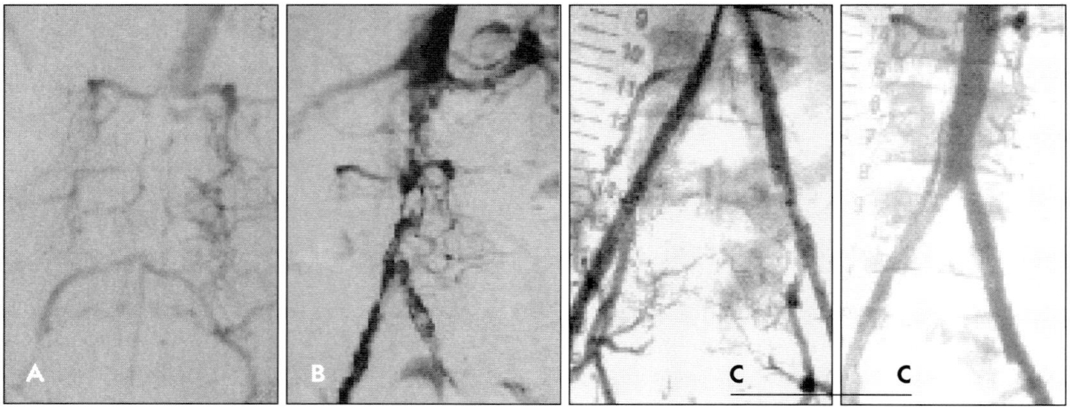

FIG. 2 Locoregional thrombolytic therapy of an acute occlusion of the abdominal aorta in a 44 year old patient. A - Aortic occlusion before treatment. B - Angiography during thrombolysis. C - Completion angiogram after thrombolysis followed by multiple stents.

Intraoperative use of fibrinolysis

This strategy is used in patients that exhibit thromboses that are very distal or not accessible to common surgical thrombus extracting techniques with a Fogarty catheter. The technique consists of the infusion of 100 000 units of urokinase diluted in 60 cc of physiological solution and then injected locally into the artery during the operation in a 20-minute time span [19]. The best indications are:

1 - the persistence of thromboembolisms after desobstruction with a Fogarty catheter,
2 - distal embolization after occlusion by popliteal aneurysm,
3 - distal embolization after angioplasty or vascular surgery,
4 - thrombosis of tibial vessels.

Our experience of intraoperative fibrinolysis initially concerned cases with residual thrombus following an unsuccessful Fogarty procedure (25 cases in 115 embolectomies carried out over the past 3 years). This population of patients is characterized by an advanced age and by arteriosclerotic lesions associated to suspected cardiogenic embolism. We have now changed our policy because of disappointing results. Following completion angiography we position a thin catheter percutaneously down to the target region and thrombolytic therapy is administered at a dosage of 100 000 U/hr. In this way we have been able to recuperate 30 % of legs to the expense of 20 % of hemorrhages, which forced us to discontinue the therapy. Interruption of the therapy has always led to amputation.

Another indication for intraoperative fibrinolytic treatment has been the absence of runoff in the course of surgical treatment for acute thrombosis of popliteal aneurysms. In five cases we performed local thrombolysis. The results, however, were disappointing and therefore we start with high-dose peroperative locoregional thrombolysis to ensure and verify the distal bed, even leaving the option to complete the whole procedure with an endoprosthesis. This attitude turns out to be obligatory in the event that the preoperative angiography reveals an absent or seriously compromised distal bed.

The use of fibrinolysis in bypass thrombosis

Locoregional thrombolysis appears to be an important effective alternative to surgical reoperation. The resolution of the thrombosis allows, when possible, to correct the lesion causing the occlusion of the graft.

Favorable results are reported in 73-79% of cases treated with urokinase or r-TPA [6,13,14,24,28]. No randomized study comparing surgery and thrombolysis has yet been published.

During the last three years we have treated 45 graft occlusions with thrombolysis (Fig. 3). Approximately half of the cases were successfully treated and subsequent correction of the lesion was either performed by PTA (45%) or surgery (55%). Unlike venous grafts, prosthetic grafts could still be treated several days after thrombosis. In 10% of cases no reason for the occlusion of the prosthetic implant was revealed.

Results

Although locoregional fibrinolytic treatment in limb ischemia is widely accepted, the case studies published are very heterogeneous and consequently the results are very variable (25%-85%). Thrombolysis seems to be more efficient in high aortoiliac occlusions than in bypass thrombosis. Various recent studies have compared the results of thrombolysis with those of surgical therapy. The only well-known trial today that addresses fibrinolysis as the initial approach in patients with recently onset arterial occlusion is the STILE study [20,29].

In patients with critical ischemia due to arterial or bypass thrombosis, existing less than seven days, thrombolytic treatment followed by angioplasty or surgical therapy has been compared to an immediate surgical approach [21,22,25,29-31]. Ischemia was resolved in 70% of the patients who underwent thrombolysis. The percentage of limb salvage in the two groups was similar (82% after 1 year) with a better overall survival rate in patients randomized to thrombolysis due to a reduction in the cardiac and respiratory hospital complications. Surgical thrombectomy was compared to intraarterial thrombolysis in patients with severe limb ischemia due to acute arterial occlusion [21]. In 60% of the patients treated with fibrinolysis a favorable result was obtained and thrombectomy provided positive results in 65%. Analysis of the experiences and the documented results allow a few considerations to be made.

1 - Thrombolysis should not be a monotherapy, but be combined with invasive or endoluminal procedures.
2 - There are no certain predictive clinical or angiographical criteria for the success of thrombolysis. The only certain criterion is represented by the passage of the metallic guide wire and positioning of the catheter inside the thrombus.
3 - The clinical situations that carry the best results are obtained in aortoiliac arterial and prosthetic occlusions and prosthetic bypasses [25,30,32-35]. Thrombolysis in saphenous vein grafts is characterized by a reduced patency with time, especially if carried out on grafts not older than one year.
4 - The best results are obtained in acute ischemia and in PTA and stenting associated thrombolysis. The results in time are very variable, since 10% to 30% of patients undergo a second vascular or endovascular procedure on the same arterial tract within five years.
5 - There are not many data regarding the intermittent jet infusion of thrombolytic agents (*pulse spray*), however, preliminary data are promising. Yusuf et al. [17] reported a reduction of lysis time in 24 treated cases. Kandarpa et al. [30], however, did not encounter any difference between the two methods.

Conclusions

Locoregional thrombolysis in acute and chronic critical ischemia has become an integral part of endovascular and surgical treatment. Despite good clinical results and manageable applications, the elevated cost, risk of complications, and intense monitoring require strict indications for this treatment. Since thrombolysis is combined with endovascular and surgical procedures, this management can only be performed in qualified and skilled centers. However, further studies are necessary to define secured protocols with respect to efficacy of the treatment and safety for the patient.

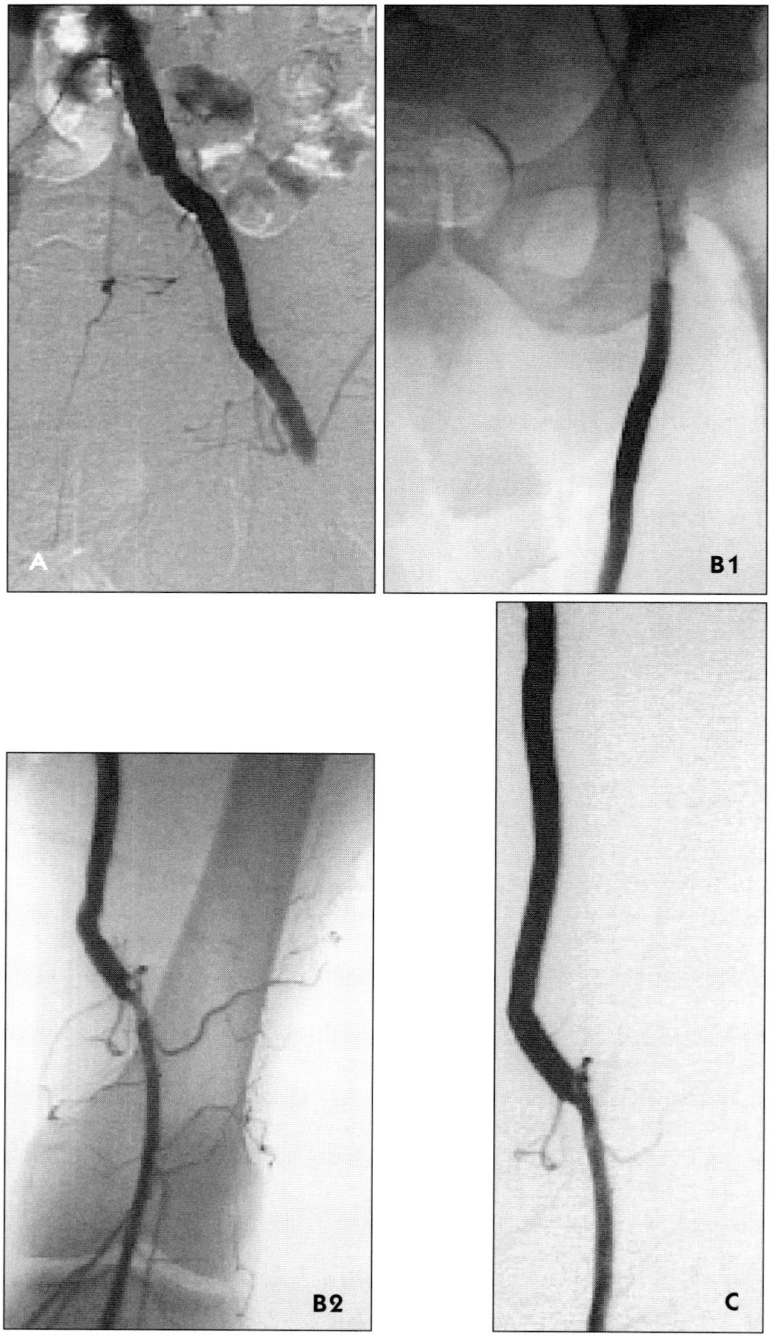

Fig. 3 Locoregional thrombolysis of an occluded femoropopliteal bypass. A - Angiography before treatment showing a complete occlusion of the common femoral artery. B1 + B2 - Angiographic control during procedure showing reopening of the bypass reopacification. C - Final result with control of the distal anastomosis of the bypass.

■ THROMBOASPIRATION (SUCTION EMBOLECTOMY)

Alternative techniques that combine the speed of the Fogarty procedure and local thrombectomy associated with minimal invasive access and less endothelial damage would avoid long-lasting procedures and related complications and significantly reduce the costs. Percutaneous thromboaspiration or suction embolectomy meets these criteria. Percutaneous aspiration of thrombotic material was first performed by Greenfield et al. [36] in a patient with a pulmonary embolism. The arterial application was described by Buxton and Mueller in 1974 [37], who successfully extracted a thrombus from the renal artery, which developed during angioplasty. Subsequently, percutaneous suction thromboembolectomy was only applied for iatrogenic acute ischemia caused by endovascular procedures. Turnipseed et al. [38] extended the indications to other arterial obstructions. Starck et al. [39] have extensively described the different techniques and indications.

TECHNIQUE

Wagner and Starck [23] described the basic equipment necessary for the procedure: an 8-9 French introducer, equipped with a removable valve (Cordis), in order to be able to easily extract trapped thrombus within the introducer; 0.035 inch hydrophillic guide wire to recanalize the occlusion (Terumo Japan); thromboaspiration catheter with a slightly tapered atraumatic tip of variable diameter (from 8 to 5 French) according to the caliber of the vessel to be treated; a 50 cc syringe.

More recently, double lumen thin wall catheters with an aspiration canal (of large caliber) and an eccentric canal (of small caliber) have been introduced that allow the passing of a 0.014 inch guide wire, which avoids the repeated trauma associated with the repositioning of the guide wire (Angiomed). Moreover, with the aim to remove thrombus stuck to the vessel walls by fragmentation, basket guide wires that can be rotated manually (Angiomed), with soft branches or 5-8 French spirals with atraumatic tip, have been developed and used by different authors.

The procedure begins with puncturing of the femoral artery either by an antegrade or retrograde access, depending on the arterial segment to treat. After having positioned the introducer and injected 5 000 units of heparin, one performs angiography for diagnostic purposes. The hydrophillic guide wire is placed beyond the occlusion and an angiographic catheter with an angulated tip (Berenstein-like) is advanced and an angiographic control is performed. The angiography catheter and its ease or difficulty in passing the occlusion gives vital information regarding the nature and degree of organization of the embolic or thrombotic material. Subsequently, the angiography catheter is removed and the suction catheter is positioned with its tip behind the head of the thrombus. The aspiration is then performed using the 50 cc syringe. If the thrombotic material is of small caliber it can be totally removed via the catheter. In most cases, however, the material adheres to the tip of the catheter, which therefore must be extracted completely, washed and reinserted to perform a new aspiration. This highlights the importance of new guiding catheters that prevent trauma linked to the repositioning of the same guide wire and catheter as described in the original technique. The number of aspirations necessary to recanalize the vessel is directly proportional to the age and length of the thrombus, even though in general three to five aspirations are sufficient to obtain a good patency of the vessel.

Some complementary techniques have been described including: removal of thrombus adhering to the vessel wall with a caliber of more than 6 mm by means of an 8 French catheter with a hockey type tip; fragmentation of old and nonaspirable thrombotic material entirely by means of a corkscrew guide wire in small arteries or basket in larger caliber vessels, or by means of segmental dilatation of a balloon; short term fibrinolysis at high dosages (200 000 to 400 000 units of urokinase in one hour); and angioplasty in case of recanalization of atheromatous vessels with multiple stenoses.

Any possible distal embolization in the course of thromboaspiration (10% in Starck's experience [23]) can be resolved by means of aspiration of the embolic material with catheters of an adequate caliber for the embolized vessel. In order to avoid distal embolization, some authors inflate the balloon of a Fogarty catheter, beyond the thrombus, inserted in the thromboaspiration catheter. At the end of the procedure an angiographic control is carried out and in case of residual stenosis, the lesion can be treated by angioplasty. Anticoagulation with heparin is considered necessary for the first two or three days after thromboaspiration and is subsequently replaced by antiplatelet drugs in cases of atherosclerotic lesions or with oral anticoagulants in case of cardiogenic embolism.

In our experience we often use a long introducer (Cook) to avoid vascular trauma due to the repeated repositioning of the aspiration catheter and to pass the guide wire under angioscopic control, both to avoid dissections and to reduce the radiation exposure. Moreover, this technique allows to directly locate the head of the thrombus and to decide where to position the catheter (Fig. 4). While the transit of the guide wire in the posterior tibial artery turned out to be relatively easy, it was often necessary to use a multipurpose type catheter to guide the wire to the anterior tibial artery.

Indications

Although thromboaspiration is a relatively easy and fast technique, very limited experience has so far appeared in the literature and therefore indications and contraindications have not completely been defined. We can assert that the main indication is acute ischemia due to embolization in healthy arteries or due to thrombosis in stenotic arteries. In these cases thromboaspiration is performed to complete the diagnostic angiography. In general the vessels to be treated are the superficial femoral, popliteal or distal arteries. In case of iliac obstructions and common femoral artery disease, the treatment of crossover with thromboaspiration is certainly more difficult as compared to conventional surgical treatment. In connection with the age and length of the thrombus the main indications are:

- presence of embolic occlusion less than 10 days without limits of length and quality of the runoff,
- thrombus on a preexisting atheromatous lesion, provided that the obstruction is less than 15 cm in length,
- residual thrombus or distal embolism due to fibrinolytic treatment,
- embolization or thrombus due to endovascular procedures,
- embolization of the inferior popliteal and distal vessels, in the course of surgical embolectomy surgery with a Fogarty catheter.

Also, in reality the contraindications are still poorly defined. An asymptomatic patient is generally considered as a contraindication to treatment with thromboaspiration. Nevertheless, the occlusion of a tibial or peroneal vessel in the course of other endovascular procedures can be a reason to try to restore the previous runoff. Long obstructions in pathological arteries represent a relative contraindication. In fact, in cases where it is impossible to execute a fibrinolysis or a surgical revascularization for limb salvage it is, however, conceivable to try suction thrombectomy. Late obstructions are also not generally suitable for treatment with thromboaspiration because of the adherence of the thrombus to the vessel wall. The use of the basket guide wire, spiral guide or angioplasty, as performed by Wagner and Starck [23], causing a mechanical fragmentation of the thrombus can enlarge the number of treatable cases also in the long-term. Obstruction of bypasses below the knee is a poor indication for both fibrinolysis treatment and thromboaspiration. Acute thrombosis of femoropopliteal aneurysms represents an indication for thrombolysis rather than thromboaspiration.

Results

Although published experience is still rather limited, the results in terms of immediate clinical and angiographic success appear promising. Complication rate is remarkably low and ranges between 4% and 13% [40]. Furthermore, complications are mostly limited to groin hematomas and false aneurysms. Angiographic success rate varies between 80% and 93%, whereas clinical results are positive in 87% of patients.

Conclusions

Fogarty embolectomy remains the most widely applied technique for acute limb ischemia. However, it should be emphasized that intimal damage is common and that crural and pedal embolectomy offers limited results. With different techniques available the following indications with subsequent treatment can be defined, underlining the additional value of their complementary application: Fogarty thrombo-embolectomy in cases of recent occlusions, above all embolic at the level of aorta, iliac and femoral

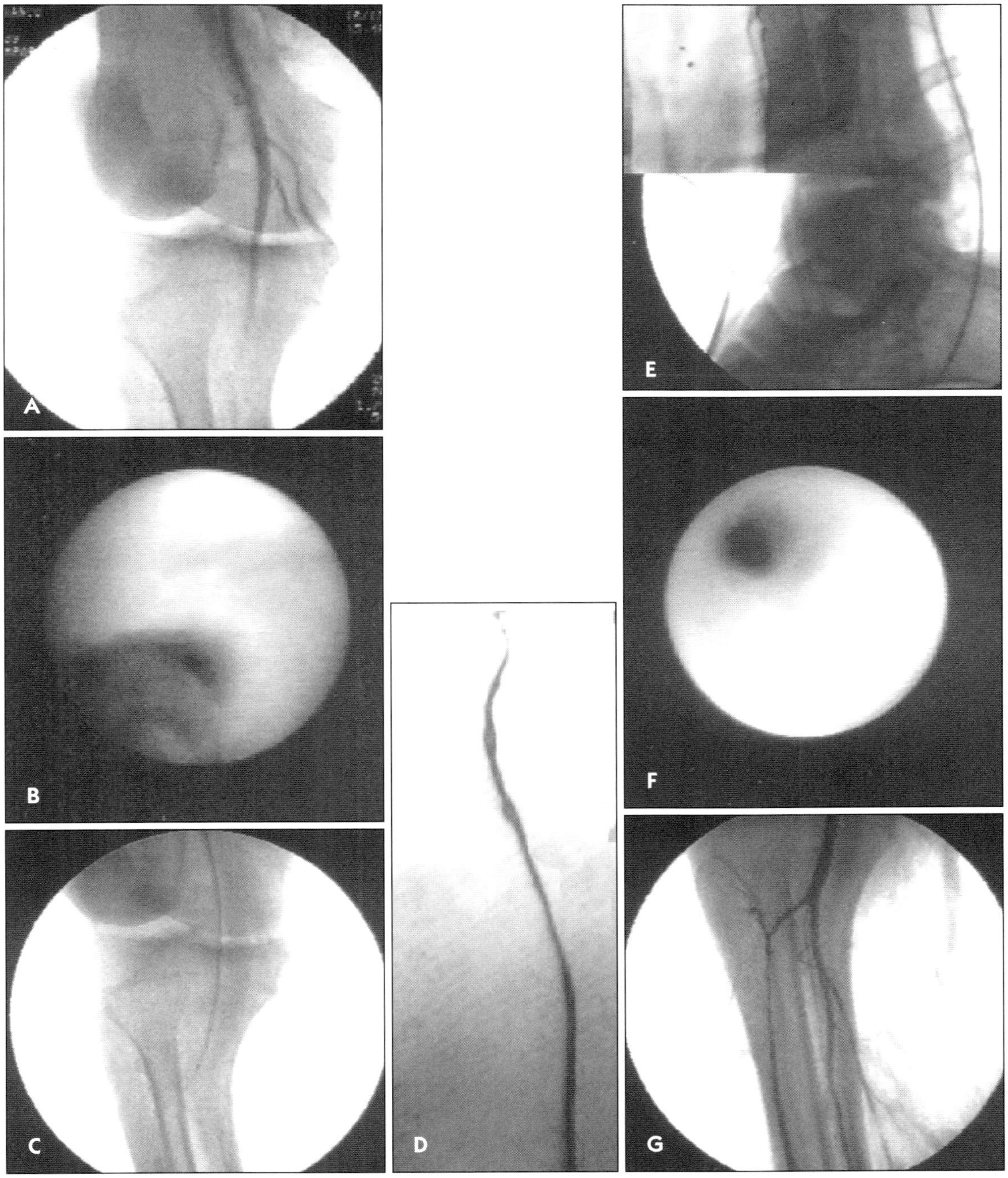

FIG. 4 Suction embolectomy for acute thrombosis of the popliteal and crural arteries. A - Occlusion of the distal popliteal artery. B - Angioscopic view of the head of the thrombus. C - Catheterization of the anterior tibial artery by the guide wire. D - Removed thrombus. E - Aspiration catheter in the posterior tibial artery. F - Angioscopic control. G - Completion angiography after suction embolectomy with complete revascularization of the popliteal and tibial vessels.

arteries; fibrinolysis in cases of long or late femoropopliteal, tibial and bypass obstructions; and thrombo-aspiration in cases of short and fresh femoral, popliteal and tibial obstructions and for cases in whom fibrinolysis is contraindicated.

As has already been mentioned the methods can, and at times must be combined in order to achieve an improvement in clinical and angiographic results. In particular, the removal of thromboembolic material with a Fogarty catheter at the femoropopliteal level can be completed by suction thromboembolectomy at the tibial level. Residual thrombus after fibrinolytic treatment can be removed by thrombo-aspiration. In the same way, thrombotic material that cannot be removed by the aspiration technique can be dissolved by means of thrombolytic agents [41]. The advantages and limits of the various methods are summarized in Table II.

TABLE II THROMBOASPIRATION VS. THROMBOLYSIS AND FOGARTY THROMBOEMBOLECTOMY

	Thrombolysis	*Thrombo-aspiration*	*Thrombo-embolectomy*
Lenght of procedure	24 - 48 h (90 min with *Pulse Spray*)	Fast	Fast
Lenght of occlusion	> 15 cm	<10 - 15 cm	Indifferent
Age of thrombus	< one month	<10 days	< 2 - 3 days
Major complications	2 - 8 %	Few	0.5 - 1%
Cost	High	Medium	Low
Management and practicality	Difficult	Medium	Easy
Contraindications	Several	Few	Few

ROTABLATOR

The results of PTA are specifically limited in coarse plaques, especially in small caliber arteries [42]. Several alternative techniques have been developed to remove the calcified material, like laser and atherectomy devices. However, clinical results have been disappointing [43,44]. There is still an interest for these devices, notably if a technique is able to reduce the saliency of a calcified plaque, without damaging the adjacent endothelium. Some authors [45] have published relevant data on this new device called Rotablator (Heart Technology, Redmond, WA).

APPARATUS

The Rotablator (Fig. 5) is a high-speed rotating device for transluminal atherectomy with a central guide wire. It is composed of an oval shaped rotating tip that is available in different sizes from 1.5 to 4.5 mm in diameter, consisting of multiple fragments of diamond sizes between 20 and 120 microns. The tip is connected to a flexible transmission stem covered by a Teflon layer that protects the vessel during rotation and forms an irrigation channel inside. A saline solution circulates to lubricate and cool the transmission wire and the tip. The stem is joined to a turbine fed by compressed gas, contained in a plastic sheath at the top of which there is a cursor that allows the operator to introduce or withdraw the tip carefully. The speed of rotation can reach 200 000 rpm and is regulated by means of the compressed gas pressure that activates the turbine. The system is completed by a stroboscopic light that measures the speed of rotation and boasts a device that can block the rotation if there is a low torsion force, in order to avoid twisting of the arteries around the catheter or tip. The guide wire for atherectomy is very thin (0.009 inches) and is fairly rigid in order to be able to sustain and limit the oscillation of the catheter and tip during rotation. The system cannot be resterilized.

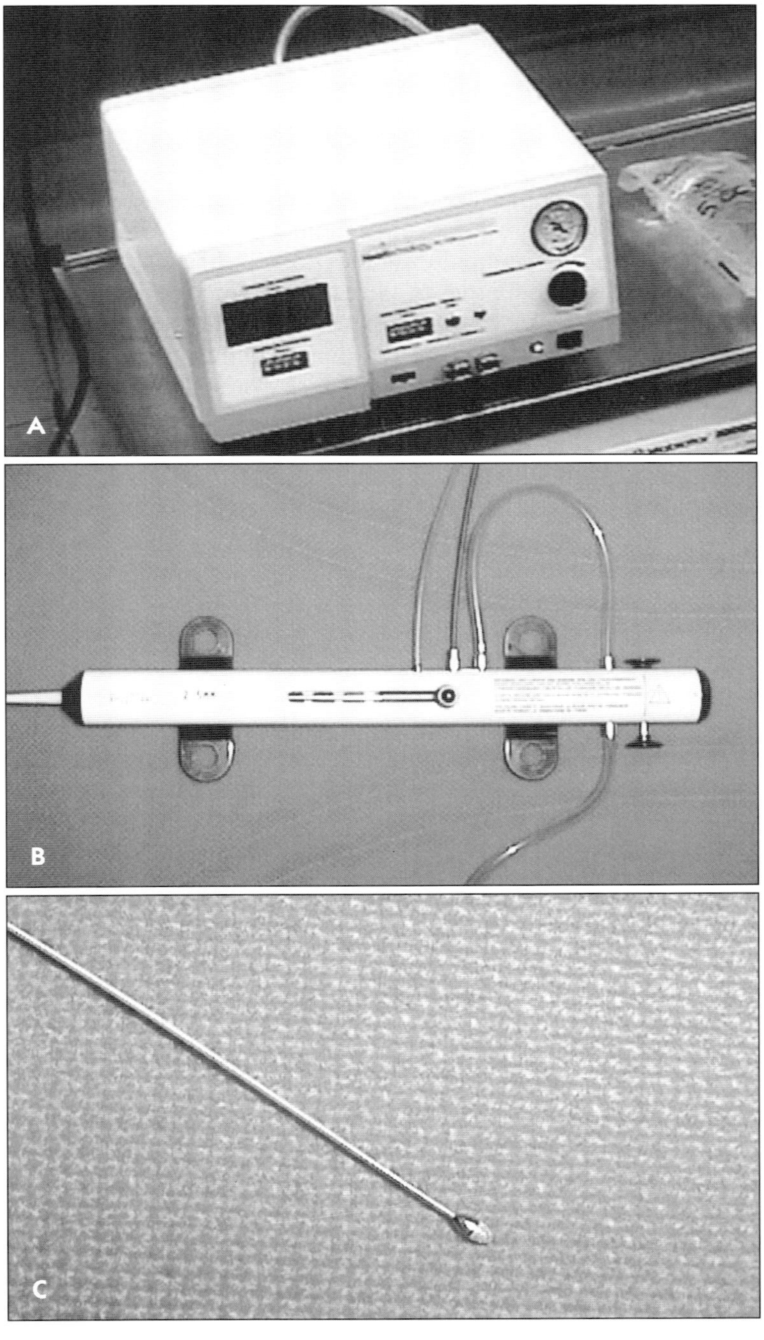

FIG. 5 The Rotablator system. A - Rotablator console. B - Rotablator catheter system. C - Oval shaped rotating tip.

The Rotablator works through the cutting action of the diamond fragments that, thanks to the high-speed rotation, are able to splinter and remove the plaque by reducing it to microparticles of a size inferior to that of the red globules that pass through the vessels, so that they do not cause obstructions. The rigid and hard tissue, unlike soft and elastic tissue, is not able to escape the action of the diamond fragments, so that the removal action works efficiently on calcified plaques while leaving the artery wall undamaged; for this reason it can also be used on eccentric plaques.

Experimental studies have shown that the arterial wall after using the Rotablator is smooth, circular and regular. The damage inflicted to the media at the treated site is minimal and the intima is fixed to underlying tunica [46], despite passage of the burr. This differs from the changes occurring after PTA, which compresses and fragmentizes the plaque. With both techniques, however, damage to the endothelium occurs, both adjacent to the lesion as well as distally.

Indications and Contraindications

The indications for the use of the Rotablator are obliterating atherosclerosis in patients with advanced stage II and critical leg ischemia [47]. Although it can also be used at the level of the iliac artery, the preferred use is in the femoral-popliteal-tibial region, particularly at bifurcations. Both stenoses and occlusions can be treated, provided that the guide wire can pass the lesion. The lesions can be long, eccentric, and preferably calcified, while elastic plaque and old thrombus do not offer optimal results. It can be used both as a primary treatment to PTA, and in order to allow recanalization of the vessel. At the level of the tibial arteries a satisfactory angiographic result can be achieved when using only the Rotablator. In the superficial femoral and in the popliteal artery, however, it must be used in combination with the balloon because of the dimension of the probes, in which case the assured removal of the plaque and the reduced insufflation pressure reduce the risk of dissection.

Based on clinical experience and nature of the equipment, the manufacturing company lists the following contraindications:

- lesions not passable with the guide wire,
- lesion extension greater than 20 cm,
- presence of thrombus,
- intimal dissection,
- previous episodes of arterial spasm

Materials and Technique

Introducer: the caliber depends on the dimension of the Rotablator tip; it can vary from 5 Fr for the 1.5 mm burr, to 14 Fr for the 4.5 mm burr. Guide wire: hydrophillic from 0.035 to 0.025 inches for recanalization; rigid from 0.009 inches for the Rotablator. Catheter: from 3 to 5 Fr for the replacement of the wire. Rotablator: available tips are 1.5, 2, 2.5, 3, 3.5, 4, and 4.5 mm. The chosen diameter is 75-85% of the diameter of the distal artery at the lesion; the optimal rotation velocity varies from 180 000 rpm for a 1.5 mm burr to 120 000 rpm for a 4.5 mm burr. Balloon: for complementary dilation, of an equal diameter to the artery.

One day before the procedure a calcium-antagonist is administered to reduce the occurrence of spasm. Percutaneous antegrade femoral access is generally performed, however, retrograde popliteal access is also possible for superficial femoral lesions and contralateral access for common femoral disease. Once the lesion is recanalized with a hydrophillic guide wire, an angiography catheter is positioned distally and the guide wire is replaced by the 0.009 inch wire. The positioning of the first guide wire can be executed under angioscopic control in order to avoid the risk of dissection. The tip of the Rotablator is placed in the target lesion and withdrawn slightly in order to allow its rotational movement. When it reaches the required velocity of 120 000 to 180 000 rpm, according to the tip caliber, it is slowly introduced with a *woodpecker* movement, by means of the cursor placed on the sheath of the turbine. One should not proceed too quickly in order to avoid blocking of the device and not maintain the position of the device for too long to avoid overheating of the vessel. After surpassing the lesion, an angiographic control should be performed to evaluate the success of the intervention (Fig. 6).

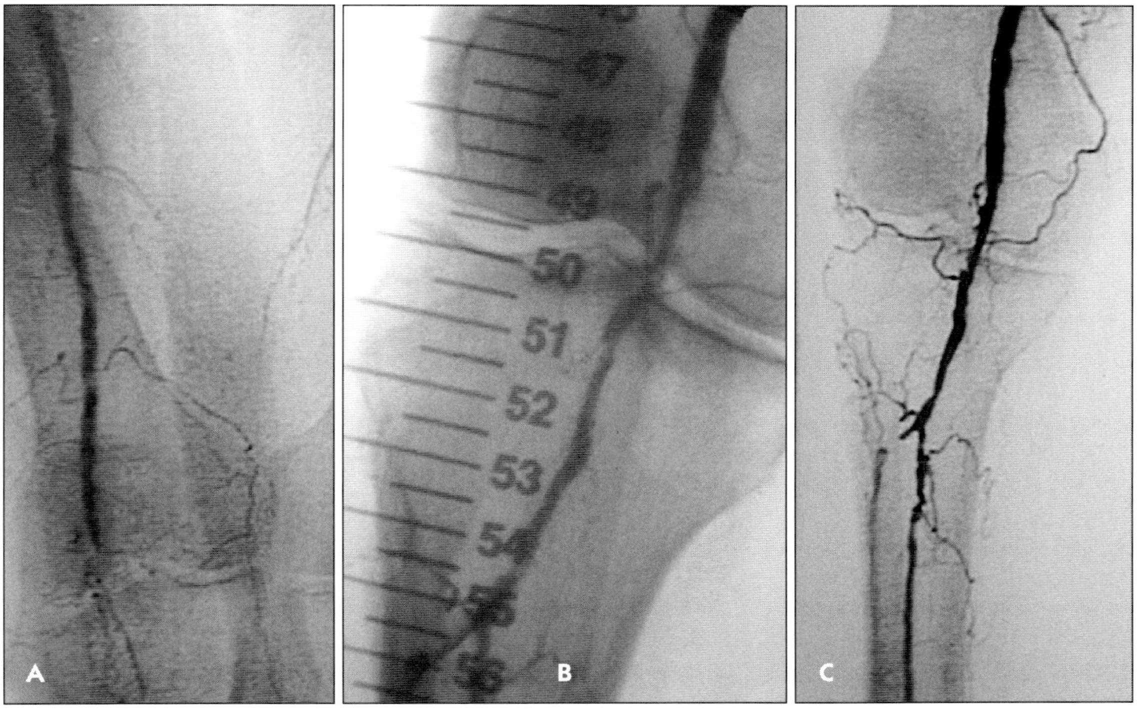

FIG. 6 Treatment by Rotablator of a long obstruction of the popliteal artery. A - Angiography before treatment. B - Intraoperative control after recanalization with Rotablator. C - Angiography after one year showing patency of popliteal artery but obstruction of the posterior and anterior tibial arteries.

Different strategies can be used:

- execute different passages through the area of the lesion to be treated with increasing probe calibers,
- use a single probe with a caliber of roughly 75-80% of the native artery,
- recanalize the vessel with a small caliber probe to remove the calcified component of the plaque and dilate the residual stenosis with a low pressure balloon,
- postoperatively a low dose of heparin is administered for 12 to 24 hours, the patient is discharged with calcium heparin for 15 days.

During the procedure the following care must be observed: do not use oxygen for feeding (risk of explosion); handle the guide with care and avoid twisting; use the device with a flow of saline solution and do not inject contrast medium in the irrigation channel of the probe. Interrupt the procedure if resistance to progression or rotation is experienced.

COMPLICATIONS

A vasospasm can occur distally at the treated site, but it can also occur 2-3 cm proximally to this point. The peripheral embolization is considered a relatively insignificant clinical event, however, some episodes of cutaneous gangrene or amputation of the leg have been reported [47]. This would appear to be a consequence of a prolonged rotation and the large dimension of the tip. A transient hemoglobinuria is probably linked to the cavitation action of the high speed rotating tip, which provokes the breakage of red globules. In all cases, however, these complications are transitory and can be resolved without consequences, even though in a small percentage a transitory renal insufficiency occurred. Early rethrombosis depends on technical failures like residual stenosis, intimal flap, recoil of the vessel, dissection, or vasospasm. The perforation of the vessel or intimal dissection is a consequence of a badly positioned guide wire.

RESULTS

Two important clinical studies are published so far with a significant number of cases and follow-up. The American Multicenter Study [47] known as CRAG *(Collaborative Rotablator Atherectomy Group)* is based on 79 cases with an average follow-up of 27 months. The reported primary patency rate was 47% at 6 months and 18.6% at 24 months. The conclusion of this study was that it is not presently advisable in the treatment of peripheral arteriopathy due to the numerous precocious failures, the high percentage of restenosis and the low long-term patency [47]. The French Singlecenter Study [48] is based on 150 patients with an average follow-up of 14.4 months. The reported primary patency rate was 76% at 4 months. The conclusion was that it holds a position in the treatment of peripheral lesions, enjoys a low percentage of complications and restenosis which is encouraging to its use, and is limited only by its high cost.

DISCUSSION

The angiographic success of removing the plaque with the Rotablator, both used singularly and in association with balloon dilation or surgical intervention, varies between 89% and 97%. The differences between the two main clinical trials [47,48] relate to the different organization of the studies. The CRAG study reported data of three different centers and different surgeons, thereby comprising learning curves of the different surgeons. Moreover, the selected patients suffered from different degrees of ischemia, critical ischemia being present in 56% of CRAG cases against only 24% in the French study. Both the complexity and the length of the lesions differed: the CRAG trial had an average length of 9 cm against 4 cm in the French study. If one limits the analysis of the results to lesions larger than 7 cm, a significant difference in the percentage of restenosis (55% versus 19% above the knee and 80% versus 18% at the distal level) is obvious.

The percentage of complications is also markedly different and this appears to be linked to the use of large caliber probes and to the length and complexity of the treated lesion, requiring long periods of rotation (Table III).

In our experience the use of the Rotablator requires a considerable learning period. The length and fragility of the guide wire, the multiple connections between the probe, console and the bag of saline solution, as well as the choice of tip caliber, are the most frequent causes of errors. In order to limit costs we tended to use a single probe at the distal level, combined with a balloon at the femoral-popliteal level.

We used the Rotablator exclusively in cases of critical ischemia. In 25% of cases the lesion could not be recanalized. In 28 cases in which removal of plaque was performed, the initial angiographic result was good in 96.4% of cases. In one patient surgical correction was necessary. Another patient underwent surgical revascularization for thrombosis during hospital stay. All patients were discharged with patent arteries and an improvement of the clinical symptomatology. The intervention was always well tolerated and only one case of minor peripheral embolism occurred (3.6%). We did not experience any cases of hemoglobinuria or transitory renal failure. This can probably be explained by: the length of the lesions treated (average 3 cm); the short rotation time (average 4 minutes); and the size of the probe used (only in 3 cases the superficial femoral artery was treated). So far we have used the Rotablator in very aged patients (average age 74.6 years),

TABLE III REPORTED COMPLICATIONS OF ROTABLATOR

Complication	*Crag [47]* %	*Henry [48]* %
Thrombosis	11.4	5.6
Embolization	10.0	1.3
Hemoglobinuria	13.0	5.0

diabetics (71.4%), or in those with a high surgical risk and low life expectancy (heart disease 57.4%, respiratory failure 46.4%, neoplasm 32.1%). In such cases, even though there tends to be a short follow-up, the possibility of limb salvage for a relatively short period becomes very significant. Patients were usually discharged with anticoagulant therapy, together with antiplatelet drugs, or otherwise with double antiplatelet (ASA, ticlopidine) treatment.

CONCLUSIONS

The Rotablator requires experience and good management in endovascular procedures. It is difficult to justify the use of this device in relation to cost, complex application and the frequent necessity of complementary techniques. Experience collected from the literature is rather limited and recent contributions to this procedure have been rather modest. Results from anecdotal reports do not allow to draw definite conclusions regarding correct indications and the best characteristics of the procedure, because the type and extension of the lesions and the quality of the runoff are not always well specified.

However, it is possible to describe the lesions that limit the results and increase the incidence and seriousness of complications. We can state that, by avoiding these indications, the complications will be limited and the percentage of initial success will improve.

In summary, the situations best suited to Rotablator treatment are:

- relatively short calcified plaques of the superficial femoral artery (necessity of complementary PTA),
- relatively lengthy calcified plaques of the popliteal and distal arteries,
- high risk of failure of traditional revascularization interventions,
- high risk of infection in the presence of necrosis or gangrene.

REFERENCES

1 Fogarty TJ, Cranley JJ, Drause RJ et al. A method for extraction of arterial emboli and thrombi. *Surg Gynecol Obstet* 1963; 116: 241-244.

2 Jivegard L, Holm J, Schersten T. Acute limb ischemia due to arterial embolism or thrombosis: influence of limb ischemia versus pre-existing cardiac disease on postoperative mortality rate. *J Cardiovasc Surg* 1988; 29: 32-36.

3 Greep JM, Aleman PJ, Jarret F et al. A combined technique for peripheral arterial embolectomy. *Arch Surg* 1972; 105: 869-874.

4 Zimmerman JJ, Cipriano PR, Hayden WG et al. Balloon embolectomy catheter used percutaneously. *Radiology* 1986; 158: 260-262.

5 Van Ommen VG. Van der Veen FH, Geskes, GG et al. Comparison of arterial wall reaction after passage of the Hydrolyser device versus a thrombectomy balloon in an animal model. *J Vasc Intervent Radiol* 1996; 7: 451-454.

6 McNamara TO, Bomberger RA, Merchant RF. Intra-arterial urokinase as the initial therapy for acutely ischemic lower limbs. *Circulation* 1991; 83 (2 Suppl): I106-109.

7 Berridge DC. Advances in thrombolytic therapy. *Br J Surg* 1994; 81: 1249-1250.

8 Fiessinger JN, Vitoux JF, Pernes JM et al. Complications of intraarterial urokinase-lys-plasminogen infusion therapy in arterial ischemia of lower limbs. *AJR Am J Roentgenol* 1986; 146: 157-159.

9 Edwards JE, Taylor LM Jr, Porter JM. Treatment of failed lower extremity bypass grafts with new autogenous vein bypass grafting. *J Vasc Surg* 1990; 11: 136-145.

10 McNamara TO, Bomberger RA. Factors affecting initial and 6 month patency rates after intraarterial thrombolysis with high dose urokinase. *Am J Surg* 1986; 152: 709-712.

11 Braithwaite BD, Petrik PV, Ritchie AW et al. Computerized angiographic analysis of the outcome of peripheral thrombolysis. *Am J Surg* 1995; 170: 131-175.

12 Dotter CT, Rosch J, Seaman AJ. Selective clot lysis with low-dose streptokinase. *Radiology* 1974; 11: 31-37.

13 Graor RA, Risius B, Young JR et al. Thrombolysis of peripheral arterial bypass graft: surgical thrombectomy compared with thrombolysis. A preliminary report. *J Vasc Surg* 1988; 7: 347-355.

14 McNamara TO, Fisher JR. Thrombolysis of peripheral arterial and graft occlusions: improved results using high-dose urokinase. *AJR Am J Roentgenol* 1985; 144: 769-775.

15 Bookstein JJ, Fellmeth B, Roberts A et al. Pulsed-spray pharmacomechanical thrombolysis: preliminary clinical results. *AJR Am J Roentgenol* 1989; 152: 1097-1100.

16 Valji K, Roberts AC, Davis GB et al. Pulsed-spray thrombolysis of arterial and bypass graft occlusions. *AJR Am J Roentgenol* 1991; 156: 617-621.

17 Yusuf SW, Whitaker SC, Gregson RH et al. Experience with pulse-spray technique in pheripheral thrombolysis. *Eur J Vasc Surg* 1994; 8: 270-275.

18 Ouriel K, Shortell CK, De Weese JA et al. A comparison of thrombolytic therapy with operative revascularization in the initial treatment of acute peripheral arterial ischemia. *J Vasc Surg* 1994; 19: 1021-1030.

19 Quiñones-Baldrich WJ, Zierler RE, Hiatt JC. Intraoperative fibrinolytic therapy; an adjunct to catheter thromboembolectomy. *J Vasc Surg* 1985; 2: 319-326.

20 Anonymous. Results of a prospective randomized trial evaluating surgery versus thrombolysis for ischemia of the lower extremity. The STILE trial. *Ann Surg* 1994; 220: 251-268.

21 Nilsson L, Albrechtsson U, Jonung T et al. Surgical treatment versus thrombolysis in acute arterial occlusion: a randomised controlled study. Eur J Vasc Surg 1992; 6: 189-193.
22 Ouriel K, Shortell CK, Azodo MW et al. Acute peripheral arterial occlusion: predictors of success in catheter-directed thrombolytic therapy. *Radiology* 1994; 193: 561-566.
23 Wagner HJ, Starck EE. Acute embolic occlusions of the infrainguinal arteries: percutaneous aspiration embolectomy in 102 patients. *Radiology* 1992; 182: 403-407.
24 Cragg AH, Smith TP, Corson JD et al. Two urokinase dose regimens in native arterial and graft occlusions: initial results of a prospective randomized clinical trial. *Radiology* 1991; 178: 681-686.
25 Valji K, Bookstein JJ, Roberts AC et al. Occluded peripheral arteries and bypass grafts: lytic stagnation as an end point for pulse-spray pharmacomechanical thrombolysis. *Radiology* 1993; 188: 389-394.
26 Kandarpa K, Goldhaber SZ, Meyerovitz MF. Pulse-spray thrombolysis: the "careful analysis". *Radiology* 1994; 193: 320-324.
27 Motarjeme AJ. PTA and thrombolysis in leg salvage. *J Endovasc Surg* 1994; 1: 81-87.
28 Parent FN III, Piotrowski JJ, Bernhard VM et al. Outcome of intra-arterial urokinase for acute vascular occlusion. *J Cardiovasc Surg* 1991; 32: 680-689.
29 Weaver FA, Comerota AJ, Youngblood M et al. Surgical revascularization versus thrombolysis for nonembolic lower extremity native artery occlusions: results of a prospective randomized trial. *J Vasc Surg* 1996; 24: 513-523.
30 Kandarpa K, Chopra PS, Arung JE et al. Intra-arterial thrombolysis of lower extremity occlusion: prospective, randomized comparison of forced periodic infusion and conventional slow continuous infusion. *Radiology* 1993; 188: 861-867.
31 Sullivan KL, Gardiner GA Jr, Kandarpa K et al. Efficacy of thrombolysis in infrainguinal bypass grafts. *Circulation* 1991; 83 (Suppl 1): I99-105.
32 Comerota AJ, Weaver FA, Hosking JD et al. Results of prospective, randomized trial of surgery versus thrombolysis for occluded lower extremity bypass grafts. *Am J Surg* 1996; 172: 105-112.
33 Knaus J, Ris HB, Do D, Stirnemann P. Intraoperative catheter thrombolysis as an adjunct to surgical revascularization for infrainguinal limb-threatening ischaemia. *Eur J Vasc Surg* 1993; 7: 507-512.
34 Anonymous. Thrombolysis in the management of lower limb peripheral arterial occlusion - A Consensus Document. Working party on thrombolysis in the management of limb ischemia. *Am J Cardiol* 1998; 81: 207-218.
35 Law MM, Gelabert HA, Colburn MD et al. Continuous post-operative intra-arterial urokinase infusion in the treatment of no reflow following revascularization of the acutely ischemic limb. *Ann Vasc Surg* 1994; 8: 66-73.
36 Greenfield LJ, Kimmel GO, McCurdy WC III. Transvenous removal of pulmonary emboli by vacuum-cup catheter technique. *J Surg Res* 1969; 9: 347-352.
37 Buxton DR Jr, Mueller CF. Removal of iatrogenic clot by transcatheter embolectomy. *Radiology* 1974; 11: 39-41.
38 Turnipseed WD, Starck EE, Dermott JC et al. Percutaneous aspiration thromboembolectomy (PAT): an alternative to surgical balloon techniques for clot retrieval. *J Vasc Surg* 1986; 3: 437-441.
39 Starck EE, McDermott JC, Crummy B et al. Percutaneous aspiration thromboembolectomy. *Radiology* 1985; 156: 61-66.
40 Wagner HJ, Starck EE, Reuter P. Long-term results of percutaneous aspiration embolectomy. *Cardiovasc Intervent Radiol* 1994; 17: 241-246.
41 Verstraete M. New developments in antiplatelet and antithrombotic therapy. *Eur Heart J* 1995; 16 (Suppl L): 16-23.
42 Samson RH, Sprayregen S, Veith FH et al. Management of angioplasty complications, unsuccessful procedures and early and late failures. *Ann Surg* 1984; 199: 234-240.
43 Cull DL, Feinberg RL, Wheeler JR et al. Experience with laser-assisted balloon angioplasty and a rotary angioplasty instrument: lessons learned. *J Vasc Surg* 1991; 14: 332-339.
44 Ahn SS, Conception B. Current status of atherectomy for peripheral arterial occlusive disease. *World J Surg* 1996; 20: 635-643.
45 Ahn SS, Auth DC, Marcus DR et al. Removal of focal atheromatous lesions by angioscopically guided high-speed rotary atherectomy. Preliminary experimental observations. *J Vasc Surg* 1988; 7: 292-300.
46 Ahn SS, Arca M, Brauel G et al. Histologic and morphologic effects of rotary atherectomy on human cadaver arteries. *Ann Vasc Surg* 1990; 4: 563-569.
47 Anonymous. Peripheral atherectomy with the Rotablator: a multicenter report. The Collaborative Rotablator Atherectomy Group (CRAG). *J Vasc Surg* 1994; 19: 509-515.
48 Henry M, Amor M, Ethevenot G, Henry I. Percutaneous peripheral rotational ablation using the Rotablator: immediate and mid-term results. Single center experience concerning 146 lesions treated. *Int Angiol* 1993; 12: 231-244.

15

TRANSLUMINAL ANGIOPLASTY COMBINED WITH SURGICAL RECONSTRUCTION FOR LOWER LIMB ISCHEMIA

JEAN-BAPTISTE RICCO, CHRISTOPHE CAMIADE
NELLA MANGIACOTTI, BERTRAND EDE, ALAIN BRANCHEREAU

During recent years several authors have shown the significance of transluminal angioplasty (TLA) in the treatment of atherosclerotic arterial disease [1]. Others have demonstrated that the results of TLA depend upon the type of lesion and the anatomic level of the affected artery [2,3]. However, the role of TLA combined with conventional revascularization remains poorly defined. TLA can indeed cause instantaneous complications (embolic events, arterial dissection) which may compromise the strategy based on successive combined techniques during the same operative procedure. For this reason, many surgeons favored a sequential approach, in which the endovascular procedure is performed prior to surgery. The improvement of endovascular techniques and the development of vascular stents in particular permitted the broadening of indications and the combination of procedures during the same session, allowing a surgical reconstruction at one level and an endovascular procedure at another level. In order to explain the principles, advantages, and drawbacks of this combined approach and to define its potential indications, we report our experience with 41 patients suffering from symptomatic multilevel arterial disease undergoing a combined endovascular and surgical procedure as a primary treatment.

■ PATIENTS

This retrospective study reports on 41 consecutive patients operated upon between January 1996 and December 1997 in two departments of vascular surgery (Hospital Jean Bernard, Academic Hospital of Poitiers and Academic Hospital of Marseille). All patients had atherosclerotic lesions at the level of the iliac arteries and the arteries of the lower limb. They all underwent an endovascular procedure combined with a conventional surgical technique during the same operative session. The group of patients included 13 women (31%) and 28 men (69%). The mean age was 70 years, ranging between 41 and 89 years. Table I depicts the commonly observed risk factors of these patients. Preoperative symptoms included rest pain in 7 patients (17%), gangrene in 10 patients (25%), severe intermittent claudication with a walking distance less than 100 meters in 12 patients (29%) and another 12 patients complained of intermittent claudication with a walking distance between 100 and 500 meters. Table II shows the different arterial lesions. Each patient had significant arterial disease, either occlusion or severe stenosis, managed by TLA and subsequent conventional surgery. In this series we only performed three recanalizations because we believe that surgical reconstruction remains the therapy of choice for arterial occlusions and that TLA is indicated for short arterial stenoses. Common femoral arterial disease was not treated with TLA but by conventional surgical techniques.

Table I RISK FACTORS

Risk	*Patients (n = 41)*	*%*
Hypertension	10	24
Diabetes	15	36
Coronary disease	19	46
Hyperlipedemia	11	27
Smoking	29	70
Renal insufficiency	9	22

Table II ANATOMIC DISTRIBUTION OF ARTERIAL LESIONS IN 41 PATIENTS AND ASSOCIATED TECHNIQUE OF RECONSTRUCTION

Artery	*Type of lesion*	*Treated lesions N*	*Surgical repair N*	*Endovascular repair N*
Common iliac	Stenosis	12	0	12
Common iliac	Occlusion	7	7	0
External iliac	Stenosis	15	0	15
External iliac	Occlusion	3	3	0
Superficial femoral	Stenosis	7	0	7
Superficial femoral	Occlusion	18	15	3
Common femoral	Occlusion/Stenosis	12	12	0
Popliteal	Stenosis	2	0	2
Popliteal	Occlusion	2	2	0
Crural	Stenosis	2	0	2
Crural	Occlusion	2	2	0
Total		**82**	**41**	**41**

Table III SURGICAL TECHNIQUES IN 41 PATIENTS

Surgical techniques		*Patients*
Aortobifemoral graft		1
Femoro-femoral cross-over		8
Femoral artery		12
- Prosthetic patch	*5*	
- Endarterectomy	*3*	
- Profunda femoral graft	*4*	
Supragenicular femoropopliteal bypass*		10
Infragenicular femoropopliteal bypass**		6
Infrapopliteal bypass***		3
Infragenicular popliteal thrombectomy		1
Total		**41**

* 9 prosthetic grafts, 1 saphenous vein graft
** 2 prosthetic grafts, 4 saphenous vein grafts
*** 3 saphenous vein grafts

SURGICAL TECHNIQUES

Table III presents the different surgical techniques. A femoro-femoral cross-over bypass was systematically performed when a unilateral iliac occlusion was associated with a normal contralateral iliac artery (n=3) or with a vessel which could be dilated during the same session (n=5). Surgical reconstruction of the common femoral artery and the femoral bifurcation (n=12) comprised five prosthetic patch plasties, three common femoral artery endarterectomies and four bypasses on the profunda femoral artery at the level of the perforating branches in patients suffering from proximal profunda femoral artery disease. Finally, 16 femoropopliteal and 3 infrapopliteal bypasses were performed for extensive occlusions in the superficial femoral artery and crural arteries.

ENDOVASCULAR TECHNIQUES

The endovascular procedures were performed in the operation theatre on a table allowing intraoperative fluoroscopy. A C-arm with image magnification was used for digital substraction angiography. The surgical field was prepared in such a way that all areas for surgical and endovascular procedures were covered. When an iliac angioplasty was planned, the abdomen was also prepared thus allowing surgical access if necessary. Conventional surgical access (vein harvesting, tunneling, arterial dissection) was performed prior to heparinization, followed by transluminal angioplasty and subsequent surgical reconstruction. This sequence was reversed if the surgical reconstruction was localized proximal to the side of angioplasty.

Endoluminal devices were introduced through the common femoral artery, except in two cases in which the popliteal artery was punctured to dilate crural vessels. The side of puncture was usually performed on a soft portion of the anterior wall of the artery, commonly situated at the level of the anastomosis of the bypass. A straight and flexible guide wire was advanced in the artery for approximately 20 cm thus allowing positioning of the introducer, of which diameter and length were selected according to the type of lesion and the characteristics of the balloon catheter used (Fig. 1). A 6 F-introducer was chosen in case of single TLA, whereas 7 F or 9 F-introducers were preferred if a stent was to be implanted. In case these endovascular techniques were performed after surgical arterial dissection, the size of the introducer was not limited by the diameter of the arterial puncture, which represents an advantage of this strategy. A first arteriography was performed by the lateral access of the introducer with a rapid image sequence. The size of the balloon was determined by measuring the caliber of the arterial segments adjacent to the stenosis. The balloon catheter was advanced over the guide wire under fluoroscopic control (Fig. 2) and TLA was performed. The guide wire remained in place until the end of the procedure and a completion arteriography was carried out. If completion angiography showed a residual stenosis or dissection, a stent was immediately implanted.

Balloon expandable stents as well as Nitinol self-expandable stents have been used in this series. An additional dilatation of the deployed stent has frequently been performed by reinflating the balloon at the proximal and distal ends of the stent. If completion angiography delineated a good result, the guide

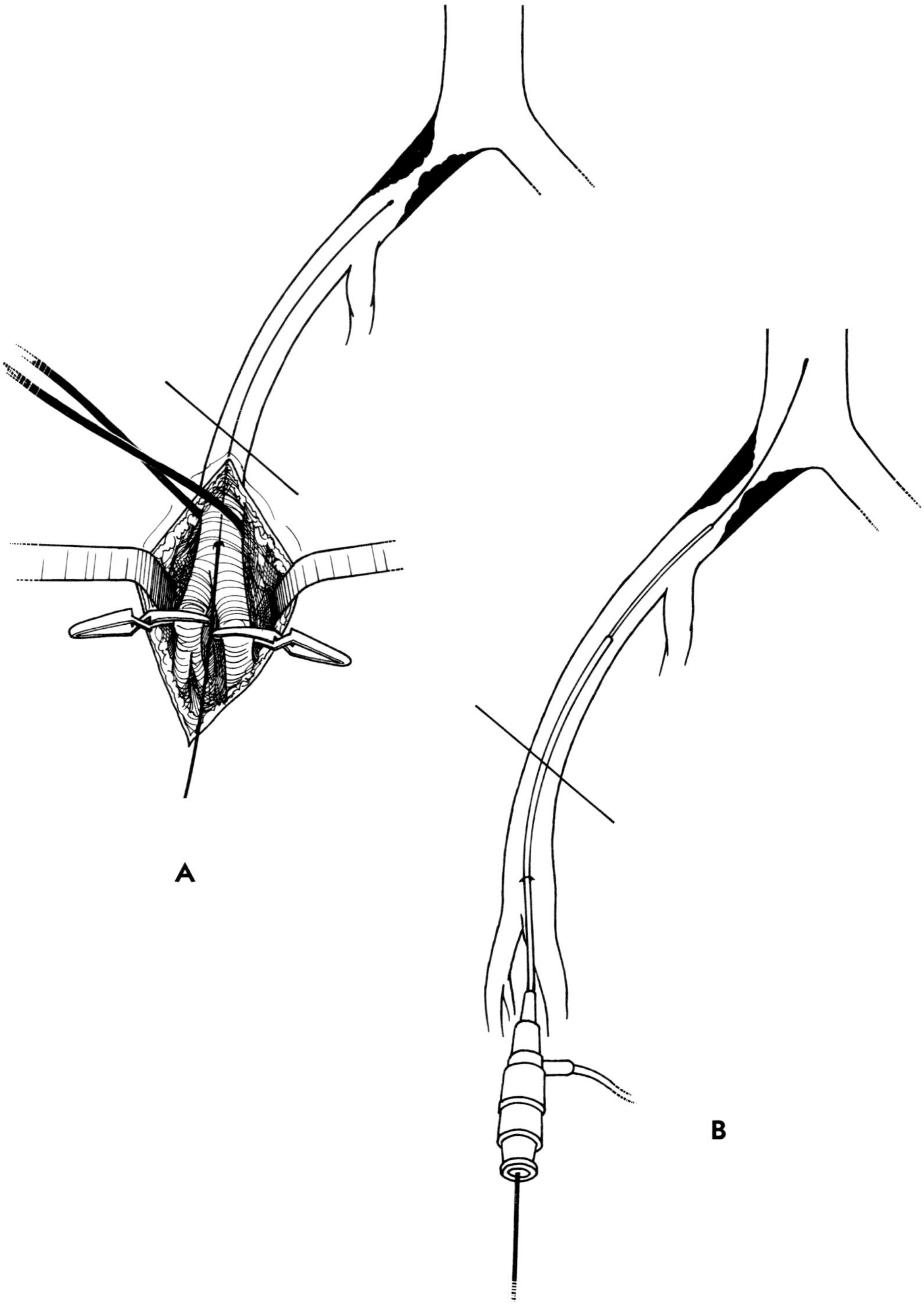

FIG. 1 A - Introduction of the guide wire and passing the lesion. B - The introducer and dilator are positioned.

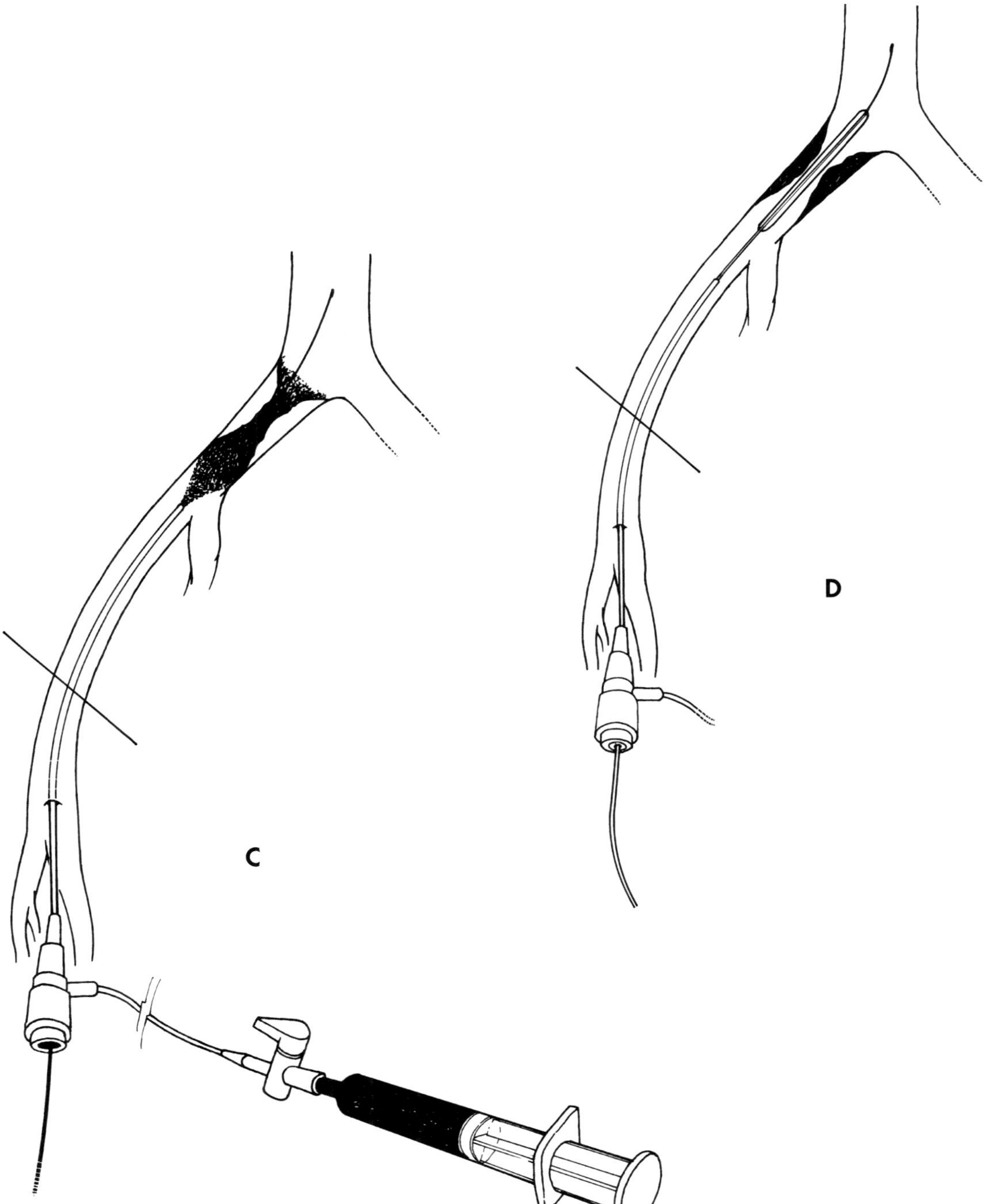

FIG. 1 C - Angiography is performed through a sideport of the introducer. D - The balloon catheter is advanced over the guide wire.

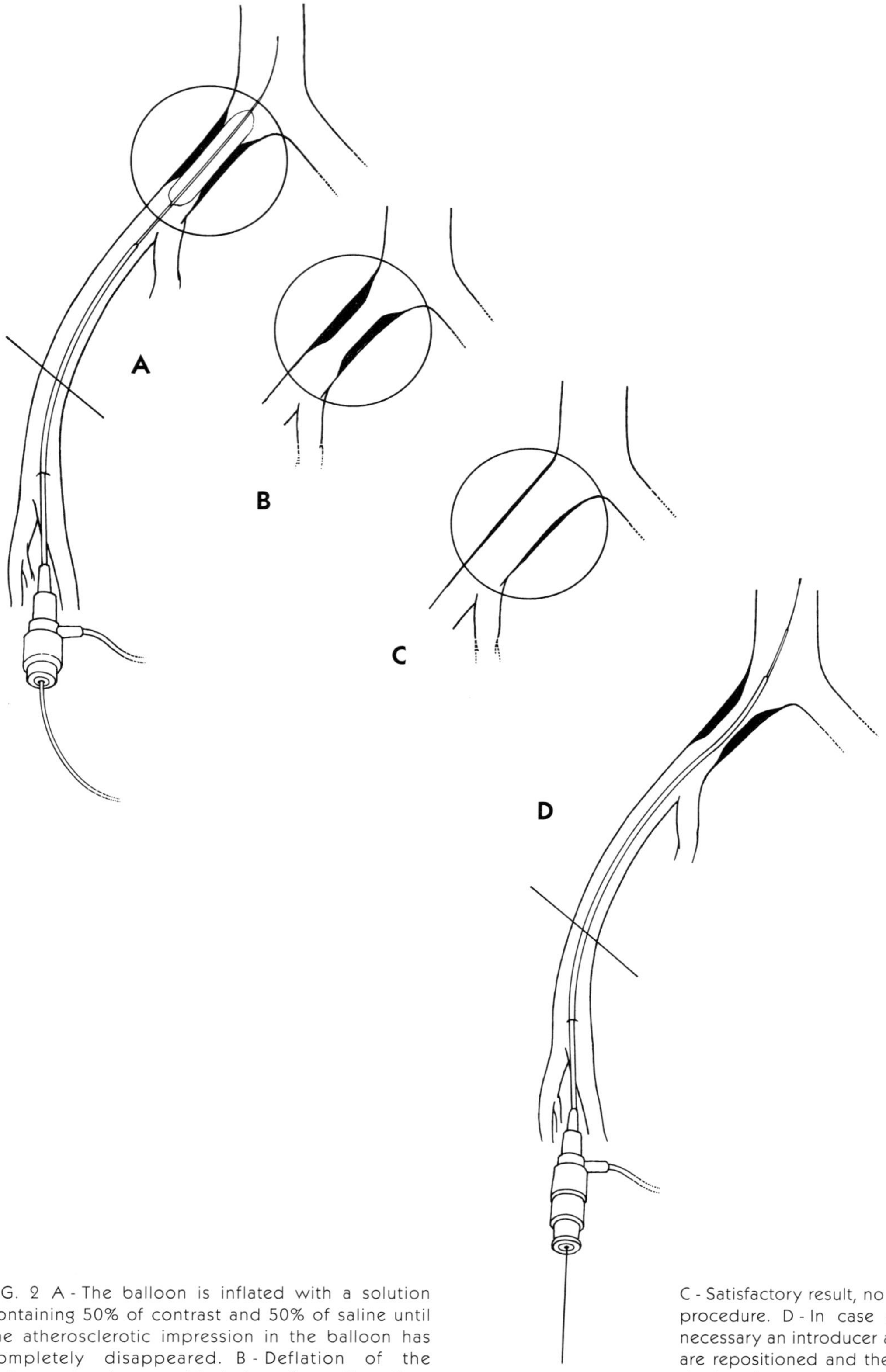

FIG. 2 A - The balloon is inflated with a solution containing 50% of contrast and 50% of saline until the atherosclerotic impression in the balloon has completely disappeared. B - Deflation of the balloon, showing a residual stenosis or dissection. C - Satisfactory result, no additional procedure. D - In case a stent is necessary an introducer and dilator are repositioned and the device is deployed.

Table IV ASSOCIATED ENDOVASCULAR TECHNIQUES

Level of arterial disease	*Angioplasty*	*Stent*
Common iliac artery*	12	7
External iliac artery*	15	3
Superficial femoral artery**	10	3
Distal popliteal/ crural artery*	4	0
Total	**41**	**13**

* All stenotic lesions
** 7 stenotic and 3 occlusive lesions

Table V SURGICAL TECHNIQUES ASSOCIATED WITH 27 ILIAC ARTERY TLA'S

Surgical procedure	*Iliac artery TLA*
Femoro-femoral cross-over bypass	5
Femoropopliteal/crural bypass	16
Common femoral and femoral bifurcation revascularization	6
Total	**27**

Table VI SURGICAL TECHNIQUES ASSOCIATED WITH 10 SUPERFICIAL FEMORAL ARTERY TLA'S

Surgical procedure	*Superficial femoral artery TLA*
Femoro-femoral cross-bypass	3
Aortobifemoral bypass	1
Common femoral and femoral bifurcation revascularization	3
Femoropopliteal/crural bypass	3
Total	**10**

TLA: transluminal angioplasty

wire and the introducer were removed and the surgical procedure was accomplished (Fig. 3). Table IV shows the details of all 41 endovascular procedures performed. Among those, 27 angioplasties (66%) have been carried out at the iliac artery level, including stent deployment in 10 cases. Ten angioplasties (24%) have been completed at the level of the superficial femoral artery with stent placement in the three cases of recanalization. Finally, four angioplasties have been performed at the level of the distal popliteal artery and crural arteries. Tables V and VI depict the surgical techniques associated with the 27 iliac artery angioplasties and the 10 superficial femoral angioplasties, respectively. Figure 4 summarizes all the combined techniques including iliac artery angioplasty.

■ ASSESSMENT AND PATIENT FOLLOW-UP

The main parameters to analyze the results were improvement of the ischemic stage and patency of the treated lesions as assessed by color duplex scanning. Primary and secondary patency rates were calculated according to the standards of the SVS-ISCVS [4]. Overall primary patency rate corresponded to the primary patency rate of all procedures performed on the same patient. Consequently, in order to have a valid overall patency rate, each patient should have a patent angioplasty and surgical reconstruction without intervention.

Survival rate and primary and secondary patency rates have been calculated by means of the Kaplan-Meier method [5], calculating the 95% confidence interval for each interval on the curve.

■ RESULTS

The mean follow-up of the study was 12 months. The mean hospital stay was 9 ± 4 days. One patient (2.4%) died within the first 30 postoperative days, due to multiorgan failure. Four other patients died within the first year (2 myocardial infarction, 1 perforation of duodenal ulcer, 1 multiorgan failure following a vascular reintervention). Four

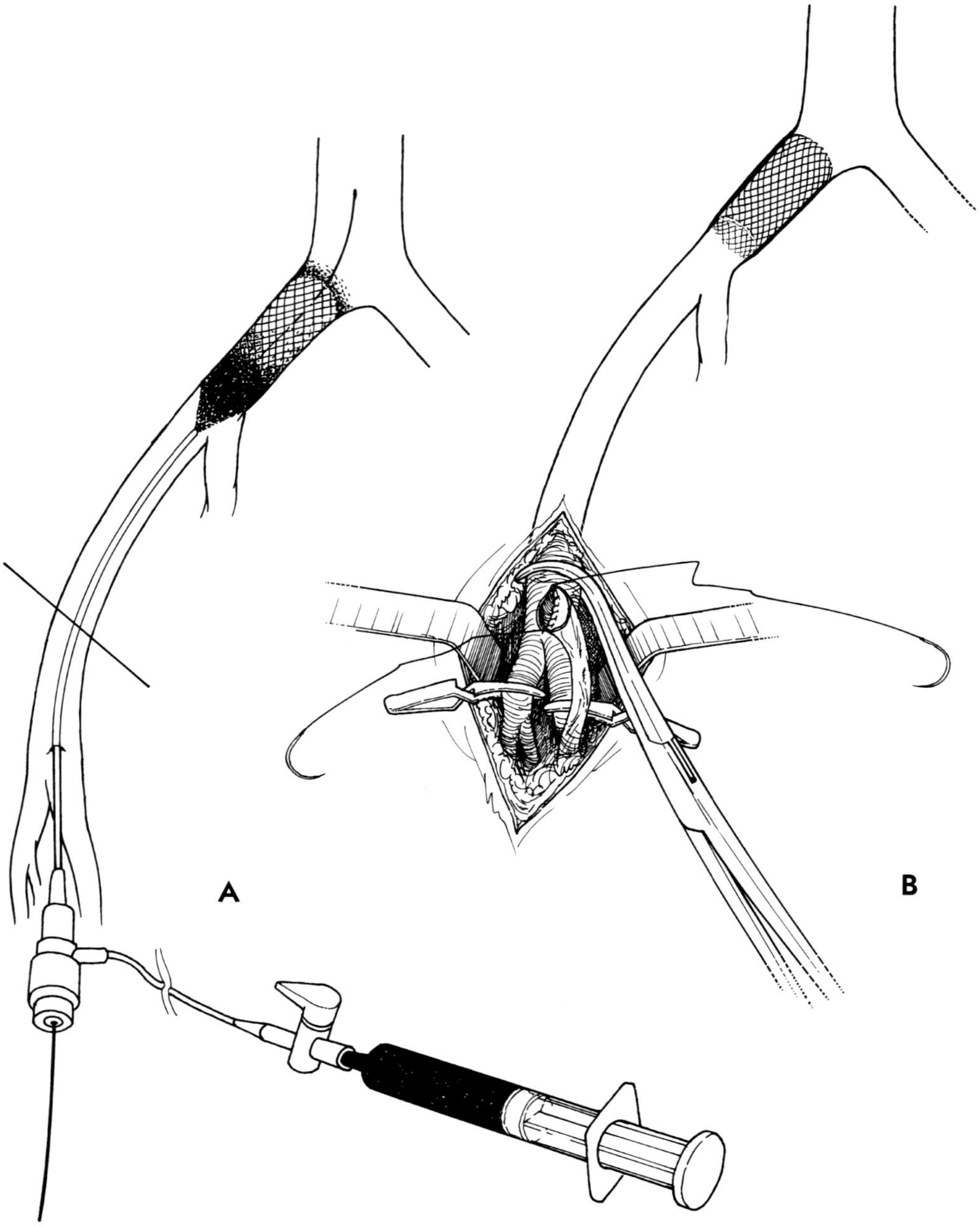

FIG. 3 A - Following placement of the stent, the balloon is deflated and completion angiography is performed. B - The introducer and guide wire are removed, the femoral artery is clamped, the arteriotomy is distended and the anastomosis performed, and the bypass is finished in a conventional way.

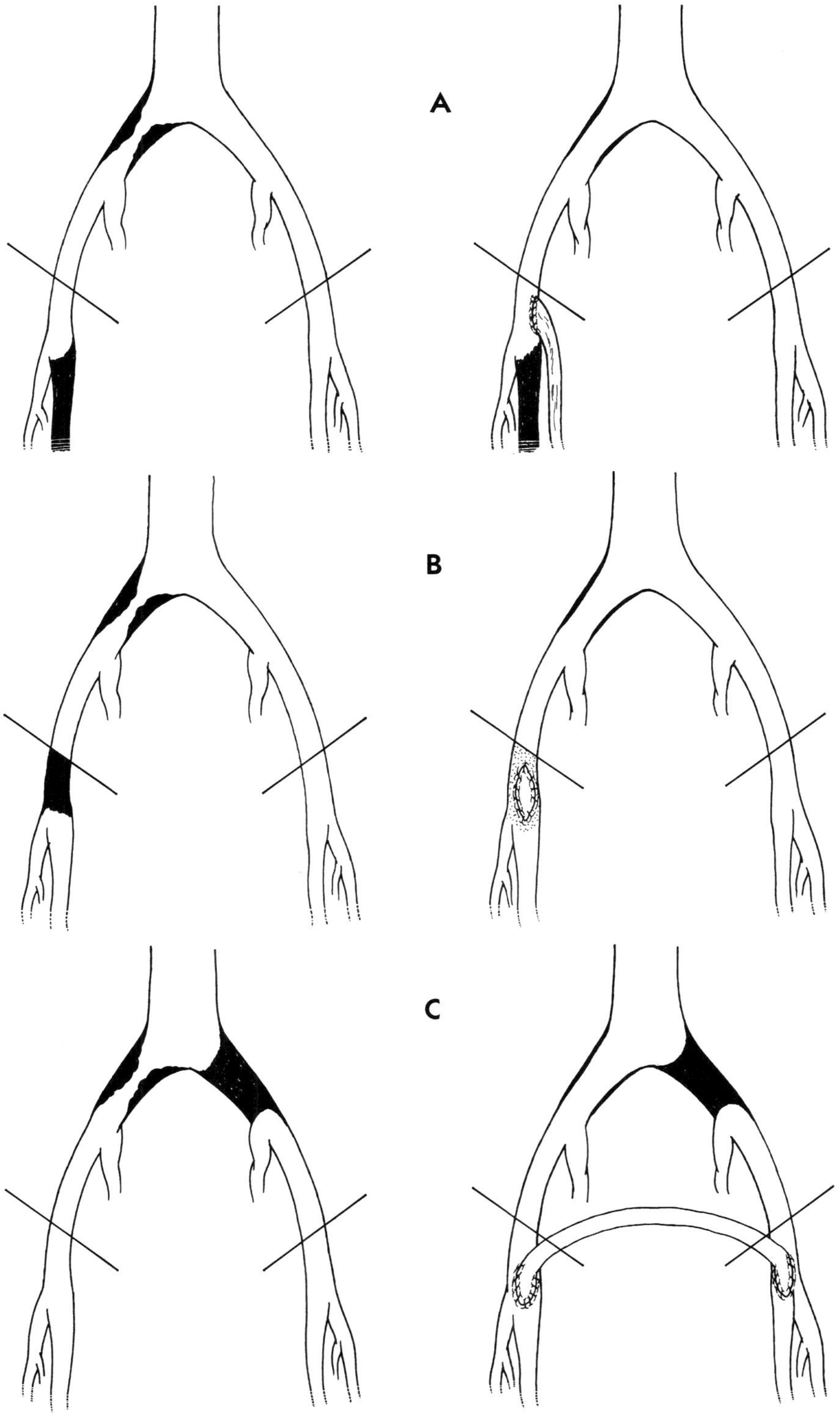

FIG. 4 Combined iliac artery TLA and conventional surgery for multilevel disease. A - Iliac artery TLA and femoropopliteal- or crural bypass (n=16). B - Iliac artery TLA and reconstruction of common femoral artery and femoral bifurcation (n=6). C - Iliac artery TLA and femoropopliteal cross-over bypass (n=5).

patients (9.7%) suffered from skin necrosis at the access site in the groin. No arterial or prosthetic infections occurred.

Survival rates at six months and one year were 97.5% ± 6% and 84% ± 14%, respectively (Fig. 5). Overall primary patency rate of the TLA and surgical procedures was 89.1% ± 10% and 66.4 ± 17% at six months and one year, respectively (Fig. 6A). Primary patency rate of the TLA's was 69.4% ± 16% (Fig. 6B) at one year and 78.4% ±16% for the surgical revascularizations (Fig. 6C). Secondary patency at six months and one year were 92.5% ± 7% and 84.7% ± 12%, respectively (Fig. 7A). Secondary patency rate of the TLA's was 87.3% ± 11% after one year (Fig. 7B) and 82% ± 9% after surgical procedures (Fig. 7C). Table VII and VIII show the primary patency rates of the bypasses and TLA's associated with the level of the lesion.

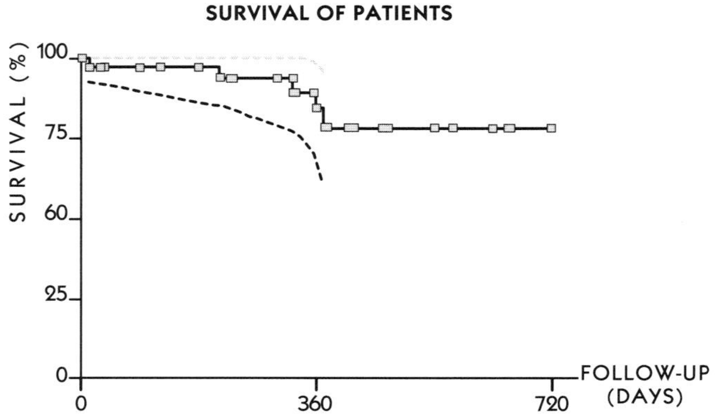

FIG. 5 Survival curve of 41 patients. Kaplan-Meier method with 95% confidence intervals. Survival rate at one year was 84% ± 14%.

Table VII SURGICAL RESULTS AT ONE YEAR ASSOCIATED WITH THE LEVEL OF ARTERIAL DISEASE (PRIMARY PATENCY)

Level of arterial disease	*N*	*Patent grafts (%)*	*Restenosis/ Occlusion*
Iliac artery	9	9 (100)	0
Common femoral/ bifurcation	12	9 (75)	3
Superficial femoral artery	16	12 (75)	4
Popliteal/ crural artery	4	3 (75)	1
Total	**41**	**33**	**8**

Table VIII PRIMARY PATENCY OF ENDOVASCULAR ANGIOPLASTY ASSOCIATED WITH THE LEVEL OF ARTERIAL DISEASE

Level of arterial disease	*N*	*Patency (%)*	*Restenosis/ Occlusion*
Common iliac artery	12	8 (66)	4
External iliac artery	15	12 (80)	3
Superficial femoral artery	10	6 (60)	4
Popliteal/ crural artery	4	2 (50)	2
Total	**41**	**28**	**13**

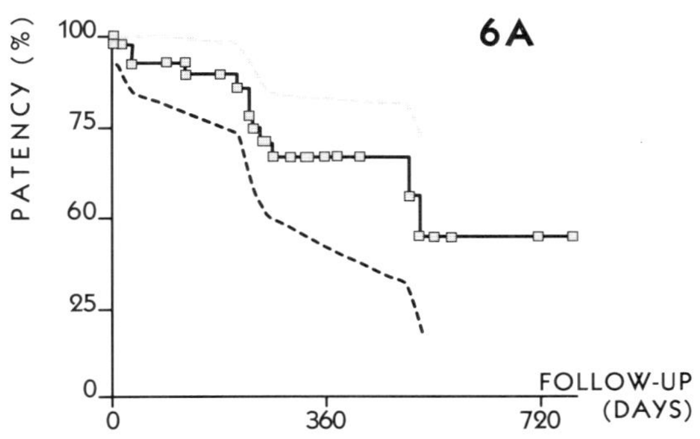

OVERALL PRIMARY PATENCY
89.1% ± 10% at 6 months and 66.4% ± 17% at one year.

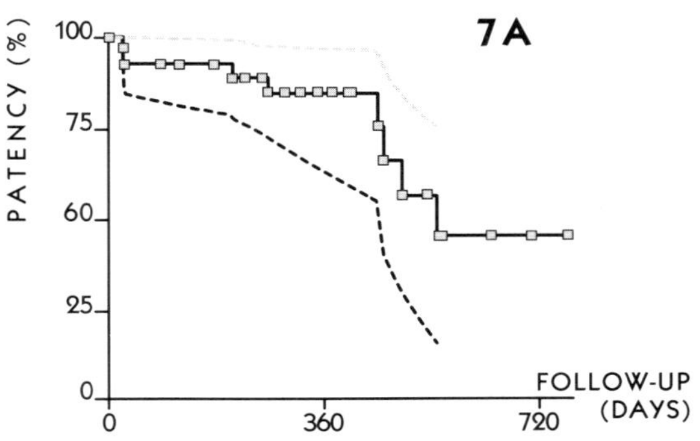

OVERALL SECONDARY PATENCY
92.5% ± 7% at 6 months and 84.7% ± 12% at one year.

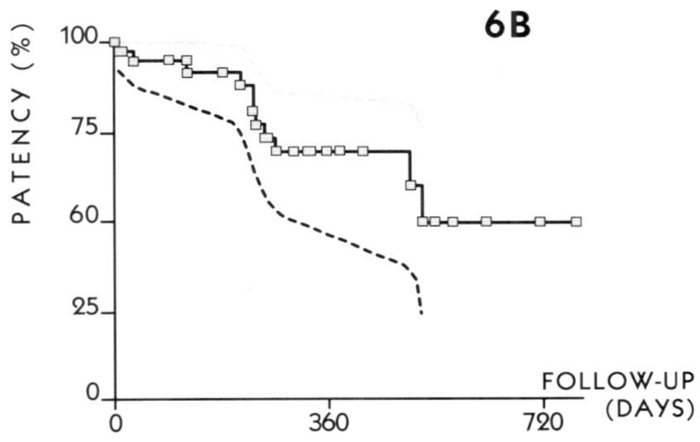

PRIMARY PATENCY OF ANGIOPLASTIES
69.4% ± 16% at one year.

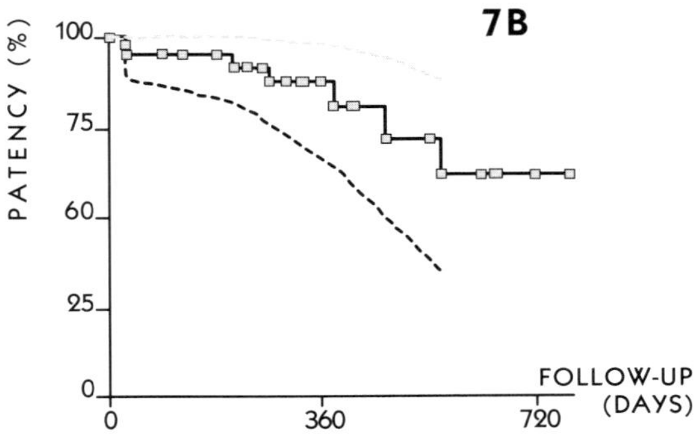

SECONDARY PATENCY OF ANGIOPLASTIES
87.3% ± 11% at one year.

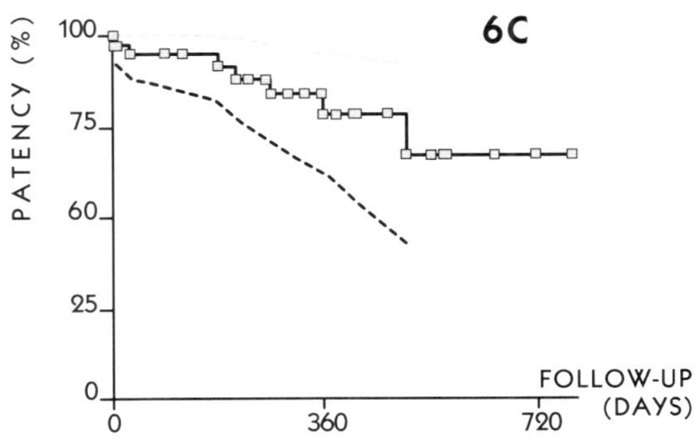

PRIMARY PATENCY
OF SURGICAL REVASCULARIZATIONS
78.4% ± 16% at one year.

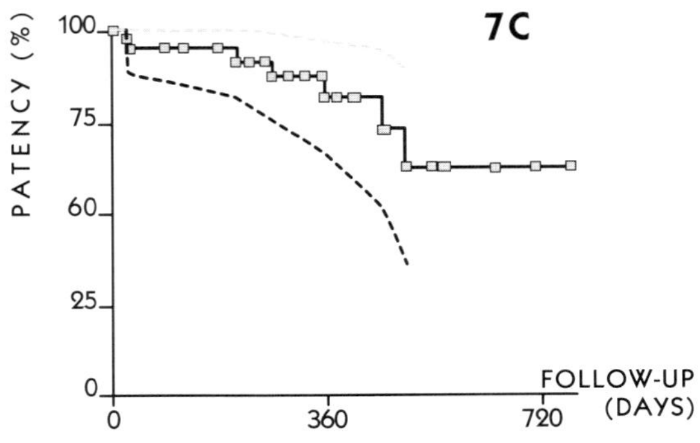

SECONDARY PATENCY
OF SURGICAL REVASCULARIZATIONS
81.9% ± 9% at one year.

FIG. 6 (A-B-C) et FIG. 7 (A-B-C) Patency rates by Kaplan-Meier method with 95% confidence intervals.

During follow-up, two patients (4.8%) underwent a major amputation. Table IX shows the evolution of the ischemic stages during time. One year after the procedure, 61% of the patients (n=25) were asymptomatic.

Table IX EVOLUTION OF CLINICAL STAGE

Clinical stage	*Preoperative clinical stage* N	(%)	*Clinical stage at 12 months* N	(%)
Asymptomatic	0	(0)	25	(61)
Intermittent claudication	24	(58)	5	(12)
Rest pain	7	(17)	4	(10)
Gangrene amputation	10	(25)	3	(7)
Deaths	0	0	4	(10)
Total	**41**		**41**	

GENERAL CONSIDERATIONS

Feasibility of the Technique

This preliminary study shows that TLA and a surgical reconstruction can be combined in the same procedure for the treatment of multilevel arterial disease causing chronic ischemia of the legs. Operative mortality of this combined treatment was low (2.4%) and hospital stay relatively short. Procedure related morbidity was limited and no arterial or prosthetic infections occurred. Secondary patency rate at one year (85%) was satisfactory and only two major amputations were performed during the follow-up of 12 months.

These encouraging results confirm the outcomes reported by others [6-8]. However, the real durability of these procedures has to be judged after a follow-up of three to five years. Although the results are preliminary, the technique is appealing. From a logistic point of view, most vascular surgical theaters are equipped with the necessary materials to perform endovascular procedures and obtain high quality angiography. The increased application of intraoperative angioplasty complementary to conventional vascular surgery is now recognized by several authors [9]. The applied techniques depended upon the level of the arterial disease and the type of the lesion. Table V, Table VI, and Figure 4 clearly show the reasons why an endovascular procedure of a short iliac stenosis (n=27) followed by a femoro-femoral crossover bypass in case of a contralateral iliac occlusion was preferred over a direct iliac recanalization. For the same reason we preferred to perform an angioplasty of short asymptomatic stenoses of the superficial femoral artery (n=10), while multilevel occlusions or stenoses were treated with a femoropopliteal bypass. Iliac artery stents have been systematically positioned in case of a dissection or residual stenoses following TLA. Among 27 dilatations, 10 stents have been implanted with an excellent final result allowing a safe distal surgical reconstruction during the same session.

TLA and Surgery for Multilevel Lesions

In general, combined transluminal angioplasty and vascular surgery offer an original management of aortoiliac lesions associated with infrainguinal vascular pathology. The classical approach preferably aims at vascular reconstruction of proximal lesions because of a higher hemodynamic efficiency. Reconstruction of proximal lesions only, however, meets several disadvantages. In the case of exercise induced ischemia, a residual claudication can remain and the patient will consider this as an insufficient treatment with regard to the functional result and improvement of quality of life. In the case of critical limb ischemia (CLI), and especially in the case of skin ulcers, hemodynamic improvement of a proximal revascularization might only relieve the ischemic pain but not heal the ischemic ulcer. Finally, a limited proximal vascular repair has an increased risk on early occlusion if femoropopliteal disease is associated with significant profunda femoral arterial lesions (Fig. 8). The combined approach allows treatment of the most complex lesions by conventional surgery and the most simple pathology by endovascular techniques.

The most frequently occurring situation, especially in CLI, includes femoropopliteal disease requiring a bypass, associated with one or more iliac artery stenoses (Fig. 9). The mean advantage of iliac artery transluminal angioplasty with or without stent is that a surgical aorto- or iliofemoral bypass can be avoided. Furthermore, in case of bilateral iliac lesions, only one accessible iliac artery for endovascular procedure is necessary followed by a femoro-femoral crossover bypass to restore the blood flow to both legs (Fig. 10).

Another possibility, however less frequent, is represented by aortoiliac lesions requiring surgical repair associated with infrainguinal pathology accessible for endovascular treatment: short superficial femoral arterial occlusion or stenosis, or popliteal stenosis. The endovascular management of these lesions combined with aortofemoral surgery improves the short- and mid-term functional result without impairing the future. In CLI and especially in the presence of skin ulcers, recanalization of femoropopliteal lesions shorter than 10 cm, or transluminal angioplasty on one or several crural artery

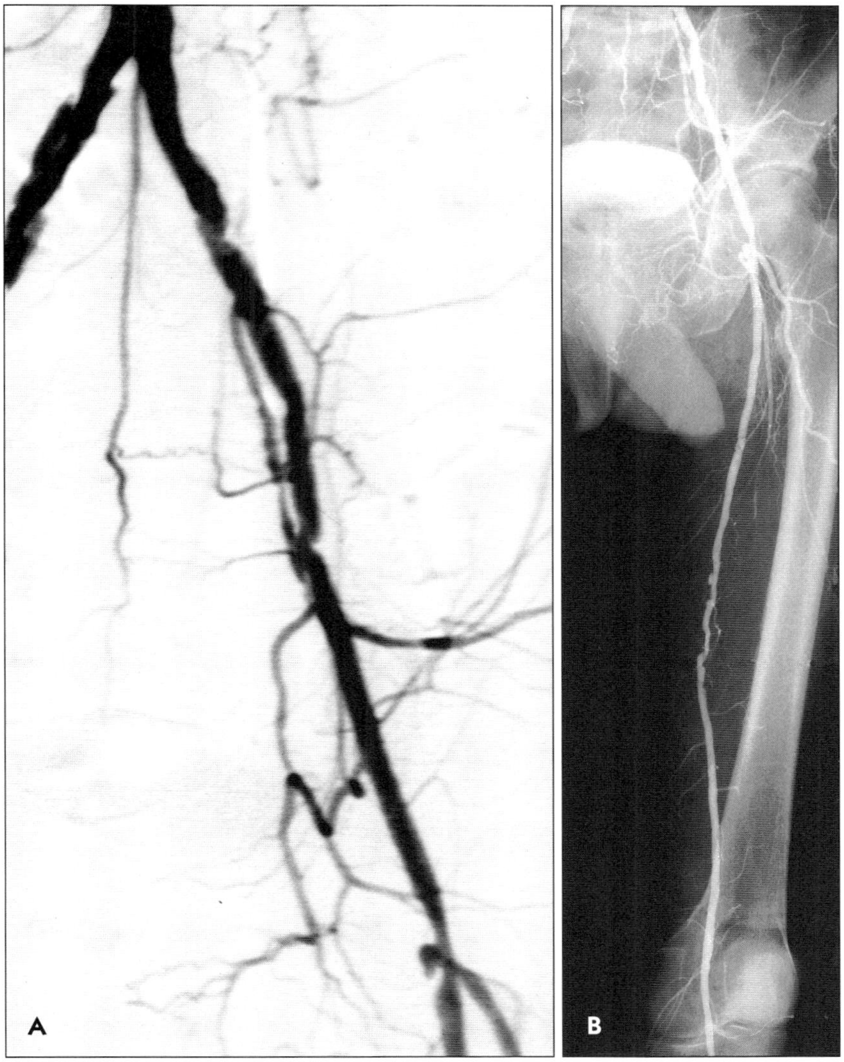

FIG. 8 Example of iliac artery disease and associated occlusive lesions in the femoropopliteal and profunda femoral arteries. An isolated aorta femoral bypass would have an increased risk for acute occlusion. Surgical reconstruction of the common femoral artery and femoral bifurcation was combined with TLA of the iliac-and superficial femoral artery.

stenoses (Fig. 11) can avoid the necessity of a femorodistal bypass and offer adequate peripheral perfusion for safer and faster wound healing.

PTA in Surgical Environment

Our current experience shows the advantages of performing endovascular procedures in the operation theatre combined with conventional surgical revascularization. During this simultaneous approach the surgeon can immediately correct insufficient results or complications of the endovascular procedure. Furthermore, a combined session allows easy endovascular access after dissection of the artery or a venous graft, avoiding difficult percutaneous puncturing in fibrous scar tissue or contralateral femoral arteries. Finally, this combined technique has allowed us to ease the treatment of progressive lesions like restenoses in distal grafts by angiography followed by an endovascular intervention or the most appropriate surgical technique.

We also believe, like others [10,11] that this combined approach offers optimal training and education for young surgeons, familiarizing them with endovascular techniques and especially the correct indications.

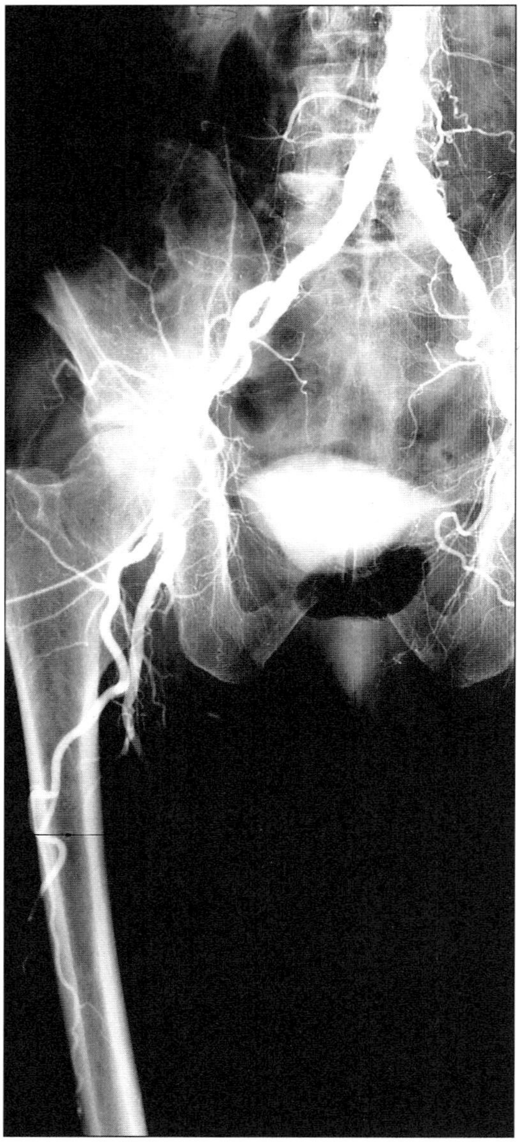

FIG. 9 Femoropopliteal occlusion associated with a proximal external iliac artery stenosis. Treatment consisted of a femoropopliteal bypass to the infragenicular popliteal artery, combined with an iliac artery TLA.

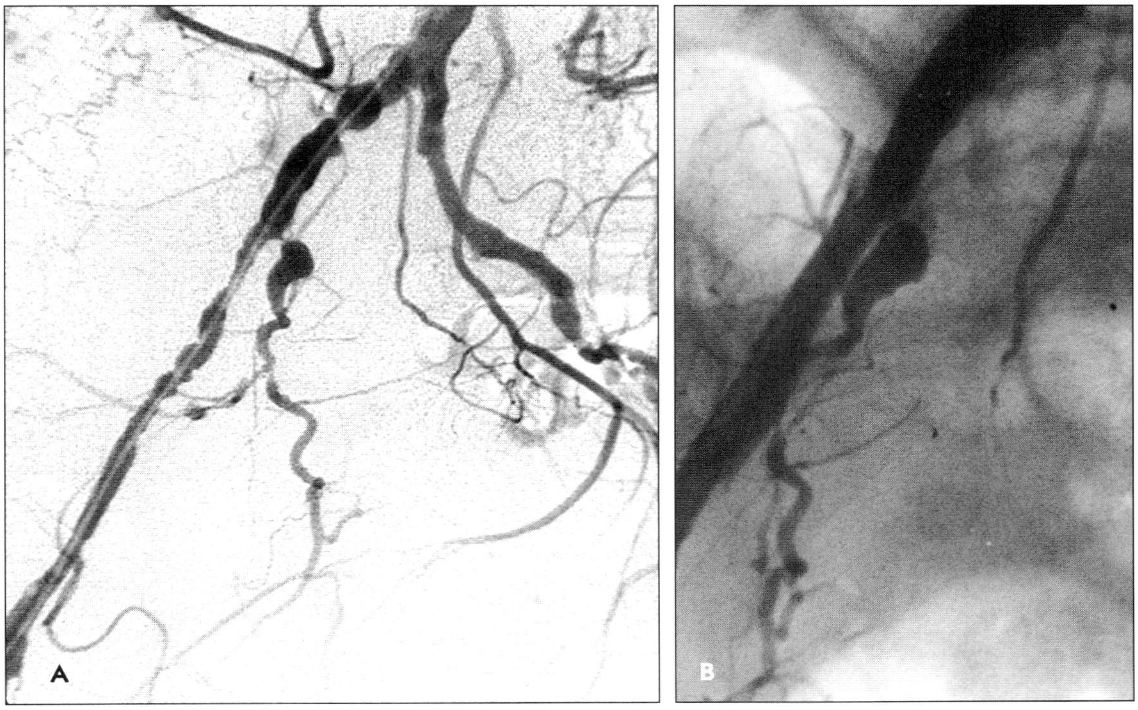

FIG. 10 Bilateral iliac artery lesions of which one is simply accessible for TLA. TLA combined with a femoro-femoral crossover bypass allows revascularization of both legs in a relatively easy way. A - Preoperative iliac artery lesions. B - Intraoperative angiographic control of the iliac artery following TLA, before crossover bypass.

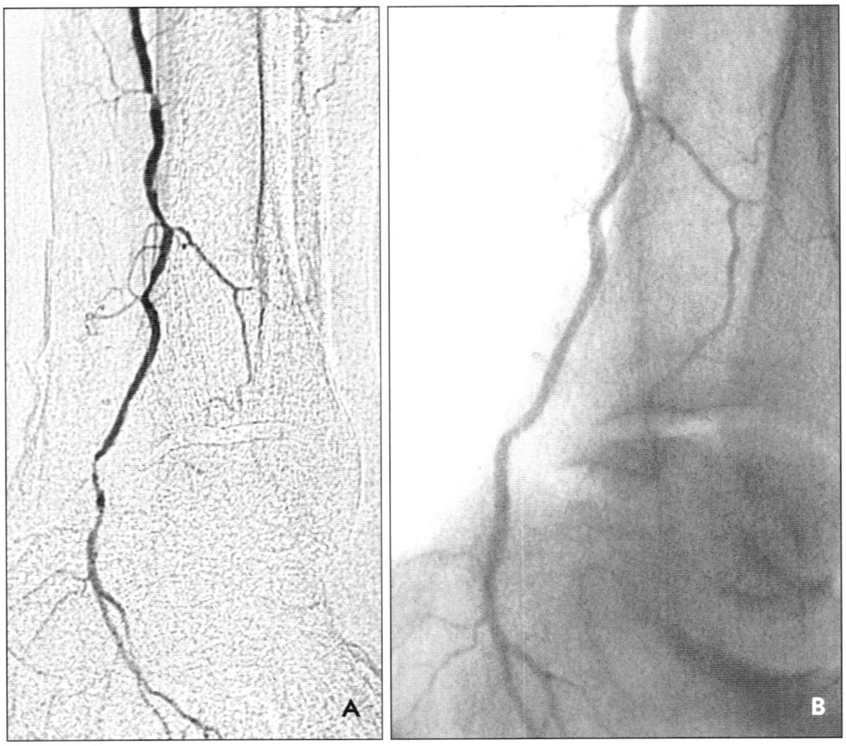

FIG. 11 Aortoiliac disease associated with severe crural artery disease (A) in a patient with CLI and skin ulcers. Aortofemoral reconstruction and posterior tibial artery TLA (B) offer an improved peripheral perfusion to heal the ulcer.

REFERENCES

1 Hunink MG, Donaldson MC, Meyerovitz MF et al. Risks and benefits of femoropopliteal percutaneous balloon angioplasty. *J Vasc Surg* 1993; 17: 183-194.

2 Blair JM, Gewertz BL, Moosa H et al. Percutaneous transluminal angioplasty versus surgery for limb threatening ischemia. *J Vasc Surg* 1989; 9: 698-703.

3 Varty K, Nydahl S, Nasim A et al. Results of surgery and angioplasty for the treatment of chronic severe lower limb ischaemia. *Eur J Vasc Endovasc Surg* 1998; 16: 159-163.

4 Rutherford RB, Baker JD, Ernst C et al. Recommended standards for reports dealing with lower extremity ischemia: revised version. *J Vasc Surg* 1997; 26: 517-538.

5 Kaplan EL, Meier P. Non parametric estimation for incomplete observations. *J Am Statist Ass* 1958; 53: 457-481.

6 Pfeiffer RJr, String ST. Adjunctive use of balloon dilatation catheter during vascular reconstructive procedures. *J Vasc Surg* 1986; 3: 841-845.

7 Kugel RD, Pereyra R. Combined femorotibial bypass and distal intraoperative transluminal angioplasty. *J Vasc Surg* 1986; 4: 533-535.

8 Schneider PA, Abcarian PW, Ogawa DY et al. Should balloon angioplasty and stents have any role in operative intervention for lower extremity ischemia ? *J Vasc Surg* 1997; 11: 574-580.

9 Calligaro KD, Dougherty MJ, Patterson DE et al. Value of an endovascular suite in the operating room. *Ann Chir Vasc* 1998; 12: 296-298.

10 Hodgson KJ, Mattos MA, Mansour A et al. Incorporation of endovascular training into a vascular fellowship program. *Am J Surg* 1995; 170: 168-173.

11 White RA, Fogarty TJ, Baker WH et al. Endovascular surgery credentialing and training for vascular surgeons. *J Vasc Surg* 1993; 17: 1095-1102.

16

CRITICAL LIMB ISCHEMIA DUE TO POPLITEAL ARTERY ANEURYSM: ENDOVASCULAR APPROACH

ANDRÉ NEVELSTEEN, HENDRIK LACROIX
LUC STOCKX, PAUWEL BERNAERTS

Although the prevalence of popliteal artery aneurysms in general population is low, they represent the most frequently encountered peripheral artery aneurysms. In a series of 1488 patients with aneurysms of the aorta and its distal branches, reported by Dent et al. popliteal artery aneurysms accounted for 84% of all peripheral aneurysms [1]. Since atherosclerosis is responsible for over 90% of the popliteal artery aneurysms, they are found most frequently in men during the sixth or seventh decade of life. Bilateral aneurysms are seen in 50% of the cases. In addition, concomitant aneurysms in the aortofemoral region are found in 40 to 80% of the patients. This high incidence of concomitant vascular disease has led some authors to refer to popliteal artery aneurysms as an index of generalized vascular disease [2].

■ NATURAL HISTORY AND CLINICAL MANIFESTATIONS

The natural history of popliteal artery aneurysms has been reviewed in several, although unfortunately exclusively retrospective, series. Gifford et al. [3] described already in 1953 a series of 100 popliteal aneurysms where, over a 46-month period, 35% of the asymptomatic aneurysms developed complications, resulting in an amputation rate of 85%. In 1986, Mellière et al. reviewed a series of 77 popliteal aneurysms of which 23 were initially treated conservatively [4]. With a mean follow-up of

48 months, six patients needed an operation for new symptoms or enlargement of the aneurysm, while amputation was performed in two others. More recently, Lowell et al. described a series of 67 popliteal artery aneurysms presenting without symptoms [5]. During follow-up, 18% of the aneurysms led to chronic or acute symptoms, resulting in three major amputations. The mean interval for the onset of symptoms was 17 months (range 4.5 to 30 months). These data are furthermore confirmed in a retrospective review of 42 patients by Dawson et al. [6]. With a mean follow-up of 6.2 years, these authors found that the cumulative risk of developing complications was 24% at 1 year rising to 68% at 5 years.

The high incidence of complications explains also why popliteal artery aneurysms have been characterized in the past as *sinister harbingers of sudden catastrophe* [3]. In fact, approximately 70% of the patients with popliteal artery aneurysms are symptomatic at initial presentation. Eight to 15% of the patients present with compression of local venous or nervous structures especially in the presence of large and saccular aneurysms. Infection or rupture has been described occasionally, occurring in less than 5% of the cases [7]. Ischemia is, however, the presenting symptom in over 80% of the patients and 20% to 40% of the patients present with acute limb threatening ischemia. The most common cause here is acute aneurysm thrombosis preceded by repetitive episodes of distal atheroembolization resulting in loss of outflow and retrograde thrombosis of the tibial arteries. Individual factors, predisposing to these thromboembolic complications have been the subject of several studies. Ramesh et al. stressed the importance of flow distortion within the aneurysm or within the popliteal artery either above or below the aneurysm [8]. Several authors [9,10] have documented the importance of distal occlusive disease and intraluminal thrombosis. On the other hand, the influence of aneurysm size on the rate of complications is still controversial. As it has been found in several series [11-13], it seems logical to suppose that the incidence of thrombosis will increase with increasing aneurysm size. However, Inahara and Toledo emphasized already in 1978 the tendency of even small aneurysms to thrombose [14]. These data are confirmed more recently by Lancashire et al. [15] and also by our own experience, in which thrombolysis identified small popliteal aneurysms (< 2.5 cm) as the cause of thrombosis in nearly 15% of the patients presenting with localized occlusion of the popliteal artery. It stresses also the fact that the incidence of complicated popliteal aneurysms might still be underestimated.

■ SURGICAL MANAGEMENT AND RESULTS

The first recorded attempt of surgery in case of a popliteal artery aneurysm goes back to Antyllus, a third century Greek who ligated both poles of a popliteal aneurysm. The aneurysm sac was drained and the wound left open to heal by secondary intention [16]. It was John Hunter who popularized in 1785 ligation of the superficial femoral artery, well above the aneurysm as the standard treatment [17]. Such ligation without opening the aneurysmal sac was accompanied by distal gangrene in over 10% of the cases but it remained the procedure of choice for more than one century. Modern treatment started in the beginning of the twentieth century when Goyanes in 1906 [18], followed by Pringle in 1912 [19] both described the reconstruction of a popliteal artery aneurysm by means of venous bypass. Synthetic graft substitution was reported for the first time by Crawford in 1958 [20], but autogenous saphenous vein bypass by the inlay technique or with exclusion of the aneurysm, as introduced in 1969 by Edwards [21] is still the operation of choice.

Over the last twenty years there have been numerous reports concerning patency rates and limb salvage after reconstruction for popliteal artery aneurysms. For asymptomatic aneurysms, patency rates range between 70% and 90% at 5 years, remaining fairly stable over the following years. The corresponding limb salvage rates are between 95% and 100% [9,13,14,22-25]. As for conventional femorodistal surgery, the results with saphenous vein bypass are certainly superior to those after prosthetic insertion. Anton et al. reported a five years patency rate of 94% after saphenous vein reconstruction, which decreased to 43% with synthetic materials [22]. These data were confirmed by Roggo et al. who reported five-year patency rates of 81%, 64% and 33% with the use of saphenous vein, PTFE and Dacron, respectively [25]. At the same time, the importance of the distal runoff with regard to graft patency remains controversial. Shortell et al. documented a 3-year patency rate of 89% in extremities with good runoff (> 2 vessels) and 40% in those with a poor (< 1 vessel) runoff [26]. Anton and coworkers reported a 10 years patency rate of 64% for grafts that regained one or both pedal pulses after reconstruction, in contrast to a 32% patency rate for limbs that did not regain palpable pulses [22]. In their series of 91 reconstructions, Poirier et al. were, however, unable to find a correlation between long-term patency and the number of patent tibial vessels [27]. They confirmed hereby the findings of Hagino et al. who also stated that the use of autologous reversed vein grafting and attention to technical details yields excellent long-term patency despite a suboptimal runoff anatomy [23].

Results of surgery are by far less encouraging in symptomatic aneurysms, especially in cases of aneurysm thrombosis and acute ischemia. Primary amputation may be the only solution in 20% to 40% of the cases [13,25,26]. Lowell et al. reported a primary amputation rate of 33% in patients presenting with acute aneurysm thrombosis [5]. Also in 1994, Varga et al. presented 62 patients with acute ischemia in a multicenter study of 200 popliteal aneurysms [28]. Three patients underwent a primary amputation, while another eight underwent a major amputation after bypass surgery (amputation rate 18%). Due to outflow impairment and the urgent nature of the procedure, the long-term results of vascular reconstruction in cases of symptomatic aneurysms are also by far inferior to asymptomatic cases. The reported 5-year patency rates range from 40% to 60%, with corresponding limb salvage usually well below 90% [12,13,22, 25,26,29].

■ ENDOVASCULAR APPROACH

Transcatheter intra-arterial fibrinolytic therapy has been shown over the years to be an effective adjunct in the management of patients with acute thromboembolic manifestations both in native arteries and prosthetic bypasses. Taking into account the dissatisfactory surgical results, it is no surprise that complementary local (pre- or perioperative) fibrinolytic therapy has also been studied as an adjunct in patients with acute ischemic complications due to a popliteal artery aneurysm. The primary purpose of thrombolysis here is to reduce distal ischemia by reopening the distal tibial vessels and even the intramuscular arteries that are not accessible to standard balloon embolectomy techniques. Thrombolysis therefore anticipates technically difficult surgery and patients who are considered inoperable may even become amenable for vascular reconstruction. In addition, long-term results of surgery may be improved because of better outflow conditions. Last but not least, preoperative fibrinolytic therapy also allows for better patient preparation and evaluation of risk factors preoperatively. Recanalization of the aneurysm itself is certainly not the primary goal, but might become more important in the future with the continuing development of prosthetic endografts.

The first report of thrombolysis in patients with a popliteal aneurysm were presented in 1984, when Schwartz et al. reported on a 61-year old male who presented with acute ischemia due to thrombosis of a popliteal artery aneurysm [30]. Angiography showed a complete occlusion of the tibial vessels. He was treated with a local catheter infusion of 5000 U/hr of streptokinase for 18 hours, after which partial clearance of the popliteal aneurysm and reopening of the tibial vessels was achieved. The following day, the patient underwent surgical repair of his popliteal aneurysm and he was discharged 3 days later from the hospital. In the same year, Taylor et al. described another three patients with acute thrombosis of a popliteal artery aneurysm and limb threatening ischemia [31]. All three were relieved of their acute ischemia by streptokinase infusion accompanied by lysis of clots in the popliteal artery outflow tract. Each patient then successfully underwent elective popliteal artery bypass. The importance of intraoperative thrombolysis was stressed in a paper presented by Thompson et al. in 1993 [32]. These authors discussed six patients, operated upon in emergency because of popliteal aneurysm thrombosis. Balloon embolectomy and bypass surgery was completed by intraoperative infusion of streptokinase (n=4) or tissue plasminogen activator (n=2) in the tibial vessels. Intraoperative angiography was performed in five patients and confirmed additional lysis in all cases.

Over the last few years, there have been at least 16 papers concerning thrombolysis of popliteal artery aneurysms [8,13,15,30-42] and evidence is accumulating that thrombolysis may indeed become an interesting adjunct. Relevant data with regard to the results in acute and limb threatening ischemia are depicted in the Table. The results need, however, to be interpreted with caution. Most of the papers are in fact anecdotal, while the larger series suffer from inhomogeneity and represent a mixture of direct surgery all or not in combination with thrombolysis without a detailed discussion. For a total of 101 procedures, preoperative thrombolysis was used in 85% of the cases. Perioperative thrombolysis is only reported by Thompson et al. [32] (n=6), Carpenter et al. [13] (n=1), and Varga et al. [28] (n=9). The fibrinolytic agent itself was mentioned in all but two papers (n=66). Streptokinase (n=30) was especially popular in the first reports. Urokinase was used on 34 occasions, while Thompson et al. [32] presented two cases where tissue plasminogen activator was used perioperatively. Complete or partial lysis was documented in 85% of the cases (n=68). The overall limb salvage rate (after additional surgery) was 91% (n=87). Differences between streptokinase and urokinase were not significant.

Although these results are certainly encouraging, objective comparison with direct surgery remains a problem in the absence of randomized studies. Hoelting et al. [39] presented a retrospective study where they analyzed their experience in 24 patients with acute ischemia from popliteal artery aneurysms. Four patients underwent primary amputation for irreversible gangrene. The others (n=20) were treated by either bypass grafting (n=11, series before 1985) or bypass grafting combined with preoperative lysis with urokinase (n=9, series after 1984). Bypass grafting alone resulted in an early amputation rate of 11% (1/11) and occlusive complications of 45% (5/11) compared to no limb loss and no bypass complication in patients who underwent surgery and preoperative lysis. These data are confirmed by Carpenter et al. [13] who, also in a retrospective non-randomized study, documented a significantly better graft patency and limb salvage rate after preoperative thrombolysis in comparison with direct emergency surgery.

Table LOCAL THROMBOLYSIS IN ACUTE ISCHEMIA DUE TO POPLITEAL ARTERY ANEURYSMS: LITERATURE REVIEW

1st author [ref.]	*year*	*N*	*Fibrinolytic agent*	*Complication rate*	*Documented lysis*	*Limb salvage*
Schwartz [30]	1984	1	Streptokinase	-	1/1	1/1
Taylor [31]	1984	3	Streptokinase	-	3/3	3/3
Fergusson [35]	1986	10	Streptokinase	-	?	7/10
Bowyer [36]	1990	9	Streptokinase	4/9*	6/9	9/9
Ramesh [8]	1993	12	NA	5/12**	7/12	NP
Thompson [32]	1993	6	Streptokinase (n = 4) TPA (n = 2)	-	5/6	6/6
Carpenter [13]	1994	7	Streptokinase (n = 2) Urokinase (n = 5)	-	7/7	7/7
Elsey [37]	1994	1	Streptokinase	-	1/1	1/1
Garramone [38]	1994	3	Urokinase	-	3/3	3/3
Hoelting [39]	1994	9	Urokinase	-	9/9	9/9
Rousseau [40]	1994	1	Urokinase	-	1/1	1/1
Varga [28]	1994	23	NA	NA	NA	16/23
Gawenda [41]	1995	14	Urokinase	NA	13/14	13/14
Debing [42]	1998	2	Urokinase	-	2/2	2/2

* bleeding n=4   ** bleeding n=3, acute limb deterioration n=2   NA: not available

In the absence of definite proof of superiority, it should also be remembered that the technique of thrombolysis itself is not without complications: minor bleeding and catheter related problems occur in up to 25% of cases [33]. The reported rates of stroke and major bleeding occur in 1% and 5%, respectively [43]. In 1994, we described also three fatal complications in a series of 50 graft occlusions treated by local thrombolysis [44]. In the series of Ramesh et al. [8], 25% of the thrombolytic procedures had to be stopped because of bleeding. Bowyer et al. [36] reported an incidence of bleeding of 44% and one of their patients died due to the consequences of hemorrhagic shock.

A further complication recently reported is that of acute deterioration of the limb during local thrombolysis. This complication appears to be more common during treatment of thrombosed popliteal aneurysms than during that of emboli or thrombosed atheromatous arteries of grafts [45]. The incidence in the series of Ramesh et al. [8] was 16%. Thompson et al. [32] and Wilkins et al. [34] have reported additional cases. The procedure had to be stopped in each case and there were at least two major amputations. The explanation for this phenomenon is probably massive thrombus destabilization followed by distal embolization. Therefore some authors advice to pass the perfusion catheter immediately through the aneurysm to direct lysis at the runoff vessels alone rather than attempting to lyze the aneurysm thrombus itself.

Time delay has been cited as a last disadvantage of thrombolysis. The mean duration of the procedures (n=6) in the series of Carpenter et al. [13] was 29 hours. Bowyer et al. [36] reported a mean duration of 32 hours in nine procedures. As for others, they however also report a marked improvement of most limbs during the first hours of treatment, which makes patients with acute ischemia also amenable to thrombolysis. Intraoperative thrombolysis at the other hand can be an attractive adjunct in patients, who require an urgent operation because of associated motor and sensory deficit. The development of newer thrombolytic agents such as (recombinant) staphylokinase might also further reduce this problem [46].

Apart from thrombolysis, the advent of stented grafts might offer an additional option in the endovascular approach of popliteal artery aneurysms. In 1994, Marin et al. [47] were the first to report about a 63-year-old man with a popliteal artery aneurysm, treated by an endoluminal stented graft constructed with two balloon-expandable stents and a 6 mm polytetrafluoroethylene (PTFE) graft. Angiography and duplex ultrasonography, performed up to 3 months after the procedure, documented graft and distal artery patency and complete aneurysmal exclusion without distal emboli. One year later, Joyce et al. [48] and Gieskes et al. [49] reported the use of the *Cragg Endopro system (MintecTM)* for popliteal artery aneurysms. The one patient described by Joyce et al. was followed for 1 year. Gieskes et al. reported another three patients, with patency and aneurysm exclusion documented up to 9 months in two cases. The third graft occluded 15 days after insertion. An additional six cases with the same device have been reported by Marcadé [50]. Again the follow-up was limited to nine months. There were no occlusions, but one patient needed a secondary intervention because of an endoleak at 6 months. In 1997, Krajcer et al. [51] described the use of a Wallstent in combination with a PTFE graft in one patient with a follow-up of 8 months. Most recently, the possible importance of combined surgical and endovascular approach, as already mentioned by Joyce et al. [48], was stressed in two case reports. Rosenthal et al. [52] described a 48-year-old man with a popliteal artery aneurysm, treated by a combination of endovascular in-situ saphenous vein bypass and transluminal antegrade coil embolization of the aneurysm. Continued patency and exclusion of the aneurysm was documented at 14 months. At the same time, Puech-Leao et al. [53] described reconstruction with greater saphenous vein and a Palmaz stent attached at its proximal end. The procedure was done through an open posterior approach to the distal popliteal artery. The stent graft was introduced in retrograde fashion to the superficial femoral artery, where the proximal stent was deployed. The saphenous vein was then conventionally anastomosed to the distal popliteal artery. Follow-up at two months showed continued exclusion of the aneurysm and normal patency of the venous graft. To our knowledge, the combination of thrombolysis and stent grafts for thrombosed popliteal artery aneurysms has been reported only by Link et al. [54]. These authors treated two patients with local thrombolysis followed by percutaneous insertion of a *Cragg Endopro stent graft (MintecTM)*. One graft occluded after 1 month, the second was still patent at 6 months. Although all these techniques seem promising, it is clear that the collected experience up to now is still premature and that additional follow-up will be needed before the use of stent grafts for popliteal artery aneurysms can be recommended for widespread use.

■ PERSONAL EXPERIENCE

Selected for this retrospective review were all patients presenting with acute ischemic popliteal aneurysm complications at the *Department of Vascular Surgery (University Clinic Gasthuisberg, Leuven, Belgium)* during the period 1989-1998. For a total of 24 patients there were 21 men and 3 women with a mean age of 69.5 years (range 54 to 92 years). One female patient presented with two consecutive episodes of acute ischemia in both legs, which results in a total of 25 aneurysms. Twenty-three of these aneurysms were thrombosed at the time of presentation. Two patients presented with limb threatening ischemia due to distal embolization from a patent popliteal aneurysm.

The diagnosis of popliteal artery aneurysm was suspected by clinical examination and/or confirmed by ultrasonography in 16 cases. Three were discovered during operative exploration, while another five were diagnosed during thrombolysis. Finally, the last aneurysm was discovered after percutaneous thrombus aspiration for localized popliteal artery occlusion.

TREATMENT

Two patients presenting with a nonviable ischemia were handled by primary above knee amputation. Another ten patients underwent immediate surgical exploration, which consisted of femorodistal bypass with exclusion of the aneurysm and thrombectomy of the tibial vessels. Perioperative thrombolysis, as described by Thompson et al. [32] was used, without any beneficial effect, in three cases (streptokinase; n=2, urokinase; n=1). Additional fasciotomy was performed in three cases.

Preoperative thrombolytic therapy was attempted in 13 cases. For reasons of analysis the series is subdivided in group 1, which includes eight patients were the diagnosis of popliteal aneurysm was already confirmed before the start of the thrombolysis (intentional thrombolysis group). All these patients showed a popliteal artery occlusion with extensive signs of distal embolization or complete occlusion of the tibial vessels. All of them were considered unreconstructable at angiography. Group 2 includes five patients, where the diagnosis of popliteal artery aneurysms was made only during thrombolysis (non-intentional thrombolysis group). They presented all with angiographic occlusion of the popliteal artery. Three of them had additional occlusions at the tibial vessels.

Thrombolytic therapy was administered percutaneously by direct intrathrombus administration in all cases. The different agents used were streptokinase (n=2), tissue plasminogen activator (n=3), alteplase (n=4) and staphylokinase (n=4). The mean duration of thrombolytic therapy was 17.1 hours (range 4 to 26 hours).

Thrombolysis was considered successful in five of the eight patients (62%) in the intentional group. Complete clearing of the popliteal artery and tibial vessels was obtained in only one case, but four patients showed partial clearing with reconstitution of previously thrombosed runoff arteries (Figs. 1 and 2). Four of them underwent additional surgery with exclusion of the aneurysm. The last patient was handled percutaneously by insertion of an endoprothesis *(Cragg Endopro System 1, MintecTM)*. Thrombolysis had to be discontinued with beneficial effect in three cases (38%) because of hematoma formation. One patient showed also ischemic deterioration. Two of these patients underwent an above knee amputation. The last limb was salvaged by urgent surgical thrombectomy and femorodistal grafting.

There were no complications in the non-intentional group (n=5). Thrombolysis resulted in complete or partial clearing of the popliteal aneurysm in all cases (Fig. 3). Additional lysis of the distal arteries was also obtained in one of the three patients presenting with associated tibial artery involvement. Percutaneous reconstruction with an endoprothesis *(Cragg Endopro System 1, MintecTM)* was performed in one case. The remaining four patients underwent additional surgical reconstruction of the aneurysm (venous bypass by the inlay technique).

RESULTS

There were no operative deaths. Operative morbidity was noted in four cases. Two patients developed a peroneal nerve palsy postoperatively. In addition to this, one of the two endoprostheses occluded after 10 days. Treatment consisted of repeated thrombolysis, but the patient ultimately needed a venous interposition. Finally, one patient of the intentional thrombolysis group thrombosed his graft after 2 days. He was handled by a below knee reconstruction. Taken into account the four amputations, performed without attempt to vascular reconstructions, the early limb salvage rate was 80% (Fig. 4).

The mean follow-up of these series was 41.4 months (range 2 to 90 months). Four patients died of unrelated causes and the survival rate at 3 years was 89%. One patient needed a secondary intervention after 7 years for local compression symptoms caused by continuing expansion of the aneurysm despite surgical ligation during the primary operation.

Late occlusion was seen in five cases (3-year patency rate 68%, Fig. 4) and resulted in two additional amputations (3-year limb salvage rate 71%, Fig. 4). Included in the occlusions was also the second endoprosthesis, which thrombosed after 1 year. Patency here has be restored by thrombolysis. Both the late amputation occurred in patients of the intentional thrombolysis group. In fact five patients (62%) of this group underwent a major amputation within 1 year after the initial event. There were no major amputations in the non-intentional thrombolysis group.

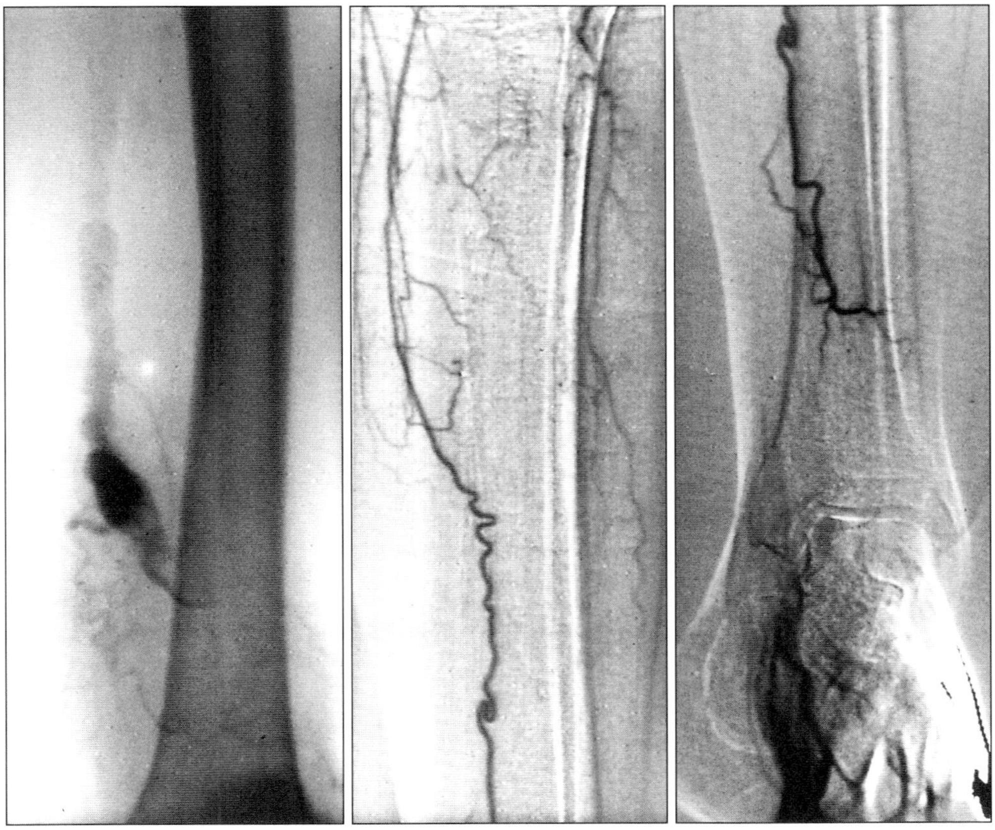

FIG. 1 Intra-arterial digital subtraction angiography: acute thrombosis of popliteal artery aneurysm with virtually no runoff.

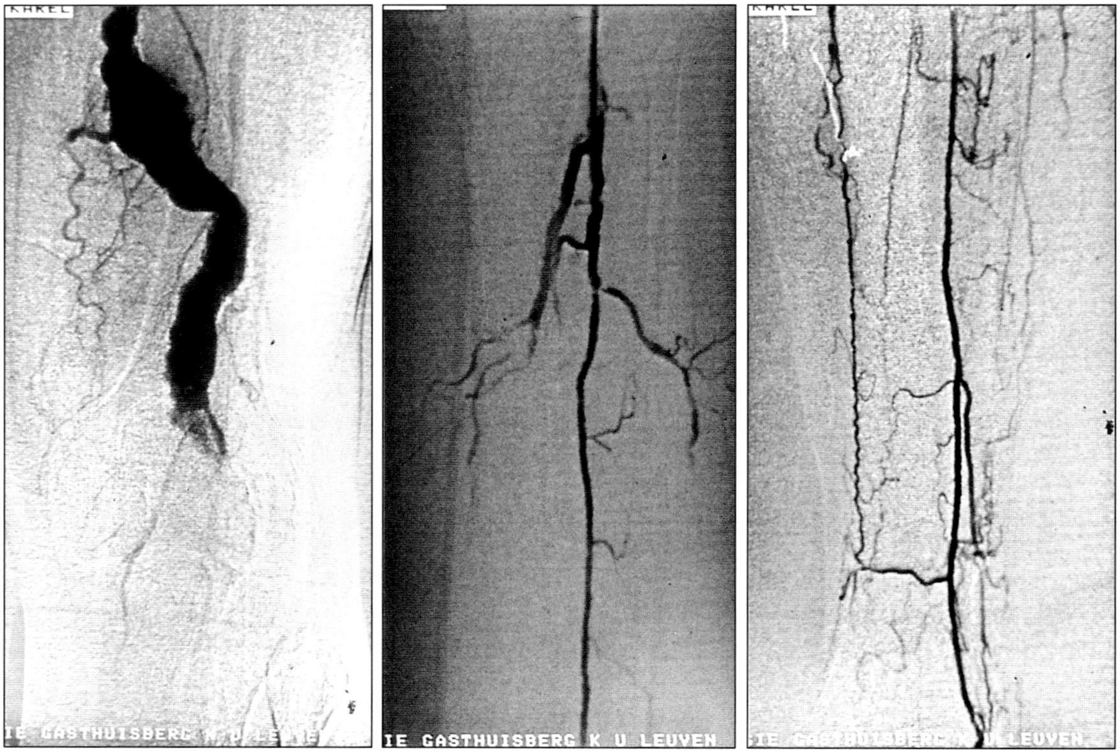

FIG. 2 Local thrombolysis with staphylokinase for 24 hours results in partial clearing of the popliteal artery and reconstitution of the peroneal artery (same patient as Fig. 1).

■ CONCLUSION

Acute limb threatening ischemia due to popliteal artery aneurysm thrombosis remains a challenge to the vascular surgeon. Despite progress in surgical techniques, the associated amputation rate is still 20% to 30%. Preoperative thrombolysis represents an attractive adjunct to surgical therapy, especially in case of distal embolization and associated occlusion of the infrapopliteal vessels. The technique has, however, its own morbidity and thrombolysis as merely diagnostic tool should be avoided. Perioperative thrombolysis may be an alternative in patients who need an emergency operation but the results are less convincing. Elective reconstruction remains the treatment of choice for most popliteal artery aneurysms and the availability of thrombolysis does not justify a conservative approach. Stent grafts for popliteal artery aneurysm reconstruction are still under investigation and should be used in selected cases only.

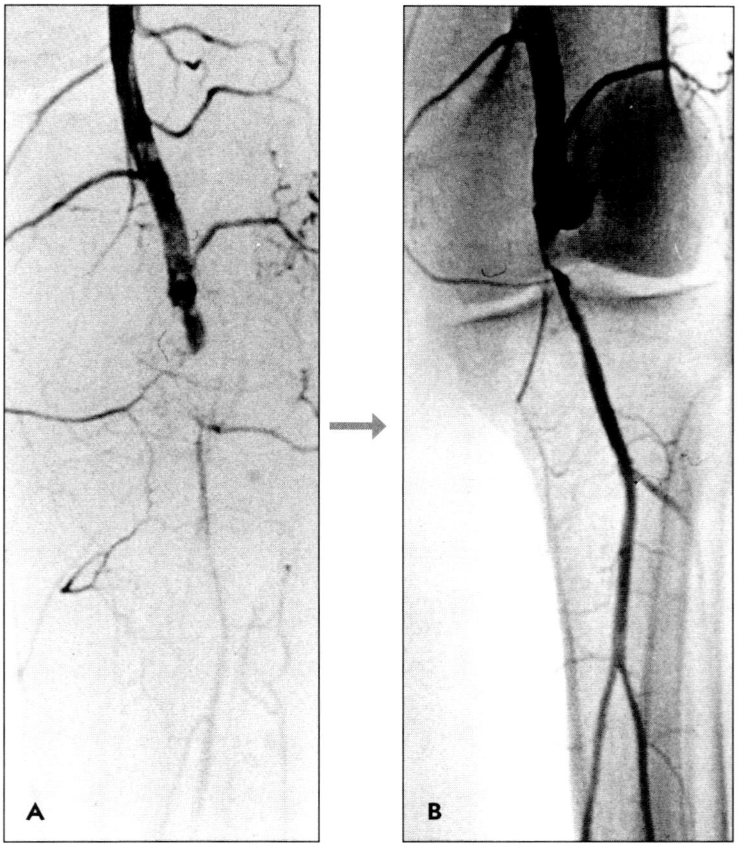

FIG. 3 A - Acute occlusion of the popliteal artery. B - Intra-arterial thrombolysis with alteplase for 24 hours unmasks a small popliteal artery aneurysm.

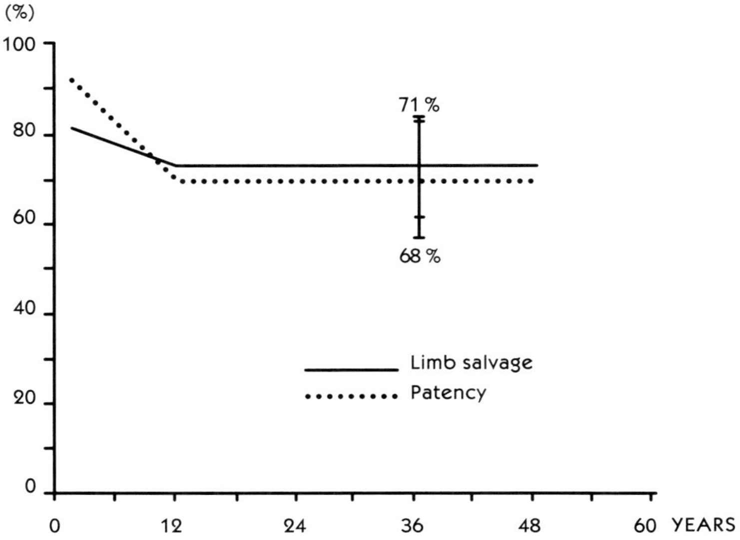

FIG. 4 Cumulative limb salvage and patency rates in 25 cases of critical limb ischemia due to popliteal artery aneurysm.

REFERENCES

1 Dent TL, Lindenauer SM, Ernst CT, Fry WJ. Multiple arteriosclerotic arterial aneurysms. *Arch Surg* 1972; 105: 338-344.
2 Cole CW, Thijssen AM, Barber GG et al. Popliteal aneurysms: an index of generalized vascular disease. *Can J Surg* 1989; 32: 65-68.
3 Gifford RW, Hines EA, Janes JM. An analysis and follow-up study of one hundred popliteal aneurysms. *Surgery* 1953; 33: 284-293.
4 Mellière D, Veit R, Becquemin JP, Etienne G. Should all spontaneous aneurysms be operated on? *J Cardiovasc Surg* 1986; 27: 273-277.
5 Lowell RC, Gloviczki P, Hallett JW, et al. Popliteal artery aneurysms: the risk of non operative management. *Ann Vasc Surg* 1994; 8: 14-23.
6 Dawson I, Sie R, van Baalen M, van Bockel JH. Asymptomatic popliteal aneurysm: elective operation versus conservative follow-up. *Br J Surg* 1994; 81: 1504-1507.
7 Sie RB, Dawson I, van Baalen JM et al. Ruptured popliteal artery aneurysm. An insidious complication. *Eur J Vasc Endovasc Surg* 1997; 13: 432-438.
8 Ramesh S, Michaels JA, Galland RB. Popliteal aneurysm: morphology and management. *Br J Surg* 1993; 80: 1531-1533.
9 Dawson I, van Bockel JH, Brand R, Terpstra JL. Popliteal artery aneurysms. Long-term follow-up of aneurysmal disease and results of surgical treatment. *J Vasc Surg* 1991; 13: 398-407.
10 Muller RC, Striffeler H., Stirnemann P. Popliteal aneurysm. *Schweiz Med Wochenstr* 1993; 123: 2390-2393.
11 Szilagyi DE, Schwartz RL, Reddy DJ. Popliteal artery aneurysms. Their natural history and management. *Arch Surg* 1981; 116: 724-728.
12 Schellack J, Smith RB III, Perdue GD. Nonoperative management of selected popliteal aneurysms. *Arch Surg* 1987; 122: 372-375.
14 Inahara T, Toledo AC. Complications and treatment of popliteal aneurysms. *Surgery* 1978; 84: 775-783.
15 Lancashire MJR, Torrie EPH, Galland RB. Popliteal aneurysms identified by intra-arterial streptokinase: a changing pattern of presentation. *Br J Surg* 1990; 77: 1388-1390.
16 Henry AK. Some surgical aspects of aneurysm. *Practioner* 1943; 150: 136-143.
17 Schechter DC, Bergan JJ. Popliteal aneurysm: a celebration of the bicentennial of John Hunter's operation. *Ann Vasc Surg* 1986; 1: 118-126.
18 Goyanes DJ. Nuevos trabajos de cirurgia vascular. Sustitución plástica de las arterias por las venas o arterioplastia venosa, applicada, como nuevo método, al tratamiento de los aneurismas. *El Siglo Medico* 1906; 53: 561-564.
19 Pringle JH. Two cases of vein grafting for the maintenance of a direct arterial circulation. *Lancet* 1913; i:1795-1796.
20 Crawford ES, DeBakey ME. Clinical use of synthetic arterial substitutes in 317 patients. *Arch Surg* 1958; 76: 261-270.
21 Edwards WS. Exclusion and saphenous vein bypass of popliteal aneurysms. *Surg Gynecol Obstet* 1969; 128: 829-830.
22 Anton GE, Hertzer NR, Beven EG et al. Surgical management of popliteal aneurysms. Trends in presentation, treatment and results from 1952 to 1984. *J Vasc Surg* 1986; 3: 125-134.
23 Hagino RT, Fujitani RM, Dawson DL et al. Does infrapopliteal arterial runoff predict success for popliteal artery aneurysmorrhaphy? *Am J Surg* 1994; 168: 652-658.
24 Schröder A, Gohlke J, Gross-Fengels W, Horstmann R. Poplitea-Aneurysmen - operative Versorgung versus konservativus Vorgehen. *Langenbecks Arch Chir* 1996; 113 Suppl II: 857-863.
25 Roggo A, Brülhart KB, Brunner U, Largiader F. Graftdurchgängigkeit und Extremitä tenerhaltung verwendeten Graftmaterials. *Helv Chir Acta* 1994; 60: 893-896.
26 Shortell CK, DeWeese JA, Ouriel K, Green RM. Popliteal artery aneurysms: a 25-year surgical experience. *J Vasc Surg* 1991; 14: 771-779.
27 Poirier NC, Verdant A, Pagé A. L'anévrisme poplité: le traitement chirurgical s'impose avant l'apparition de complications. *Ann Chir* 1996; 50: 613-618.
28 Varga ZA, Locke-Edmunds JC, Baird RN. A multicenter study of popliteal aneurysms. Joint Vascular Research Group. *J Vasc Surg* 1994; 20: 171-177.
29 Farina C, Cavallaro A, Schultz RD et al. Popliteal aneurysms. *Surg Gynecol Obstet* 1989; 169: 7-13.
30 Schwartz W, Berkowitz H, Taormina V, Gatti J. The preoperative use of intra-arterial thrombolysis for a thrombosed popliteal artery aneurysm. *J Cardiovasc Surg* 1984; 25: 465-468.
31 Taylor LM Jr, Porter JM, Baur GM et al. Intra-arterial streptokinase infusion for acute popliteal and tibial artery occlusion. *Am J Surg* 1984; 147: 583-588.
32 Thompson JF, Beard J, Scott DJA, Earnshaw JJ. Intraoperative thrombolysis in the management of thrombosed popliteal aneurysm. *Br J Surg* 1993; 80: 858-859.
33 Browse DJ, Torrie EPH, Galland RB. Early results and 1-year follow-up after intra-arterial thrombolysis. *Br J Surg* 1993; 80: 194-197.
34 Wilkins DC, Bliss BP, Wells IP. Conservative management of asymptomatic popliteal aneurysm. *Br J Surg* 1992; 79: 710.
35 Ferguson LJ, Faris I, Robertson A et al. Intra-arterial streptokinase therapy to relieve acute limb ischemia. *J Vasc Surg* 1986; 4: 205-210.
36 Bowyer RC, Cawthorn SJ, Walker WJ, Giddings AE. Conservative management of asymptomatic popliteal aneurysm. *Br J Surg* 1990; 77: 1132-1135.
37 Elsey JK, Rosenthal D. The use of adjunctive thrombolytic therapy in the management of acute popliteal aneurysm thrombosis. *Am Surg* 1994; 60: 942-945.
38 Garramone RR Jr, Gallagher JJ Jr, Drezner AD. Intra-arterial thrombolytic therapy in the initial management of thrombosed popliteal artery aneurysms. *Ann Vasc Surg* 1994; 8: 363-366.
39 Hoelting T, Paetz B, Richter GM, Allenberg JR. The value of preoperative lytic therapy in limb-threatening acute ischemia from popliteal artery aneurysm. *Am J Surg* 1994; 168: 227-231.
40 Rousseau H, Soula P, Joffre F. Anévrisme poplité et ischémie de jambe: thrombolyse première. *J Mal Vasc* 1994; 19 Suppl A: 154-157.
41 Gawenda M, Sorgatz S, Müller U et al. The thrombosed popliteal aneurysm with distal arterial occlusion - successful therapy by interdisciplinary management. *Thorac Cardiovasc Surgeon* 1995; 43: 112-116.
42 Debing E, Van den Brande P, van Tussenbroek F et al. Intra-arterial thrombolysis followed by elective surgery for thrombo-embolic popliteal aneurysms. *Acta Chir Belg* 1999; In press.
43 Berridge DC, Makin GS, Hopkinson BR. Local low dose intra-arterial thrombolytic therapy: the risk of stroke and major haemorrhage. *Br J Surg* 1989; 76: 1230-1233.
44 Lacroix H, Suy R, Nevelsteen A et al. Local thrombolysis for occluded arterial grafts. *J Cardiovasc Surg* 1994; 35: 187-191.
45 Galland RB, Earnshaw JJ, Baird RN et al. Acute limb deterioration during intra-arterial thrombolysis. *Br J Surg* 1993; 80: 1118-1120.
46 Vanderschueren S. Thrombolytic properties of recombinant staphylokinase in experimental models and clinical trials. *Doctoral thesis* - Leuven University Press, 1996.
47 Marin ML, Veith FJ, Panetta TF et al. Transfemoral endoluminal graft repair of popliteal artery aneurysm. *J Vasc Surg* 1994; 19: 754-757.

48 Joyce WP, McGrath F, Leahy AL et al. A safe combined surgical/radiological approach to endoluminal graft stenting of a popliteal aneurysm. *Eur J Vasc Endovasc Surg* 1995; 10: 489-491.
49 Gieskes L, Rousseau H, Otal P et al. Traitement percutané par endoprothèse couverte des anévrysmes poplités: expérience clinique préliminaire. *J Mal Vasc* 1995; 20: 264-267.
50 Marcadé JP. Stent graft for popliteal aneurysms. Six cases with cragg Endo-Pro System 1 Mintec. *J Cardiovasc Surg* 1996; 37: 41-44.
51 Krajcer Z, Dietrich EB. Successful endoluminal repair of arterial aneurysms by Wallstent prosthesis and PTFE graft: preliminary results with a new technique. *J Endovasc Surg* 1997; 4: 80-87.
52 Rosenthal D. Atkins CP, Shuler FW et al. Popliteal artery aneurysm treated with a minimally invasive endovascular approach: an initial report. *J Endovasc Surg* 1998; 5: 60-63.
53 Puech-Leao P, Kaufmann P, Wolosker N, Anacleto AM. Endovascular grafting of a popliteal aneurysm using the saphenous vein. *J Endovasc Surg* 1998; 5: 64-70.
54 Link J, Müller-Hülsbeck S, Brossman J et al. Erste ergebnisse der perkutanen Therapie von poplitealaneurysmen mit Stents. *Rofo Fortschr Röntgenstr* 1996; 164: 244-248.

17

FEMOROPOPLITEAL RECONSTRUCTION

WILLIAM P. PAASKE

The concept of chronic limb ischemia based on atherosclerotic occlusive arterial disease is, by tradition, divided into two main categories.

In functional ischemia the defining symptom is intermittent claudication, where blood flow is adequate at rest for normal cellular function, but during muscular exercise perfusion becomes insufficient for oxygen and substrate delivery and removal of metabolites. The normal perfusion coefficient in leg skeletal muscle at rest is 2-4 ml/100g/min, but during maximum exercise a blood flow of 250 ml/100g/min may be necessary [1]. Assessment of claudication distance is difficult since it is spuriously estimated, inaccurately reported, falsely recorded, inappropriately measured, and usually misinterpreted [2]. In claudicants, exercise to the onset of calf pain followed by rest results in an ischemia/reperfusion insult with free radical formation, neutrophil activation, and systemic vascular endothelial damage [3]. Also, these patients have impaired endogenous fibrinolytic activity, and one source of the increased coagulation activity is probably the ischemic leg itself [4], together with other coagulation disorders [5].

In the other class, the clinical picture is dominated by ischemic rest pain, and, in more advanced cases, ischemic necrosis in the form of skin ulceration and/or gangrene associated with other manifestations of trophic disturbances in the peripheral tissues such as atrophy of skin, adipose tissue and muscles, or destruction of normal nerve function causing a variety of neurological disturbances.

The term critical leg ischemia is widely used in the literature, but exact definitions are highly problematic [6]. The term subcritical ischemia has been proposed to describe patients with rest pain, an intact foot and a segmental systolic ankle blood pressure above 40 mmHg. The term critical ischemia should be used restrictively to characterize patients with rest pain and either tissue loss or an ankle pressure below 40 mmHg [7]. The critical level for adequate tissue perfusion at rest is low, around 0.25 ml/100g/min, and the ischemic microcirculatory bed is characterized by maximum arteriolar dilatation and extremely long transit times for corpuscular elements and plasma through the capillaries. Restoration of blood flow to normal values at rest and during exercise does not revert the pathophysiological changes. This can be experienced when a distal ischemic ulcer fails to heal after successful revascularization of the profunda femoris artery; a segmental systolic digital blood pressure over 30 mmHg is necessary for healing of distal ulcers in addition to blood flow increase [8].

Renormalization of blood flow in atherosclerotic patients does not revert disturbances in red blood cell aggregation, hematocrit fraction, and plasma viscosity; not only abnormal endothelial cell activation [9] but also leukocyte and platelet activations remain after surgery.

■ EPIDEMIOLOGY

The Danish Vascular Registry *(Karbase)* collects information on activity within all ten departments performing arterial vascular surgery in Denmark (population 5.1 million). A total of 6329 procedures including 4261 primary arterial reconstructions were reported in 1997 (corresponding to 83.5 primary arterial procedures per 100 000 inhabitants). This figure includes 698 primary femoropopliteal operative reconstructions (698/6329 = 11.0% = 13.7 procedures per 100.000) made on all indications in the above knee (384 procedures = 55.0%) and below knee (314 = 45.0%) positions. The number of femoropopliteal reconstructions has increased by a factor of 2.4 since 1983 [10]. The Danish reporting system does not have *critical ischemia* as an independent indication variable for entry point but applies the formats *functional ischemia, rest pain,* and *gangrene + ulceration.*

Table I shows the number and type of grafts related to the stage of ischemia.

In Sweden, 12.4% of operations in all anatomical positions for critical ischemia were redo procedures [11]. Table II shows the data or proximal and distal femoropopliteal bypasses for functional ischemia, rest pain and skin lesions.

Table I ARTERIAL FEMOROPOPLITEAL RECONSTRUCTIONS IN RELATION TO BYPASS TYPE AND INDICATION ACCORDING TO THE DANISH VASCULAR REGISTRY *(Karbase)*

	*Proximal FPB**	*Distal FPB**	*Incidence***
Claudication	146	36	**3.6**
Rest pain	45	57	**2**
Ulceration or gangrene	77	96	**3.4**

* Femoropopliteal bypass ** Per 100 000 per year

Table II ARTERIAL FEMOROPOPLITEAL RECONSTRUCTION TECHNIQUES IRRESPECTIVE OF THE INDICATION, IN RELATION TO THE LOCALIZATION ACCORDING TO THE SWEDISH VASCULAR REGISTRY *(Swedvasc* [11]*)*

Technique	*Proximal FPB**	*Distal FPB**
Reversed vein	1	3
In-situ vein	37	146
Dacron	3	0
PTFE	226	25
Other	1	15

In 1983 reversed vein and prosthesis accounted for 50% each [10], but, as indicated, reversed vein is currently almost never used in the above as well as the below knee position, and the in-situ technique is gaining increased popularity, also for the above knee position.

In Sweden, 48.1% of all vascular surgical activity (17 000 procedures recorded by Swedvasc, the Swedish Vascular Registry, 1987 - 1993) was performed on the indication chronic ischemia [11], and the proportion of PTA compared to open surgery has been around 50% during the last decade in the femoropopliteal localization with the percentage of PTA for chronic leg ischemia being around 46% with an increasing trend the last 5 years [11]. The number of procedures per 100 000 inhabitants for intermittent claudication varied between the six health care regions from 11.9 (Stockholm) to 30.6 (South) and for critical limb ischemia from 11.7 (North) to 37.0 (South) [11]. Also, the ratio between procedures for intermittent claudication/critical limb ischemia varied between 0.62 and 1.23.

For Finland, information is available through the Finvasc Registry: 2753 surgical reconstructions were recorded for chronic ischemia/claudication between 1991 and 1994, and 2296 for chronic ischemia/critical ischemia; 1761 operations were made for chronic critical limb ischemia, 10.3% with femoropopliteal endarterectomy and 38.1% with femoropopliteal bypasses [12]. Of the femoropopliteal reconstructions, 40% were in-situ vein, 26% were ex-situ vein, 7% were Dacron, 23% were PTFE, and 4 were combinations.

In the United Kingdom, 63% of limbs of patients with chronic critical leg ischemia admitted to hospitals were treated with revascularization (with a 75% chance of limb salvage, overall limb salvage 68%), 16% with primary major amputation, 8% with medical or minor surgical treatment, and 13% with no treatment [13]. Limb threatening ischemia accounts for 50 000 admissions annually to hospitals in the UK [14,15].

■ THE PATIENT

The symptoms of the leg are manifestations of generalized atherosclerotic disease in patients approaching the end of life at a time when functional status is frequently declining rapidly and interval survival is short [16]. The mean age of a patient treated for intermittent claudication is 65.7 years (SD ± 10.4) and for critical ischemia 74.4 years (SD ± 9.4) [11]. Women are generally 4 years older than men are before surgery for limb salvage [17]. The cumulative progression from intermittent claudication to rest pain and/or gangrene for newly diagnosed claudicants is low; 7.5% the first year and 2.2% for each of the following years [18], and many patients with critical ischemia do not have previous intermittent claudication. The prevalence of claudication in the populations of Northern Europe is around 1%, in men aged 55-74 around 4.5% [19,20], and at the age of 70 years the prevalence is 9% in men and 3% in women [21]. A study from Iceland has indicated a peculiar decline of prevalence in that country [22]. Within ten years of onset of intermittent claudication, 43% develop congestive heart failure, 21% stroke, and 24% cardiac failure; in the USA survival following onset of intermittent claudication is only two-thirds of the general population (60% died within 10 years) [23]. Approximately 1% of men over 55 years of age will develop critical limb ischemia [6]. The incidence is estimated at 400-1000 per million population, and the prevalence in the UK population is 1 in 2500 [13].

The yearly mortality is around 20% in patients with critical ischemia [11]. Half of the nondiabetics and 55% of those alive after 5 years will have undergone amputation [18]; most patients who have infrainguinal bypass surgery for limb salvage require ongoing treatment and have persistent or recurrent symptoms until their death [16]. In a Norwegian study, the 5-year observed survival rate for all patients operated for femoropopliteal disease was 59% (expected survival 78%) indicating a double risk of death [24,25]. It should be kept in mind that up to 50% of patients with critical limb ischemia have atherosclerotic changes at multiple levels [26]. The 1-year mortality percentages were found to be identical at 22.9% after surgery and percutaneous transluminal angioplasty (PTA) [11]. For patients less than 76 years old the percentage of death was 17.6 within 1 year after surgery as well as for PTA; for patients 76 years old and older the percentages were 26.6% for surgery and 25.8% for PTA [11]. A long-term study with follow up from 10-14 years of 50 in-situ grafts and 50 reversed vein grafts implanted in the femoropopliteal level (75% below knee) provided an actuarial survival at 10 years of 30.8% in the in-situ group and 29.5% in the reversed vein group [27].

■ FUNDAMENTAL REPORTING PROBLEMS

A huge number of reports on femoropopliteal reconstructions have been published in the national and international literatures. As emphasized by Cheshire and Wolfe, many of these communications are difficult to interpret, and indeed meaningless, since information on essential points is inadequate or lacking, and only few are conducted as prospective, randomized studies [15]. Referral patterns to various individual surgeons and institutions vary, and indications are rarely constant within the time horizon of the study due to development of newer techniques and change of attitudes. The case mix is almost never described; proper definition of intention to treat and the fate of conservatively treated (and thus excluded) patients are missing. The reporting format may be awkward with no clear grouping of patients with respect to indication or even bypass type or exact localization. Risk factors are often summarily described, and adjunctive measures are rarely reported (anticoagulation, intraoperative Dextran, type of anesthesia). Besides, surveillance programs have been shown to influence ultimate outcome, but precise information on postoperative handling strategies is often lacking. New treatment options are reported without properly conducted protocol formats.

The *Ad Hoc* documents for reporting [28-31] developed by the scientific societies should be welcomed as timely initiatives to increase scientific quality and, ultimately, provide the foundation for evidence based patient care.

■ INDICATIONS FOR TREATMENT

An indication for treatment of chronic ischemic symptoms that can be ascribed to atherosclerotic arterial disease in the femoropopliteal segment should always include considerations on general and specific risk assessments. This must be balanced towards the motivation of the patient and the benefits and expected outcome of the various treatment modalities considering the expected life style and life span.

For intermittent claudication, the indication is practically always relative, and a period of conservative treatment must be recommended unless weighty social factors are decisive. *Stop smoking and keep walking* [3,32,33] should be encouraged together with weight loss in adipose patients. Conservative treatment is the primary treatment of choice; if this fails, then PTA is preferred [19]. Acetylsalicylic acid should be given to reduce risk of postoperative myocardial infarction and stroke in these stigmatized patients [34]. Young patients (under the age of 40 years) present a special risk group with high mortality (mostly due to atherosclerotic disease) and a high risk of failure of initial arterial reconstructions. This means that a liberal attitude towards reconstructive surgery, particularly in claudication, is not warranted [35].

An Italian study [36] comprising only 14 femoropopliteal bypasses (the rest femorodistal) showed no statistically significant difference in a group of patients over 80 years of age as compared to a matched group of patients below that age with respect to mortality rate, graft patency and limb salvage. An American study comprising 88 patients over 80 years of age yielded a perioperative mortality rate of 6% and a mean hospitalization duration of 18 days, but, even though the expected survival is lower, results were satisfactory with respect to graft patency, limb salvage, and function [37].

Presence of ischemic rest pain, ulceration and/or gangrene usually presents absolute indication for vascular surgical treatment even in older patients [36,37], but primary amputation should be considered for old, debilitated patients with very short life expectancy, for bedridden patients who, for various reasons, cannot be expected to become mobile, and for patients in whom the expected outcome of a reconstruction is highly problematic. The results of infrainguinal bypass surgery are very poor for patients on dialysis. A Finnish study [38] including six femoropopliteal grafts showed that none of the patients in dialysis was alive with salvaged leg 4 months after revascularization, but the results were acceptable in patients not requiring dialysis (1-year patency: 81%, leg salvage 79%). In an American study of 44 patients on maintenance dialysis [39], the 30-day morbidity rate was 39%, and perioperative mortality was 9%. The patient survival rate was 52% at 2 years. This is in contrast to the marvelous results obtained by another group [40] with a 60% patency and a 72% limb salvage rate at 2 years (369 legs treated).

If conservative strategy fails in disabling intermittent claudication, or if the patient has ischemic rest pain, ulceration and/or gangrene, full angiographic visualization of aorto-iliac-infrainguinal segments (including foot arteries, at least to the arches) should be undertaken to determine run-in to the groin, to identify lesions in the femoropopliteal segment, and to estimate runoff in the infrapopliteal segments. It is a matter of concern that decisions on treatment and strategy are generally taken on the basis of symptoms and signs in combination with (at best) two plane angiographic visualizations, usually provided by digital subtraction techniques (DSA). This means that morphology, and not function, may play a too decisive element for indication of treatment. Magnetic resonance angiography (MRA) provides images of blood flow velocities within the arteries. Interpretation of the reconstructed images can be problematic, but certain evidence points towards the fact that MRA may be more suited for visualization of crural arteries than DSA [41]. The technique will, at a later stage, be able to provide three dimensional anatomical visualization as well as functional information on the hemodynamic influence of individual lesions [42].

In some centers, preoperative evaluation is made exclusively with duplex ultrasound. The technique is dependent on highly skilled operators who are able to obtain excellent results [43-45].

In order to perform systematic and methodological analyses, the Ad Hoc scoring according to SVS/ ISCVS [28] may be applied and incorporated into departmental databases as part of a strategy to make quality control and auditing an integral part of today's practice. The immediate clinical usefulness was shown in a recent study comprising 120 patients with 132 primary femoropopliteal vein bypass procedures, 32 for intermittent claudication, and 100 for critical leg ischemia. The patients with a SVS/ ISCVS score in the highest quartile had 8.7 times higher risk for immediate graft occlusion, and multivariate analysis showed that the score was predictive of immediate patency and leg salvage [46]. Further, evidence from a prospective study of 54 patients indicated that, contrary to intraoperative hemodynamic test results with a range of examinations, the SVS/ISCVS runoff score was a good predictor of early graft failure [47].

■ TREATMENT OPTIONS

ENDOVASCULAR PROCEDURES

Although prospective, randomized series on PTA of the superficial femoral artery are few [48,49] most surgeons would agree to initial balloon dilatation of at least shorter stenoses and occlusions in this segment [50,51] but be less enthusiastic with respect to dilatation of the popliteal artery, especially below the knee. The interesting PTA technique with creation of a subintimal plane of cleavage by catheter dissection has given excellent results in the institution of invention [52,53]. PTFE endoprostheses with distal stenting have been inserted in patients with occlusion of the superficial femoral artery and open popliteal artery above the knee on a non-randomized basis with a primary patency rate of 75.1% and an assisted primary patency rate of 79.1% [53], but like other endovascular treatments such as atherectomy and laser assisted ablation, properly conducted studies are necessary before their usefulness can be definitely assessed. The literature up to now seems to be entirely anecdotal on these subjects. The efficacy of the new Hemobahn prosthesis (Gore-Tex®) is not yet documented.

ENDARTERECTOMY

Operative removal of lesions in the superficial femoral artery is occasionally reported. A retrospective series [55,56] comprising 231 consecutive cases in 197 patients over 10 years described the results of semiclosed endarterectomy of the superficial femoral artery for occlusion and various degrees of stenosis. The mortality was 0.8% and the complication rate 10%. The 5-year cumulative primary patency was 75% in patients with disabling claudication, 61% in patients with rest pain, and 45% in those with gangrene. Using a modified ring stripper in another series, closed endarterectomy of the superficial femoral artery resulted in a primary and secondary patency of 86% in 103 limbs [57].

The efficacy of PTA (62 limbs) and endarterectomy (41 limbs) was examined for short occlusion in the femoropopliteal segment with duplex and angiographic follow-up [58]. The 3-year primary patency in patients with endarterectomy was 87%, whereas PTA resulted in a patency of 44%, but redo procedures were necessary in 17% of the endarterectomy patients and in 39% of the PTA patients. The patency after redo procedures (tertiary patency) was 94% for the endarterectomy group and 74% in the PTA group, but this difference was not significant. Besides, the expenses associated with the surgical procedure were comparable to those of an endovascular procedure if costs were expressed as a cost-to-patency ratio. Local endarterectomy of the superficial femoral artery with a Vollmar ring stripper applied antegradely from the groin followed by one or more self-expanding coated stents and a 5-mm, thin-walled PTFE endoluminal graft has been reported in an anecdotal study comprising four patients. Three reconstructions were open after 6 months [59]. The need for randomized, prospective series is obvious.

Bypass Procedures

A wide range of options is available.

Reversed autologous greater saphenous vein

An adequate ipsilateral great saphenous vein will be available for at least 55% of the planned operations [60]. Published patency rates vary much, and methodological problems are inherent to several studies [15]. It would seem reasonable to expect a cumulative patency at two years in the range of 80% for above knee and 60%-70% for below knee femoropopliteal bypasses [60-62]. Ten-year secondary patencies of around 70% have been reported even for below knee bypasses [27,63]. Endoscopic harvesting of the vein has been attempted [64], but it is not documented that this technique is advantageous.

Non-reversed autologous greater saphenous vein

This technique applies excision of the best available appropriate length segment of the vein, valve lysis with Mill's valvulotome using antegrade distension with a suitable solution, and translocation to the anastomoses sites [65]. The main advantage seems to be a better match between graft and artery at the anastomoses. The results are similar to those obtained by the in-situ technique.

In-situ autologous greater saphenous vein

The largest (American) published series comprises 3148 autogenous vein bypasses over 20 years with 31% above the knee and 69% below [66]. The in-situ technique was used in 75% of cases, and the overall cumulative patency rates were 91% at 1 year, 81% at 5 years, and 70% at 10 years; limb salvage was 90% at 10 years using the in-situ technique. The Dutch BOA study recently reported a 2-year cumulative patency for femoropopliteal in-situ grafts of 81.4% [62]. The 10 years results of a European randomized prospective study [27] on femoropopliteal bypasses (75% at the below knee popliteal artery, 97% for chronic critical ischemia) gave an actuarial primary patency of 41.7%, a secondary patency of 64.8%, and an actuarial limb salvage rate of 73.5%. Another European prospective and randomized study provided a 63.5% primary patency rate at 5 years [63]. Once again, discrepancy is apparent between American and European studies.

Does vein size matter? This was examined systematically in 50 in-situ and 50 reversed vein reconstructions [27]. The actuarial secondary patency at 10 years in the in-situ graft group was 37.5% for bypasses using veins with a diameter of 4 mm or less and 80% for veins with a diameter of 4 mm or more ($p < 0.05$). In the reversed graft group the patency rate was 71.2% for veins greater than 4 mm and 65.5% for veins smaller than 4 mm (n.s.). Another study [63] on in-situ grafts showed better results with veins larger than 4 mm (see below).

Other veins

The ipsilateral lesser saphenous vein may be a principal alternative to ipsilateral greater saphenous vein [67]. In the femoropopliteal segment, arm veins are rarely necessary but may be useful. In an American study of 250 arm veins in 224 patients including 41 femoropopliteal reconstructions, arm vein (cephalic, cephalic + basilic, basilic) was used when the ipsilateral greater saphenous vein was not available [68]. Angioscopy was carried out routinely to observe valve lysis, remove abnormalities and select the optimal vein segments. The marvelous 30-day (early) patency was 94.8% with a cumulative primary patency at 1 year of 70.6% (secondary patency 76.9%) and a limb salvage rate of 88.2%. Vein splicing and configuration did not affect patency rates, but vein quality and repeat operations did. The upper arm basilic-cephalic loop can sometimes be harvested in sufficient length to reach down to midtibial levels [69]; straightening of the median cubital vein and valvulotomy does not seem to affect patency. Conduits constructed of the proximal part of the superficial femoral artery after eversion endarterectomy plus arm vein or greater saphenous vein can be used with an expected 3 years primary patency rate of 60% and a secondary patency rate of 72%; limb salvage is reported to be 66% [70].

The use of the superficial femoral, popliteal, and deep femoral veins for femoropopliteal bypass has been suggested [71], but randomized, prospective series are lacking, and the method has not gained common acceptance. Creation of an adjunct arteriovenous fistula at the distal anastomosis has been proposed to be of some benefit when the distal anastomosis is placed at crural artery level [72,73], but reports are lacking for the popliteal level.

A Taylor patch [74], Miller cuff [75], St Mary's booth [76] or other methods for interposition of a small vein graft has been suggested to improve patency [77], and hemodynamic studies have provided the theoretical background for their use [78]. A non-randomized prospective study using historical controls examined the influence of distal interposition vein cuff in (among other procedures) 20 femoropopliteal PTFE bypasses with a cumulative 2-year patency of 65% as compared to the historic controls from the same institution with 46% patency at 2 years [79]. The use of a vein cuff does not seem to be of importance for PTFE grafts in the above knee position, but at the below knee popliteal artery it might be useful [80-82]. A multicenter randomized prospective study by the UK Joint Vascular Research Group [83] showed that there was no improvement in the patency rate with the use of a distal anastomosis vein cuff in femoral above knee popliteal PTFE bypass grafts, but there was a statistically significant advantage when PTFE bypass grafts were anastomosed to the popliteal artery below the knee (median follow-up 617 days, two years patency rate 72% and 70%, respectively, for cuffed and non-cuffed above knee bypasses, below knee 52% and 29%). Instead of a cuff, a composite reconstruction may be contemplated using proximal PTFE and distal reversed vein, preferably greater saphenous vein, in legs with good runoff [84], and splicing suitable vein into a vein graft might be indicated [85].

Polytetrafluoroethylene (PTFE)

The various commercially available products are not identical but seem to be equally clinically applicable. The reported patency rates vary, also between USA and Europe, but overall 5-year patencies of around 50%-60% can be expected in the above knee femoropopliteal segment. In a series of 322 reconstructions with PTFE, femoropopliteal bypasses made for severe claudication had an actuarial primary patency of 63% compared with 38% for limb salvage, but the actuarial limb salvage in the latter group was 66% [86]. Two-thirds of all failures occurred within 12 months. For above knee PTFE bypasses the overall 5- and 7-year actuarial patency rates were 58.3% [87]. When implanted for claudication, a 57% primary patency at 5 years has been reported [88]. In a series with mixed indications for above knee bypasses, the primary patency was 75% at 1 year and 54% at 5 years with corresponding limb salvages of 88% and 70% [89]. A study of 73 secondary PTFE femoropopliteal bypasses [90] performed after a failed ipsilateral bypass showed a primary life table graft patency at 4 years of 38%, and, after an aggressive policy, the overall secondary patency was 55% at 4 years with a limb salvage rate of 74%. In 438 procedures, above knee with PTFE resulted in a primary patency of 74% after 1 year, 56% after 3 years, and 50% after 5 years [91]. A recent report from a prospective randomized investigation published a primary patency rate of 57% for PTFE in above knee position at 3 years [92]. External support with rings does not seem to influence patency [93]. With respect to PTFE for below knee reconstructions, see below.

Composite grafts consisting of a proximal segment of PTFE and a distal vein graft segment (crossing the knee joint) can in some cases be a reasonable solution but the literature is not convincing, since randomized, prospective series are lacking. A few reports indicate good patencies superior to those of PTFE grafts [94]. Composite grafts should be tried for limb salvage when autogenous vein is not available [95].

Dacron

As with PTFE, several products with different characteristics are available. A report from Japan [96] provided a cumulative patency rate for femoropopliteal bypass (all indications) of over 80%. An American study on 85 PTFE grafts and 38 knitted, non-velour Dacron grafts performed by a single surgeon showed better results for Dacron [97]. A larger American multicenter study (see below) gave a primary patency rate of 62% at 3 years [92]. A retrospective analysis on 200 femoropopliteal bypasses with externally supported Dacron grafts with a mean follow-up time of 5 years had a primary patency of 71% at for the above knee position and 65% for the below knee grafts. Limb salvage rate was 79% at 5 years [98]. Overall, patency rates for Dacron seem to be of the same magnitude as for PTFE for the above knee position.

Other graft types

Homografts [99] and allografts [100-102] are occasionally used on a semi-experimental basis. Evidence suggests that use should presently be deferred [103]. Human umbilical vein (HUV): in a Norwegian study on femoropopliteal bypasses there was no difference between reversed homologous vein and HUV in the below knee position [104]. The overall patency (25% above knee) was 50.6% at 5 years. Another study on 211 femoropopliteal bypasses with HUV gave a 1-year cumulative patency rate of 70%, a 5-year value of 45%, and at 10 years patency was 26%. The patency was slightly worse at 1 and 3 years on the indication limb salvage than on claudication. The aneurysm formation incidence was 3.3% [105]. In 87 HUV grafts included in a wider, prospective trial, the cumulative patencies were 68%, 63%, and 57% at 1, 2, and 3 years, respectively [106]. An older study reported a 74% 1-year patency for below knee bypasses (4 years: 42%) [107].

Sheep (ovine) collagen prostheses: With the Omniflow graft in the above knee position, a 3-year patency of 62% was found when runoff was estimated to be good, and 44% with poor peripheral vessels. Infection rate was zero, and only 3/274 prostheses developed an aneurysm [108]. Another study of femoropopliteal bypasses in the above knee segment provided a cumulative patency of 56% and below knee of 50% [109]. A non-randomized series on 29 patients with 28 femoropopliteal bypasses above knee and one below (indications: claudication 27, critical ischemia 2) showed a primary cumulative patency rate above knee at 3 years of 83.7% and a secondary patency rate of 91.2%. Aneurysmatic change was not observed [110]. More information is necessary.

■ WHICH BYPASS?

Should Vein be Preferred to Prosthetic Material in the Above Knee Position?

This subject is still a matter of debate [111]. In a meta-analysis from 1989, it was shown that the differences in occlusion rate in the first year between veins and prosthetic grafts were related to the length of the graft (smallest for above knee) and that vein should be the first choice for all infrainguinal bypasses [112].

The American 6-year, prospective, multicenter, randomized comparison of autologous saphenous vein and PTFE [113] in 845 infrainguinal bypass operation (485 to popliteal artery, 360 to infrapopliteal arteries) yielded, among other results, life table primary patency rates for randomized PTFE and randomized saphenous vein to the popliteal artery at 4 years of 68% for vein and 47% for PTFE (not significant). Above knee, 4-year patency was 61% for vein and 38% for PTFE, but the difference was not statistically significant. Below knee, however, 4-year patency was 76% for vein and 11% for PTFE (significant). The complex results failed to support routine preferential use of PTFE grafts for femoropopliteal bypasses, and it was concluded that vein should be used for infrapopliteal grafts. Another study on 290 above knee bypasses with a median follow-up of 20 months did not show any difference between vein (reversed and in-situ) and PTFE with respect to up to 6-year patencies, and there was no difference in the amputation level (63 limbs) with respect to graft type [114]. It has been claimed that the saphenous vein should be spared for later distal bypass or coronary artery bypass grafting. It appears that only around 10% will at a later stage need the greater saphenous vein, and, besides, secondary reconstruction with vein following a prosthetic graft seems to fare better than the opposite policy; in one series only 5.8% needed the vein for coronary artery grafting [115], and generally the need later is low [116]. The large retrospective series from Vienna on 442 patients observed over 20 years showed only a small probability of secondary below knee repair in patients with previous primary above knee bypass, so a policy to spare the saphenous vein could not be recommended from these data [117].

An analysis of a database of infrainguinal and coronary artery bypass over a 15-year period at a single American institution showed a need for subsequent coronary artery bypass of 2% at 5 years and 3% at 10 years with a cumulative probability of requiring a subsequent infrainguinal bypass of 27% at 5 years and 32% at 10 years. After primary infrainguinal bypass, additional procedures using vein may rise to 1/4 to 1/3 of patients [118].

Is In-Situ Better than Reversed Greater Saphenous Vein?

An English study comprised 123 reversed vein and 103 in-situ femoropopliteal bypasses [63]: cumulative patency rates were not significantly different between reversed and in-situ grafts at any time up to 6 years after operation with primary patency rates of 62.4% for reversed and 63.5% for in-situ vein at 5 years. Small veins (< 4 mm) had a patency of 36%, whereas larger veins (> 4 mm) had a 75.9% patency. Patency rates of small veins employed for in-situ and reversed vein were similar. A large American study showed a 5-year limb salvage rate of 82% for non-reversed greater saphenous vein and 90% for in-situ vein made for limb salvage (n.s.) [65]. Data derived from the Dutch BOA study have recently shown that the 2-year cumulative patency for femoropopliteal in-situ grafts was 81.4% and for reversed veins 83.4%; additional hazard ratio analysis showed a slightly favorable trend for reversed grafts in this position [62].

Should PTFE be Preferred to Dacron?

In a series from a single surgeon, knitted non-velour Dacron was significantly better than PTFE for femoropopliteal reconstructions at 5 years [97]. A randomized, prospective series from eight centers in the United States comprised 244 patients receiving either knitted Dacron coated with collagen or PTFE to the above knee popliteal artery [92]. The survival rate of the patients was 77% at 3 years, which is noteworthy. The primary patency rate for Dacron was 62% and for PTFE 58%. The secondary patency rate at 3 years was 75% in both groups. Grafts with a diameter of 7 and 8 mm fared better than those of 5-6 mm.

A prospective, randomized multicenter study from the Netherlands comprising 228 patients showed no significant difference in patency rates for knitted Dacron or thin walled PTFE (all patients were on oral anticoagulants). The primary patency at 3 years was 55% for PTFE, and 62% for Dacron; primary assisted patency was 64% versus 70%; secondary patency was 71% versus 76% [119]. It is obvious that there cannot be much difference between PTFE and Dacron in the above knee position, and the choice of material must be based on other criteria than patency, for example structural stability and resistance to infection.

PRINCIPLES OF TREATMENT

Isolated Lesions in the Common Femoral Artery

A peculiar, localized stenosis, or even occlusion, may occasionally be encountered in the common femoral artery just below the inguinal ligament or at the division of the common femoral artery into the superficial femoral/profunda femoral arteries. The other arteries are usually remarkably free of disease. The atherosclerotic process is generally very hard and looks like a coral reef. It is not suited for PTA, and endarterectomy, which can be done under local anesthesia, is the treatment of choice.

Isolated Lesions in the Profunda Femoral Artery

Isolated occlusions are rare, and usually well-developed collaterals are present if the occlusion is located distally to the first muscular branch. Stenoses are practically always seen in combination with changes in the common femoral or the superficial femoral arteries. Interventions to correct isolated profunda femoral stenoses or occlusions are not indicated except, possibly, in very rare cases of femoral claudication. PTA is difficult, primary endarterectomy is the preferred treatment. Profundaplasty cannot be expected to be followed by healing of distal ischemic ulcerations.

Synchronous Lesions in the Common Femoral Artery and in the Profunda Femoral Artery

The lesions in the common femoral artery should be removed with endarterectomy, and run in to the profunda femoral artery should be secured. By a separate incision in the wall of the profunda, local endarterectomy can be made, or, in selected cases, intraoperative PTA of stenoses in the second or third segment (antegrade ring stripping may be better).

Synchronous Lesions in the Common Femoral Artery and in the Superficial Femoral Artery

If hemodynamically significant stenoses are present at the same time in the common femoral artery and in the superficial femoral artery, local endarterectomy should be made in the common femoral artery with creation of unrestricted inflow to both the profunda and the superficial femoral arteries. This procedure will be adequate for some claudicants, but in other cases with claudication, and in most cases with critical ischemia, it is also necessary to dilate the stenoses in the superficial femoral artery (if lesions are relatively short) or to perform bypass grafting.

Lesions of the Superficial Femoral Artery

Reconstruction may be performed in claudicants as well as in patients with critical ischemia. If lesions are relatively short, PTA of the superficial femoral artery should be the preferred initial treatment. In some centers, even completely occluded superficial femoral arteries are treated by PTA, but usually a femoropopliteal bypass will be made to the above knee popliteal artery. Occasionally, reports on superficial femoral artery endarterectomy appears; older publications report inferior results, but newer series seem to have better results.

In case the above knee popliteal artery is of bad quality but the below knee popliteal artery is of good quality a bypass is indicated to the infragenicular popliteal artery. Generally, reconstructions with distal anastomosis below the knee joint should not be undertaken for intermittent claudication, but the operation is indicated for critical ischemia.

Whenever the above and below knee popliteal artery is diseased but a good crural runoff is available, reconstruction should be reserved for patients with critical ischemia only. Bypass to crural arteries should be made, preferably to the tibioperoneal trunk; if this is not possible, then to the posterior tibial artery, the peroneal artery and the anterior tibial artery, in that order of preference.

If both the popliteal and crural arteries are severely diseased, reconstruction is definitely only indicated in patients with critical ischemia. Bypass to crural or pedal arteries, if possible. If not, consider revascularization of the popliteal blind segment or geniculate arteries, or amputation.

Isolated Blind Popliteal Segment

Some data are available to indicate that this may be tried as a last resort with the greater saphenous vein as a bypass [120], but larger series are lacking. Best results are obtained with autogenous vein but PTFE plus Taylor vein patch should be contemplated if the saphenous or other veins are not available, since it has been shown that comparable 3-year patency rates can be expected for PTFE to isolated popliteal segments and to below knee popliteal arteries with patent tibial and peroneal run off [121]. A study of 217 femoropopliteal bypasses in 207 patients over 10 years with 15% reversed saphenous vein and 85% with PTFE [122] gave primary 5-year patency rates of 79% for vein and for PTFE 56%; limb salvage rate was 78%; mortality (30 days) was 10%, 5-year patient survival 38%. It must be noted that sequential extension to an infrapopliteal artery was required in up to 20% of cases. A recent study [123] on 20 patients with a variety of techniques to revascularize perigenicular arteries in critical limb ischemia gave an assisted primary patency rate of 85% at 1 year and a limb salvage rate of 90% (postoperative mortality 15%). Consequently, it seems as if revascularization of an isolated popliteal segment could be attempted as an alternative to amputation.

■ RISK AND COMPLICATIONS

Results of surgery are generally believed to be identical in the sexes [124], but multivariate analysis has shown that female gender is associated with a 2.4 fold decreased secondary graft patency; on the other hand the cumulative 3-year limb salvage rate was identical in men and women [125]. A systematic Finnish study (207 patients, 218 reconstructions) focused on risk factors for results in the femoropopliteal segments, and showed that increased severity of ischemia was associated with decreasing leg salvage and survival rates. However, sex, age, and diabetes did not influence patency or leg salvage; diabetes was associated with a decreased survival [126]. When analyzed separately, age above 70 years was associated with a smaller chance of being alive with a functioning graft. In a study on PTFE grafts, smoking, ankle systolic pressure of less than 50 mm Hg, presentation with rest pain or tissue necrosis and single vessel runoff were identified as having a significant adverse effect on graft patency, but multivariate analysis showed that smoking was the only significant adverse variable [127].

Although a femoropopliteal reconstruction can usually be performed under epidural or spinal anesthesia with small incisions, the operation is not without local and general risks. For example, in a typical report, 13% minor wound problems occurred, 4.2% of the operations required reoperations due to bleeding, a few per cent suffered from acute myocardial infarction, and an occasional stroke was seen [60]. In another series, 2.2% cardiac complications occurred (all nonfatal: arrhythmia, congestive heart failure, myocardial infarction), 0.4% developed moderate lymphocele, and 2.2% graft infection. Wound morbidity rate (infection, disruption, delayed healing) was 3.9%, including 3.5% mild infection [92]. The 30-day mortality rate varies from zero [92] in USA to 3%-5% in Finland and Norway [12-104]. In Denmark, all departments had an early mortality of less than 2% for femoropopliteal reconstructions [128]. In Sweden, the 30 day mortality for above knee femoropopliteal bypass on all indications was 2.5% and for below knee 2.9% in 1996.

■ SURVEILLANCE

Distinction is necessary between postoperative (first day), early (< 30 days) and late (> 30 days) thrombosis. The reasons for postoperative graft thrombosis are multiple and cannot always be determined with confidence (hypercoagulable blood, relative general dehydration, intra- and postoperative hypotension, embolization, technical difficulties, vein quality and properties, hemodynamic failure, etc.). In the longer term, grafts fail mainly due to progression of the atherosclerosis compromising inflow or runoff, neointimal hyperplasia or pseudoaneurysm formation at the anastomoses, and, for the vein grafts - especially the in-situ grafts - graft body stenosis. The Dacron grafts all dilate and elongate with time [129], but non-anastomotic aneurysms are very rare and occur occasionally in the non-anastomotic graft segments [96]. In a series with duplex scanning or arteriography [130] with a mean follow up of 4.8 years, 52% of the 150 patients previously subjected to vascular reconstructions had signs of progression of atherosclerosis. Especially, 30% of the superficial femoral arteries showed progression and one-third of the patients with stenosis greater than 50% at baseline became occluded. The progression was more pronounced in patients requiring femoropopliteal than suprainguinal surgery. In a series of 91 failing PTFE grafts the predominant reason for failure was 43 inflow lesions and 83 outflow lesions, with ten lesions in the body of the graft and only 8 at the anastomoses [131]. An aggressive policy for surgical revision of threatened vein grafts is warranted (inflow procedures, vein patch angioplasty, interposition vein grafts, vein graft extensions to new distal site). Early intervention is unlikely to be required in grafts with a low peripheral resistance and high blood flow (as measured intraoperatively) [132]. A study from Edinburgh [133] seems to indicate that stenoses prone to occlude can be identified (by velocity ratio > 3), and that they progress more rapidly than moderate stenoses, and that there is only a brief window of opportunity to treat before occlusion.

In the case of surveillance of synthetic grafts, some evidence is conflicting; the results of one study did not point towards benefit of a surveillance program because a significant number of grafts failed within the first 30 days and most of these were not predicted by the program [134]. On the other hand, another investigation on surveillance of PTFE grafts found it necessary [131] since it enhanced the 5-year patency and limb salvage rates.

An analysis of 6649 grafts distributed between surveillance and non-surveillance showed that the number of deaths and occluded grafts was greater in the non-surveillance group, whereas perioperative occlusion rates were not. The number of amputations did not differ between groups. It was concluded that patency seemed to be improved as a result of surveillance, but limb salvage was not [135]. On the other hand, a surveillance program with duplex is likely to detect an incidence of vein graft stenoses of 21.5% within one year in addition to approximately the same incidence of graft threatening inflow and outflow problems (including progression of atherosclerosis) [136]. Since about 20% of the stenoses in vein grafts occur more than 12 months after surgery, surveillance programs, at least for vein grafts, should continue indefinitely [137]. The problem of the possible beneficial effects of surveillance on vein graft patency is still unsettled [138]. A recent study from the U. K. has documented that there has been a considerable increase in workload due to modern maintenance of infrainguinal bypass grafts [139], for example: 116 procedures were necessary in 38 below knee femoropopliteal grafts and 58 interventions in the group of 34 above knee femoropopliteal grafts. Radiological intervention accounted for almost 70% of the workload [139].

In the next decade, surveillance with automated non-invasive magnetic resonance (MR) techniques will become clinically feasible, even on a larger scale, for routine purposes. MR imaging and advanced postprocessing of blood flow velocity data will be able to provide the surgeon with exact information, not only on morphology, but also on a variety of other essential hemodynamic and biomechanical variables. Our current research projects [42,140,141] are attacking these complex problems, and the present state of development is indicated in figures 1-4.

ACKNOWLEDGEMENTS
I thank my clinical assistant, Sten Oyre, MD, PhD-student, for having made the magnetic resonance examinations, the original postprocessing analyses and the illustrations specifically for this chapter.
Consultant Vascular Surgeon Leif Panduro Jensen, MD, Chairman of the Board of The Danish Vascular Registry (Karbase), is gratefully acknowledged for having provided me, in October 1998, with the information quoted in the text on activity in Denmark for 1997.

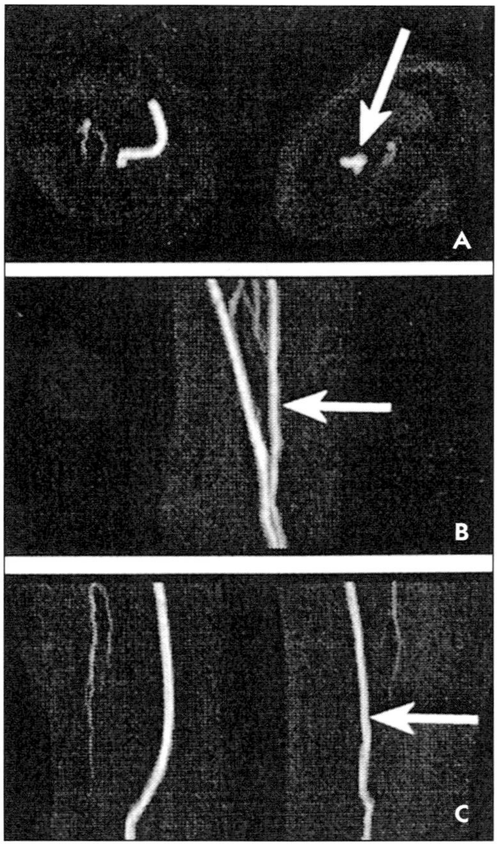

FIG. 1 The MIPs (maximal intensity projections) from the ECG (electrocardiogram, R-wave) gated magnetic resonance angiography image of both legs. The patient had an Atrium (PTFE) graft inserted in the right leg 2 days previously (left on a and c), and another Atrium graft implanted into the left leg 4.5 months ago. A - Is the projection in the feet-head direction. B - In the left-right. C - In the anterior-posterior directions.
The arrows indicate the femoropopliteal graft in the left leg on all three images. In b) and c) the arrows also show the position of the plane for the transversal blood flow measurement image. The MR investigations were performed on a Philips Gyroscan ACS-NT scanner.
All illustrations: Courtesy of Sten Oyre, MD.

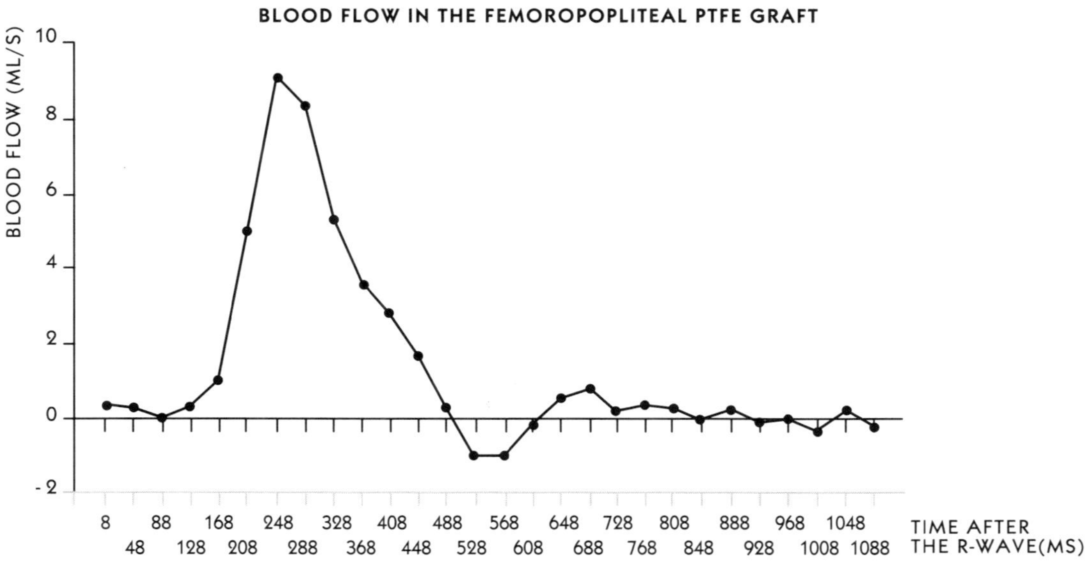

FIG. 2 This figure demonstrates the blood flow within the graft of the left leg as a function of time within the cardiac cycle. The averaged blood flow was 79 ml/min based on measurements in 28 heart phases; heart rate was 50 beats/min.

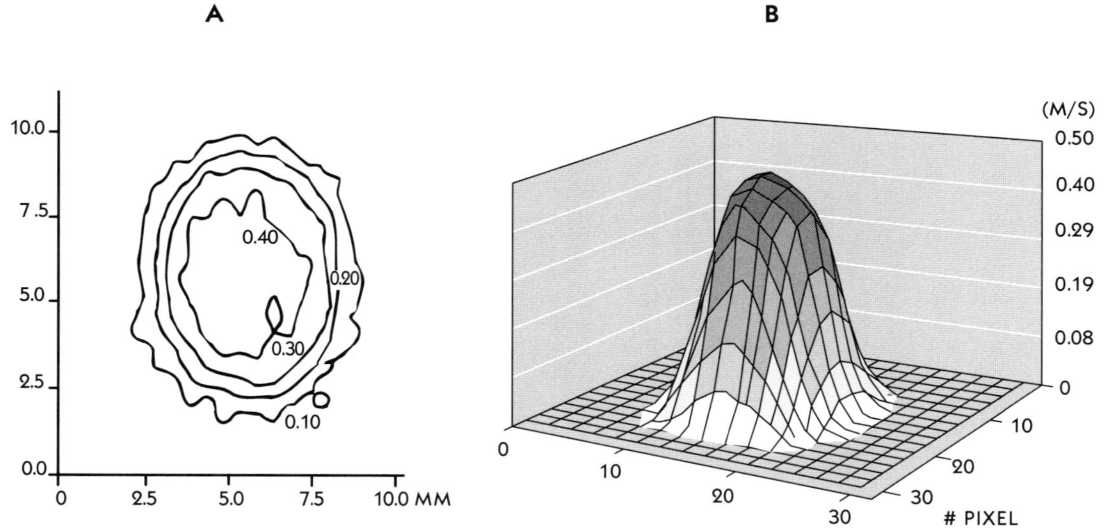

FIG. 3 The velocity profile from peak systole within the graft in the left leg is visualized as A - a contour plot (velocity [m/s]) and as B - a three-dimensional surface plot. Note the paraboloid shape of the profile.

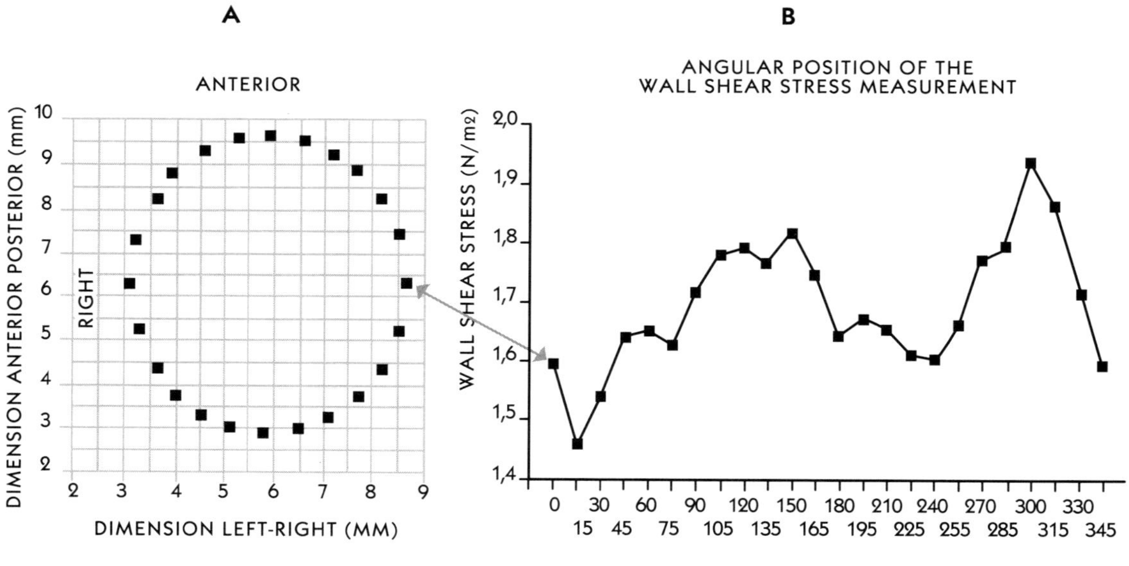

FIG. 4 A - The edge detection (the exact localization of the graft wall) in 24 evenly distributed edge points from peak systole, and B - the corresponding circumferential wall shear stress in these 24 edge points of the graft in the left leg. The arrow indicates two corresponding points, i.e., the zero degree point is at the left position of the graft. The degrees are numbered anti-clockwise. The mean diameter of the graft was found to be 6.28 mm (factory specification: 6 mm) in peak systole (area 30.96 mm^2) by the three dimensional paraboloid method. The matrix in a) was equivalent to the pixel size of 0.39 x 0.39 mm^2.

REFERENCES

1 Andersen P, Saltin B. Maximal perfusion of skeletal muscle in man. *J Physiol (Lond)* 1985; 366: 233-249.

2 Watson CJ, Phillips D, Hands L et al. Claudication distance is poorly estimated and inappropriately measured. *Br J Surg* 1997; 84: 1107-1109.

3 Tisi PV, Shearman CP. The evidence for exercise-induced inflammation in intermittent claudication: should we encourage patients to stop walking? *Eur J Vasc Endovasc Surg* 1998; 15: 7-17.

4 Killewich LA, Gardner AW, Macko RF et al. Progressive intermittent claudication is associated with impaired fibrinolysis. *J Vasc Surg* 1998; 27: 645-650.

5 Ray SA, Rowley MR, Loh A et al. Hypercoagulable states in patients with leg ischaemia. *Br J Surg* 1994; 81: 811-814.

6 Second European Consensus Document on chronic critical leg ischemia. *Eur J Vasc Surg* 1992; 6 Suppl A: 1-32.

7 Wolfe JH, Wyatt MG. Critical and subcritical ischaemia. *Eur J Vasc Endovasc Surg* 1997; 13: 578-582.

8 Holstein P, Lassen NA. Healing of ulcers on the feet correlated with distal blood pressure measurements in occlusive arterial disease. *Acta Orthop Scand* 1980; 51: 995-1006.

9 Hunt BJ, Jurd KM. Endothelial cell activation. A central pathophysiological process [editorial]. *Br Med J* 1998; 316: 1328-1329.

10 Nielsen OM, Jensen LP, Schroeder TV. The Nordic experience of preference for above knee femoropopliteal bypass. In: Greenhalgh RM, Fowkes FGR, eds. *Trials and tribulations of vascular surgery. Evidence-based vascular surgery.* London, WB Saunders, 1996; pp. 325-331.

11 Bergqvist D, Troëng T, Elfstrom J et al. Auditing surgical outcome: ten years with the Swedish Vascular Registry - Swedvasc. The Steering Committee of Swedvasc. *Eur J Surg* 1998; Supp 581: 3-48.

12 Kantonen I, Lepäntalo M, Luther M et al. Factors affecting the results of surgery for chronic critical leg ischemia - a nationwide survey. Finvasc Study Group. *J Vasc Surg* 1998; 27: 940-947.

13 Critical limb ischaemia: management and outcome. Report of a national survey. The Vascular Surgical Society of Great Britain and Ireland. *Eur J Vasc Endovasc Surg* 1995; 10: 108-113.

14 Anonymous. Amputation or arterial reconstruction? [editorial]. *Lancet* 1992; 339: 900-901.

15 Cheshire NJ, Wolfe JHN. Does distal revascularization for limb salvage work? In: Greenhalgh RM, Fowkes FGR, eds. *Trials and tribulations of vascular surgery. Evidence-based vascular surgery.* London, WB Saunders, 1996: pp. 353-364.

16 Nicoloff AD, Taylor LM Jr., McLafferty RB et al. Patient recovery after infrainguinal bypass grafting for limb salvage *J Vasc Surg* 1998; 27: 256-263.

17 Belkin M, Conte MS, Donaldson MC et al. The impact of gender on the results of arterial bypass with in-situ greater saphenous vein. *Am J Surg* 1995; 170: 97-102.

18 Jelnes R, Gaardsting O, Hougaard Jensen K et al. Fate in intermittent claudication: outcome and risk factors. *Br Med J (Clin Res Ed)* 1986; 293: 1137-1140.

19 Van der Heijden FH, Eikelboom BC, Banga JD et al. Management of superficial femoral artery occlusive disease. *Br J Surg* 1993; 80: 959-963.

20 Dormandy J, Mahir M, Ascady G et al. Fate of the patient with chronic leg ischemia. A review article. *J Cardiovasc Surg (Torino)* 1989; 30: 50-57.

21 Agner E. Natural history of angina pectoris, possible previous myocardial infarction and intermittent claudication during the eighth decade. A longitudinal epidemiologic study. *Acta Med Scand* 1981; 210: 271-276.

22 Ingolfsson IO, Sigurdsson G, Sigvaldason H et al. A marked decline in the prevalence and incidence of intermittent claudication in Icelandic men 1968-1986: a strong relationship to smoking and serum cholesterol - the Reykjavik Study. *J Clin Epidemiol* 1994; 47: 1237-1243.

23 Kannel WB. The demographics of claudication and the aging of the American population. *Vasc Med* 1996; 1: 60-64.

24 Aune S, Trippestad A. Relative mortality of patients operated for femoropopliteal occlusive disease. *Eur J Vasc Surg* 1994; 8: 188-192.

25 Aune S, Trippestad A. Femoropopliteal reconstruction for obliterative arteriosclerosis. A 6-year surgical material. *Tidsskr Nor Laegeforen* 1993; 113: 1458-1461.

26 Veith FJ, Gupta SK, Wengerter KR et al. Changing arteriosclerotic disease patterns and management strategies in lower-limb-threatening ischemia. *Ann Surg* 1990; 212: 402-412.

27 Watelet J, Soury P, Menard JF et al. Femoropopliteal bypass: in-situ or reversed vein grafts? Ten-year results of a randomized prospective study. *Ann Vasc Surg* 1997; 11: 510-519.

28 Anonymous. Suggested standards for reports dealing with lower extremity ischemia. Prepared by the Ad Hoc Committee on Reporting Standards, Society for Vascular Surgery/North American Chapter, International Society for Cardiovascular Surgery. *J Vasc Surg* 1986; 4: 80-94.

29 Ahn SS, Rutherford RB, Johnston KW et al. Reporting standards for infrarenal endovascular abdominal aortic aneurysm repair. Ad Hoc Committee for Standardized Reporting Practices in Vascular Surgery of The Society for Vascular Surgery/International Society for Cardiovascular Surgery. *J Vasc Surg* 1997; 25: 405-410.

30 Porter JM, Moneta GL. Reporting standards in venous disease: an update. International Consensus Committee on Chronic Venous Disease. *J Vasc Surg* 1995; 21: 635-645.

31 Baker JD, Rutherford RB, Bernstein EF et al. Suggested standards for reports dealing with cerebrovascular disease. Subcommittee on Reporting Standards for Cerebrovascular Disease, Ad Hoc Committee on Reporting Standards, Society for Vascular Surgery/North American Chapter, International Society for Cardiovascular Surgery. *J Vasc Surg* 1988; 8: 721-729.

32 Housley E. Treating claudication in five words [editorial]. *Br Med J (Clin Res Ed)* 1988; 296: 1483-1484.

33 Powell JT. Vascular damage from smoking: disease mechanisms at the arterial wall. *Vasc Med* 1998; 3: 21-28.

34 McCollum C, Alexander C, Kenchington G et al. Antiplatelet drugs in femoropopliteal vein bypasses: a multicenter trial. *J Vasc Surg* 1991; 13: 150-161.

35 Van Goor H, Boontje AH. Results of vascular reconstructions for atherosclerotic arterial occlusive disease of the lower limbs in young adults. *Eur J Vasc Endovasc Surg* 1995; 10: 323-326.

36 Gouny P, Bertrand P, Decaix B et al. Distal bypass for limb salvage: comparative study in patients below and above 80 years of age. *J Cardiovasc Surg* 1994; 35: 419-424.

37 Nehler MR, Moneta GL, Edwards JM et al. Surgery for chronic lower extremity ischemia in patients eighty or more years of age: operative results and assessment of postoperative independence. *J Vasc Surg* 1993; 18: 618-624.

38 Peltonen S, Biancari F, Lindgren L et al. Outcome of infrainguinal bypass surgery for critical leg ischaemia in patients with chronic renal failure. *Eur J Vasc Endovasc Surg* 1998; 15: 122-127.

39 Baele HR, Piotrowski JJ, Yuhas J et al. Infrainguinal bypass in patients with end-stage renal disease. *Surgery* 1995; 117: 319-324.

40 Isiklar MH, Kulbaski M, MacDonald MJ et al. Infrainguinal bypass in end-stage renal disease: when is it justified? *Semin Vasc Surg* 1997; 10: 42-48.
41 Hertz SM, Baum RA, Owen RS et al. Comparison of magnetic resonance angiography and contrast arteriography in peripheral arterial stenosis. *Am J Surg* 1993; 166: 112-116.
42 Oyre S, Ringgaard S, Kozerke S et al. Accurate noninvasive quantitation of blood flow, cross-sectional lumen vessel area and wall shear stress by three-dimensional paraboloid modeling of magnetic resonance imaging velocity data. *J Am Coll Cardiol* 1998; 32: 128-134.
43 Koelemay MJ, Legemate DA, de Vos H et al. Can cruropedal colour duplex scanning and pulse generated runoff replace angiography in candidates for distal bypass surger? *Eur J Vasc Endovasc Surg* 1998; 16: 13-18.
44 Koelemay MJ, den Hartog D, Prins MH et al. Diagnosis of arterial disease of the lower extremities with duplex ultrasonography. *Br J Surg* 1996; 83: 404-409.
45 Sensier Y, Fishwick G, Owen R et al. A comparison between colour duplex ultrasonography and arteriography for imaging infrapopliteal arterial lesions. *Eur J Vasc Endovasc Surg* 1998; 15: 44-50.
46 Alback A, Biancari F, Saarinen O et al. Prediction of the immediate outcome of femoropopliteal saphenous vein bypass by angiographic runoff score. *Eur J Vasc Endovasc Surg* 1998; 15: 220-224.
47 Blankensteijn JD, Gertler JP, Brewster DC et al. Intraoperative determinants of infrainguinal bypass graft patency: a prospective study. *Eur J Vasc Endovasc Surg* 1995; 9: 375-382.
48 Whyman MR, Fowkes FG, Kerracher EM et al. Is intermittent claudication improved by percutaneous transluminal angioplasty? A randomized controlled trial. *J Vasc Surg* 1997; 26: 551-557.
49 Whyman MR, Fowkes FG, Kerracher EM et al. Randomised controlled trial of percutaneous transluminal angioplasty for intermittent claudication. *Eur J Vasc Endovasc Surg* 1996; 12: 167-172.
50 Hunink MG, Wong JB, Donaldson MC et al. Revascularization for femoropopliteal disease. A decision and cost- effectiveness analysis. *JAMA* 1995; 274: 165-171.
51 Alback A, Biancari F, Schmidt S et al. Haemodynamic results of femoropopliteal percutaneous transluminal angioplasty. *Eur J Vasc Endovasc Surg* 1998; 16: 7-12.
52 Nydahl S, Hartshorne T, Bell PR et al. Subintimal angioplasty of infrapopliteal occlusions in critically ischaemic limbs. *Eur J Vasc Endovasc Surg* 1997; 14: 212-216.
53 Varty K, Nydahl S, Nasim A et al. Results of surgery and angioplasty for the treatment of chronic severe lower limb ischaemia. *Eur J Vasc Endovasc Surg* 1998; 16: 159-163.
54 Casselman F, Van Elst F, Spoelstra H. Femoropopliteal endobypass: a feasibility study in 41 patients. *Acta Chir Belg* 1997; 97: 23-26.
55 Van der Heijden FH, Eikelboom BC, van Reedt Dortland RW et al. Endarterectomy of the superficial femoral artery: a procedure worth reconsidering. *Eur J Vasc Surg* 1992; 6: 651-658.
56 Van der Heijden FH, Eikelboom BC, van Reedt Dortland RW et al. Long-term results of semiclosed endarterectomy of the superficial femoral artery and the outcome of failed reconstructions. *J Vasc Surg* 1993; 18: 271-279.
57 Moll FL, Ho GH. Closed superficial femoral artery endarterectomy: a 2-year follow-up. *Cardiovasc Surg* 1997; 5: 398-400.
58 Vroegindeweij D, Idu M, Buth J et al. The cost-effectiveness of treatment of short occlusive lesions in the femoropopliteal artery: balloon angioplasty versus endarterectomy. *Eur J Vasc Endovasc Surg* 1995; 10: 40-50.
59 Lo Scudo LS, Mascellari L, Dompe G et al. Thromboendarterectomy of the superficial femoral artery with a femoropopliteal endoprosthesis. Technical note on a preliminary experience. *Minerva Cardioangiol* 1997; 45: 541-546.
60 Taylor LM Jr., Edwards JM, Porter JM. Present status of reversed vein bypass grafting: five-year results of a modern series. *J Vasc Surg* 1990; 11: 193-205.
61 Gentile AT, Lee RW, Moneta GL et al. Results of bypass to the popliteal and tibial arteries with alternative sources of autogenous vein. *J Vasc Surg* 1996; 23: 272-279.
62 Lawson JA, Tangelder MJD, Algra A et al. The myth of the in-situ graft: superiority in infrainguinal venous bypass procedures? Results from a large prospective comparative study. *Eur J Vasc Endovasc Surg* 1999; in press.
63 Moody AP, Edwards PR, Harris PL. in-situ versus reversed femoropopliteal vein grafts: long-term follow-up of a prospective, randomized trial. *Br J Surg* 1992; 79: 750-752.
64 Johnson PR, Tan SL, Chin AK. Endoscopic femoral-popliteal/distal bypass grafting: a preliminary report. *J Am Coll Surg* 1998; 186: 331-336.
65 Belkin M, Knox J, Donaldson MC et al. Infrainguinal arterial reconstruction with nonreversed greater saphenous vein. *J Vasc Surg* 1996; 24: 957-962.
66 Shah DM, Darling RC, Chang BB et al. Long-term results of in-situ saphenous vein bypass. Analysis of 2058 cases. *Ann Surg* 1995; 222: 438-446.
67 Chang BB, Paty PS, Shah DM et al. The lesser saphenous vein: an underappreciated source of autogenous vein. *J Vasc Surg* 1992; 15: 152-156.
68 Hölzenbein TJ, Pomposelli FB Jr, Miller A et al. Results of a policy with arm veins used as the first alternative to an unavailable ipsilateral greater saphenous vein for infrainguinal bypass. *J Vasc Surg* 1996; 23: 130-140.
69 Holzenbein TJ, Pomposelli FB Jr, Miller A et al. The upper arm basilic-cephalic loop for distal bypass grafting: technical considerations and follow-up. *J Vasc Surg* 1995; 21: 586-592.
70 Taylor SM, Langan EM, Snyder BA et al. Superficial femoral artery eversion endarterectomy: a useful adjunct for infrainguinal bypass in the presence of limited autogenous vein. *J Vasc Surg* 1997; 26: 439-445.
71 Schulman ML, Badhey MR, Yatco R et al. An 11-year experience with deep leg veins as femoropopliteal bypass grafts. *Arch Surg* 1986; 121: 1010-1015.
72 Wijesinghe LD, Beardsmore DM, Scott DJ. Polytetrafluoroethylene (PTFE) femorodistal grafts with a distal vein cuff for critical ischaemia. *Eur J Vasc Endovasc Surg* 1998; 15: 449-453.
73 Ascer E, Pollina RM, Gennaro M et al. Noninvasive predictors of patency for infrapopliteal PTFE bypasses with combined arteriovenous fistula and vein interposition technique. *Am J Surg* 1995; 170: 103-105.
74 Taylor RS, Loh A, McFarland RJ et al. Improved technique for polytetrafluoroethylene bypass grafting: long-term results using anastomotic vein patches. *Br J Surg* 1992; 79: 348-354.
75 Miller JH, Foreman RK, Ferguson L et al. Interposition vein cuff for anastomosis of prosthesis to small artery. *Aust N Z J Surg* 1984; 54: 283-285.
76 Wolfe JH, Tyrrell MR. Improving the patency of prosthetic grafts with vein cuffs. *Semin Vasc Surg* 1995; 8: 246-252.
77 Tyrrell MR, Chester JF, Vipond MN et al. Experimental evidence to support the use of interposition vein collars/patches in distal PTFE anastomoses. *Eur J Vasc Surg* 1990; 4: 95-101.
78 Wijesinghe LD, Smye SW, Scott DJ. Impedance index measurements of in vitro PTFE end-to-side anastomoses: effect of angle and Miller cuff. *Eur J Vasc Endovasc Surg* 1998; 16: 65-70.
79 Pappas PJ, Hobson RW, Meyers MG et al. Patency of infrainguinal polytetrafluoroethylene bypass grafts with distal interposition vein cuffs. *Cardiovasc Surg* 1998; 6: 19-26.
80 Raptis S, Miller JH. Influence of a vein cuff on polytetrafluoroethylene grafts for primary femoropopliteal bypass. *Br J Surg* 1995; 82: 487-491.

81 Jakobsen HL, Baekgaard N, Christoffersen JK. Below-knee popliteal and distal bypass with PTFE and vein cuff. *Eur J Vasc Endovasc Surg* 1998; 15: 327-330.
82 Neville RF, Attinger C, Sidawy AN. Prosthetic bypass with a distal vein patch for limb salvage. *Am J Surg* 1997; 174: 173-176.
83 Stonebridge PA, Prescott RJ, Ruckley CV. Randomized trial comparing infrainguinal polytetrafluoroethylene bypass grafting with and without vein interposition cuff at the distal anastomosis. The Joint Vascular Research Group. *J Vasc Surg* 1997; 26: 543-550.
84 Karacagil S, Holmberg A, Narbani A et al. Composite polytetrafluroethylene/vein bypass grafts: conventional distal vein segment or vein cuff? *Eur J Vasc Endovasc Surg* 1996; 12: 337-341.
85 Chang BB, Darling RC, Bock DE et al. The use of spliced vein bypasses for infrainguinal arterial reconstruction. *J Vasc Surg* 1995; 21: 403-410.
86 Quinones-Baldrich WJ, Prego AA, Ucelay-Gomez R et al. Long-term results of infrainguinal revascularization with polytetrafluoroethylene: a ten-year experience. *J Vasc Surg* 1992; 16: 209-217.
87 Sterpetti AV, Schultz RD, Feldhaus RJ et al. Seven-year experience with polytetrafluoroethylene as above-knee femoropopliteal bypass graft. Is is worthwhile to preserve the autologous saphenous vein? *J Vasc Surg* 1985; 2: 907-912.
88 Prendiville EJ, Yeager A, O'Donnell TF Jr. et al. Long-term results with the above-knee popliteal expanded polytetrafluoroethylene graft. *J Vasc Surg* 1990; 11: 517-524.
89 Patterson RB, Fowl RJ, Kempczinski RF et al. Preferential use of ePTFE for above-knee femoropopliteal bypass grafts [see comments]. *Ann Vasc Surg* 1990; 4: 338-343.
90 Yang PM, Wengerter KR, Veith FJ et al. Value and limitations of secondary femoropopliteal bypasses with polytetrafluoroethylene. *J Vasc Surg* 1991; 14: 292-298.
91 Woratyla SP, Darling RC, Chang BB et al. The performance of femoropopliteal bypasses using polytetrafluoroethylene above the knee versus autogenous vein below the knee. *Am J Surg* 1997; 174: 169-172.
92 Abbott WM, Green RM, Matsumoto T et al. Prosthetic above-knee femoropopliteal bypass grafting: results of a multicenter randomized prospective trial. Above-knee femoropopliteal Study Group. *J Vasc Surg* 1997; 25: 19-28.
93 Gupta SK, Veith FJ, Kram HB et al. Prospective, randomized comparison of ringed and nonringed polytetrafluoroethylene femoropopliteal bypass grafts: a preliminary report. *J Vasc Surg* 1991; 13: 163-172.
94 Benedetti-Valentini F, Gossetti B, Irace I et al. Composite grafts for critical ischaemia. *Cardiovasc Surg* 1996; 4: 372-376.
95 Chang JB, Stein TA. The long-term value of composite grafts for limb salvage. *J Vasc Surg* 1995; 22: 25-31.
96 Orii M, Shirasugi N, Yamazaki M, Akiyama Y. Pseudo-aneurysm caused by disruption of an externally supported knitted Dacron graft for femoropopliteal bypass. *Tokai J Exp Clin Med* 1995; 20: 241-244.
97 Pevec WC, Darling RC, L'Italien GJ et al. Femoropopliteal reconstruction with knitted, nonvelour Dacron versus expanded polytetrafluoroethylene. *J Vasc Surg* 1992; 16: 60-65.
98 El-Massry S, Saad E, Sauvage LR et al. Femoropopliteal bypass with externally supported knitted Dacron grafts: a follow-up of 200 grafts for one to twelve years. *J Vasc Surg* 1994; 19: 487-494.
99 De Leersnijder D, Willocx P, Van Marck E, Vanmaele R. Venous homografts in infra-inguinal procedures: an eight years experience. *J Cardiovasc Surg* 1992; 33: 633-640.
100 Leseche G, Penna C, Bouttier S et al. Femorodistal bypass using cryopreserved venous allografts for limb salvage. *Ann Vasc Surg* 1997; 11: 230-236.
101 Rebane E, Tikko H, Tunder E et al. Venous allografts for infrainguinal vascular bypass. *Cardiovasc Surg* 1997; 5: 21-25.
102 Walker PJ, Mitchell RS, McFadden PM et al. Early experience with cryopreserved saphenous vein allografts as a conduit for complex limb-salvage procedures. *J Vasc Surg* 1993; 18: 561-568.
103 Gournier JP, Favre JP, Gay JL et al. Cryopreserved arterial allografts for limb salvage in the absence of suitable saphenous vein: two-year results in 20 cases. *Ann Vasc Surg* 1995; 9 Suppl: S7-14.
104 Florenes T, Zasada W, Bille S et al. Femoropopliteal bypass in arteriosclerosis. *Tidsskr Nor Laegeforen* 1994; 114: 3193-3194.
105 Jarrett F, Mahood BA. Long-term results of femoropopliteal bypass with stabilized human umbilical vein. *Am J Surg* 1994; 168: 111-114.
106 Myers KA. Semi-closed, ex-situ, non-reversed or reversed autogenous vein grafting. A technique for femoro-distal arterial bypass. *J Cardiovasc Surg* 1991; 32: 110-116.
107 Eickhoff JH, Broome A, Ericsson BF et al. Four years' results of a prospective, randomized clinical trial comparing polytetrafluoroethylene and modified human umbilical vein for below-knee femoropopliteal bypass. *J Vasc Surg* 1987; 6: 506-511.
108 Koch G, Gutschi S, Pascher O et al. Analysis of 274 omniflow vascular prostheses implanted over an eight- year period. *Aust N Z J Surg* 1997; 67: 637-639.
109 Koch G, Gutschi S, Pascher O et al. Femoropopliteal vascular replacement: vein, ePTFE or ovine collagen? *Zentralbl Chir* 1996; 121: 761-767.
110 Yoshida H, Sasajima T, Goh K et al. Early results of a reinforced biosynthetic ovine collagen vascular prosthesis for small arterial reconstruction. *Surg Today* 1996; 26: 262-266.
111 Panayiotopoulos YP, Taylor PR. A paper for debate: vein versus PTFE for critical limb ischaemia-an unfair comparison? *Eur J Vasc Endovasc Surg* 1997; 14: 191-194.
112 Michaels JA. Choice of material for above-knee femoropopliteal bypass graft. *Br J Surg* 1989; 76: 7-14.
113 Veith FJ, Gupta SK, Ascer E et al. Six-year prospective multicenter randomized comparison of autologous saphenous vein and expanded polytetrafluoroethylene grafts in infrainguinal arterial reconstructions. *J Vasc Surg* 1986; 3: 104-114.
114 John TG, Stonebridge PA, Kelman J et al. Above-knee femoropopliteal bypass grafts and the consequences of graft failure. *Ann R Coll Surg Engl* 1993; 75: 257-260.
115 Quinones-Baldrich WJ, Busuttil RW, Baker JD et al. Is the preferential use of polytetrafluoroethylene grafts for femoropopliteal bypass justified? *J Vasc Surg* 1988; 8: 219-228.
116 Wilson YG, Wyatt MG, Currie IC et al. Preferential use of vein for above-knee femoropopliteal grafts. *Eur J Vasc Endovasc Surg* 1995; 10: 220-225.
117 Berlakovich GA, Herbst F, Mittlbock M et al. The choice of material for above-knee femoropopliteal bypass. A 20-year experience. *Arch Surg* 1994; 129: 297-302.
118 Poletti LF, Matsuura JH, Dattilo JB et al. Should vein be saved for future operations? A 15-year review of infrainguinal bypasses and the subsequent need for autogenous vein. *Ann Vasc Surg* 1998; 12: 143-147.
119 Van Det RJ, Bertelink BP, van Baal JG et al. PTFE or Dacron for femoro-popliteal above knee bypass: a multicenter randomized prospective trial. *Eur J Vasc Endovasc Surg* 1999; in press.
120 Mosimann UP, Stirnemann P. Chronic critical leg ischemia: revascularization of the isolated popliteal segment in comparison with femoro-distal bypass. *VASA J Vasc Dis* 1995; 24: 49-55.
121 Loh A, Chester JF, Taylor RS. PTFE bypass grafting to isolated popliteal segments in critical limb ischaemia. *Eur J Vasc Surg* 1993; 7: 26-30.

122 Kram HB, Gupta SK, Veith FJ et al. Late results of two hundred seventeen femoropopliteal bypasses to isolated popliteal artery segments. *J Vasc Surg* 1991; 14: 386-390.

123 Barral X, Salari GR, Toursarkissian B et al. Bypass to the perigeniculate collateral vessels. A useful technique for limb salvage: preliminary report on 22 patients. *J Vasc Surg* 1998; 27: 928-935.

124 Harris EJ Jr, Taylor LM Jr, Moneta GL et al. Outcome of infrainguinal arterial reconstruction in women. *J Vasc Surg* 1993; 18: 627-634.

125 Magnant JG, Cronenwett JL, Walsh DB et al. Surgical treatment of infrainguinal arterial occlusive disease in women. *J Vasc Surg* 1993; 17: 67-76.

126 Luther M, Lepäntalo M. Femoropopliteal reconstructions. Influence of risk factors on outcome. *Ann Chir Gynaecol* 1996; 85: 238-246.

127 Sayers RD, Thompson MM, Dunlop P et al. The fate of infrainguinal PTFE grafts and an analysis of factors affecting outcome. *Eur J Vasc Surg* 1994; 8: 607-610.

128 Nielsen OM, Perko M, Crone PO et al. Femoro-popliteal bypass surgery in Denmark 1983-1987. *Ugeskr Laeger* 1990; 152: 1980-1982.

129 Riepe G, Loos J, Imig H et al. Long-term in vivo alterations of polyester vascular grafts in humans. *Eur J Vasc Endovasc Surg* 1997; 13: 540-548.

130 McLafferty RB, Moneta GL, Masser PA et al. Progression of atherosclerosis in arteries distal to lower extremity revascularizations. *J Vasc Surg* 1995; 22: 450-455.

131 Sanchez LA, Suggs WD, Veith FJ et al. Is surveillance to detect failing polytetrafluoroethylene bypasses worthwhile? Twelve-year experience with ninety-one grafts. *J Vasc Surg* 1993; 18: 981-989.

132 Mercer KG, Scott DJA, Turton EPL et al. Can intra-operative flow measurements identify grafts for intensive duplex surveillance? *Eur J Vasc Endovasc Surg* 1999; in press.

133 Whyman MR, Ruckley CV, Fowkes FG. A prospective study of the natural history of femoropopliteal artery stenosis using duplex ultrasound. *Eur J Vasc Surg* 1993; 7: 444-447.

134 Dunlop P, Sayers RD, Naylor AR et al. The effect of a surveillance programme on the patency of synthetic infrainguinal bypass grafts. *Eur J Vasc Endovasc Surg* 1996; 11: 441-445.

135 Golledge J, Beattie DK, Greenhalgh RM et al. Have the results of infrainguinal bypass improved with the widespread utilisation of postoperative surveillance? *Eur J Vasc Endovasc Surg* 1996; 11: 388-392.

136 Wilson YG, Davies AH, Currie IC et al. Vein graft stenosis: incidence and intervention. *Eur J Vasc Endovasc Surg* 1996; 11: 164-169.

137 Dunlop P, Hartshorne T, Bolia A et al. The long-term outcome of infrainguinal vein graft surveillance. *Eur J Vasc Endovasc Surg* 1995; 10: 352-355.

138 Bergqvist D, Karacagil S, Lindblad B. Is there evidence in favour of vein graft surveillance? A randomised trial. In: Greenhalgh RM, Fowkes FGR, eds. *Trials and tribulations of vascular surgery. Evidence-based vascular surgery.* London, WB Saunders, 1996: pp. 289-304.

139 Loftus IM, Reid A, Thompson MM et al. The increasing workload required to maintain infrainguinal bypass graft patency. *Eur J Vasc Endovasc Surg* 1998; 15: 337-341.

140 Oyre S, Pedersen EM, Ringgaard S et al. In vivo wall shear stress measured by magnetic resonance velocity mapping in the normal human abdominal aorta. *Eur J Vasc Endovasc Surg* 1997; 13: 263-271.

141 Laustsen J, Paaske WP, Oyre S et al. Dynamic quantification, visualisation and animation of blood velocities and flows in infrarenal aortic aneurysms in vivo by three-dimensional MR phase velocity encoding. *Eur J Vasc Endovasc Surg* 1995; 9: 383-388.

18

FEMOROCRURAL BYPASS FOR LIMB SALVAGE: REAL INDICATIONS AND RESULTS

JOSE A. GONZALEZ-FAJARDO, CARLOS VAQUERO

The results of femorocrural bypass for limb salvage have traditionally been assessed by graft patency rate, limb salvage rate, and patient survival rate [1-11]. Recently, increasing emphasis has been placed on the importance of maintaining or improving the quality of life for the patient [12-15], with the most cost-effective treatment [16-19]. Modern surgical techniques have resulted in an aggressive approach of distal arterial reconstruction with high patency rates and significantly reduced amputation rates [3,4,20,21]. However, such operations often require multiple reinterventions to attempt to preserve patency or treat wound complications [4,22,23]. This means that for many patients, once limb-threatening ischemia occurs, the need for treatment is life-long, and the functional outcome and quality of life might be of much greater importance to the patient than the technical result [24]. Successful revascularization and limb salvage with subsequent deterioration in the lifestyle of the individual should temper contemporary enthusiasm for limb salvage at all costs [13,25]. Because there may be a group of patients who could benefit more from primary amputation rather than femorocrural grafting, it is important to identify preoperatively those patients most likely to benefit from arterial reconstruction, as well as those in whom the procedure is not likely to be successful [26-29].

This chapter reviews the experience of long femorocrural bypasses for critical lower limb ischemia. The analysis was undertaken to determine factors that affect graft patency, limb salvage, quality of life, and the hospital costs of these procedures.

■ GENERAL CONSIDERATIONS

There are enormous discrepancies between the reported results of femorocrural reconstructions from different centers. Much of this is due to the inappropriate inclusion of patients with non-critical limb ischemia. Other times, the definition of *distal* or *infrapopliteal* is also controversial. Unfortunately, even standardized methods of reporting have not been uniformly adopted, and the results are sometimes reported prematurely with an insufficient number of grafts at risk. Consequently, the patency rate may be compromised by a large standard error. In this review we have only considered those series that report well documented information about patients with critical limb ischemia and treatment by femorotibial or femoroperoneal bypass. In addition, we report our clinical experience with 235 bypass grafts to crural vessels performed in 195 patients with rest pain or gangrene between 1987 and 1997. Saphenous vein grafts were used in 165 cases and 70 prosthetic grafts (PTFE 6 mm) were implanted either because of absence of saphenous vein or because the presence of unsuitable autologous vein of sufficient quality to provide a long femorotibial bypass. Adjunctive vein cuffs were created in 42 PTFE prostheses. It is important to remember that comparison of patencies between reported series is affected by demographic characteristics. In addition, primary patency underscores the performance of a graft in a particular location, whereas secondary patency denotes the determination of the surgeons caring for the patient to accomplish limb salvage. Because of this, only actuarial primary patency curves were computed in our analysis using the Kaplan-Meier method and compared with the Mantel-Haenszel log-rank test. Multivariate Cox regression analysis was developed to identify independent risk factors affecting graft patency, limb salvage and mortality.

■ GRAFT PATENCY RELATED TO GRAFT MATERIAL

Autogenous Vein

Multiple reports over the last decade have documented the clear superiority of autogenous vein grafts over prosthetic reconstruction to the crural vessels (Table I). Primary and secondary cumulative patencies for autologous vein bypass ranged from 36% to 85%, and from 64% to 85%, respectively, after 3 years of follow-up. At 5 years, primary patency rates ranged from 32% to 76%, and secondary patency rates ranged from 73% to 81%. These results show that reversed and in-situ vein grafts are equally effective for femorocrural bypass. The impact of vein graft surveillance is reflected in some modern series with differences between primary and primary assisted patency rate [8,26]. In patients with unavailable greater saphenous vein (Table II), alternative sources including arm and lesser saphenous vein have been used with acceptable results [5], although the five-year patency rate has been significantly inferior (52% vs. 67%, in Taylor's series) [20]. In the absence of sufficient length of saphenous vein for conventional femorotibial bypass, a shorter length of residual vein originating from a patent distal superficial femoral artery or popliteal artery has provided satisfactory patency rates [38,39].

Prosthetic Bypass

PTFE grafts have been regarded as a poor alternative at the more distal site (Table III). In the most impressive series of 211 tibial bypass grafts for limb salvage with ringed PTFE prostheses, Schweiger et al. [2] reported a primary patency rate of 36.6% at 2 years and 22.6% at 5 years. The secondary patency rates were 45.4% and 25.4%, respectively. Two techniques using a segment of vein, the Miller cuff [23,50,51] and the Taylor patch [52], have been credited with surprisingly good results in the inventor's own series. High success rates were reported by Taylor et al. [52] in a series of 83 tibioperoneal PTFE grafts. They used distal anastomotic vein patches and achieved a primary patency rate at 1, 3, and 5 years of 74%, 58%, and 54%, respectively (Table IV). Likewise, the use of an interposition vein collar has been advocated to improve the patency of femorocrural bypasses. Recently, Wijesinghe et al. [53] reported a primary and secondary cumulative patency rate of 64% and 66% after 1 year; and 51% and 52%, respectively, after 2 years of follow-up. The value of an arteriovenous fistula (AVF) at the distal

Table I PATENCY AND LIMB SALVAGE RATES OF AUTOLOGOUS VEIN BYPASSES TO THE CRURAL VESSELS

1st author [ref.]	*Year*	*Graft N*	*CLI %*	*Crural Bypass %*	*1 yr %*	*2 yrs %*	*3 yrs %*	*4 yrs %*	*5 yrs %*
Veith [1]	1986	106	76.1	100					
Primary patency rate					-	-	-	49	-
Limb salvage rate					-	-	-	61	-
O'Mara [30]	1987	182	100	100					
>80 yrs		*40*		(vein 77.5)					
Primary patency rate					88	-	56	-	-
Limb salvage rate					91		81	-	-
<80 yrs		*142*							
Primary patency rate					86	-	85	-	-
Limb salvage rate					89	-	89	-	-
Rutherford [28]	1988	77	98	100					
Primary patency rate					85	85	78	-	-
Shah [9]	1988								
Diabetic		*387*							
Primary patency rate					85	81	77	74	74
Limb salvage rate					96	-	86	-	72
Non-Diabetic		*294*							
Primary patency rate					87	84	81	76	76
Limb salvage rate					99	-	94	-	70
Feinberg [7]	1990	57	96.5	100					
Primary patency rate					82	64	-	-	-
Limb salvage rate					71	-	-	-	-
Londrey [6]	1991	175	92	88					
Primary patency rate					70	65	63	63	63
Limb salvage rate					-	-	82	-	-
Donaldson [8]	1991	191	68	100					
Primary patency rate					-	-	-	-	63
Primary assisted					-	-	-	-	71
Secondary patency rate					81	81	78	78	78
Limb salvage rate					-	-	-	-	85
Taylor [20]	1991	203	100	100					
Primary patency rate					88	-	70	-	67 saphenous 52 other veins
Limb salvage rate					94	-	84	-	84
Bergamini [31]	1991	361	93	68					
Primary patency rate					73	70	66	63	63
Secondary patency rate					90	87	85	81	81
Limb salvage rate					94	91	90	89	86
Cheshire [17]	1992	89	100	100					
Primary patency rate					-	-	54	-	-
Secondary patency rate					-	-	72	-	-
Tordoir [27]	1993	141	100	78.7					
Primary patency rate				(Vein 82.2)	67	-	55	-	-
Secondary patency rate					75	-	64	-	-
Limb salvage rate					-	-	87	-	-

CLI : critical limb ischemia

(Table I continues on next page)

Table I PATENCY AND LIMB SALVAGE RATES OF AUTOLOGOUS VEIN BYPASSES TO THE CRURAL VESSELS

1st author [ref.]	*Year*	*Graft N*	*CLI %*	*Crural bypass %*	*1 yr %*	*2 yrs %*	*3 yrs %*	*4 yrs %*	*5 yrs %*
Morris [23]	1993	106	96	100					
Primary patency rate					52	-	36	-	32
Limb salvage rate					74	-	65	-	-
Sayers [26]	1993	78	62	77					
Primary patency rate					33	-	29	-	-
Primary Assisted					63	-	57	-	-
Secondary patency rate					63	-	64	-	-
Limb salvage rate					-	-	67	-	-
Schneider [11]	1993	256	100	79.3					
Primary patency rate					-	-	61	-	-
Primary Assisted					-	-	75	-	-
Secondary patency rate					-	-	79	-	-
Limb salvage rate					-	-	87	-	-
Chang & Stein [32]	1995	114	100	100					
Primary patency rate					79	75	75	68	68
Secondary patency rate					83	81	79	73	73
Limb salvage rate					-	-	-	-	86 at 8 years
Luther [33]	1997	187	100	100					
Primary patency rate					57	53	47	-	-
Limb salvage rate					81	-	71	-	-
Panayiotopoulos [16]	1997	109	100	82.5					
Primary patency rate				(Vein 55)	-	-	27	-	-
Primary Assisted					-	-	31	-	-
Secondary patency rate					-	-	45	-	-
Limb salvage rate					-	-	53	-	-
Current series	1998	165	100	100					
Primary patency rate					78	68	64	62	57
Limb salvage rate					82	75	72	70	70

CLI: critical limb ischemia

anastomosis as a means of achieving improved PTFE patency to tibial arteries remains yet unproven and contradictory. Dardik et al. [54] reported a 2-year patency rate of 44% in 69 patients with AVF's in prosthetic or umbilical vein grafts to the infrapopliteal arteries. Paty et al. [55] achieved a 1-year patency rate of 67% and limb salvage rate of 75% with PTFE crural bypasses, and Jacobs et al. [56,57] showed similar short-term results. Other authors have not been able to demonstrate that the arteriovenous fistula confers any advantage [43]. The recommendation of using prosthetic grafts at crural level with an adjunctive procedure (AVF, vein angioplasty, or a vein cuff) has not been evidenced by prospective randomized trials performed in this setting.

Composite Grafts

In contrast to autogenous vein and prosthetic reconstructions, less data are available to document the patency of composite grafts to the tibial level (Table V). Primary cumulative rates ranged from 28% to 72% after 1 year; and from 18% to 54% after 3 years of follow-up. Recently, Karacagil et al. [58] reported

Table II PATENCY AND LIMB SALVAGE RATES OF AUTOLOGOUS ARM OR LESSER SAPHENOUS VEIN BYPASSES TO THE CRURAL VESSELS

1st author [ref.]	*Year*	*Graft N*	*CLI %*	*Crural bypass %*	*1 yr %*	*2 yrs %*	*3 yrs %*	*4 yrs %*	*5 yrs %*
Harward [34]	1992	43	93	79					
Primary patency rate					67	58	49	-	-
Seconday patency rate					74	64	64	-	-
Limb salvage rate					74	-	63	-	-
Hölzenbein [35]	1996	250	99.2	40					
Primary patency rate					70.6	-	49.5	-	-
Seconday patency rate					76.9	-	52.8	-	-
Limb salvage rate					88.2	-	-	80.4	-
Sesto [36]	1992	35	71	71.4					
Primary patency rate					49	49	40	-	-
Secondary patency rate					68	58	44	-	-
Limb salvage rate					82	-	82	-	-
Chalmers [37]	1993	42	67	64					
Primary patency rate					51	46	34	-	-
Secondary patency rate					85	85	76	-	-
Londrey [5]	1994		88	76					
Single segment		*169*							
Primary patency rate					-	-	-	-	-
Secondary patency rate					78	-	60	-	52
Multisegment		*88*							
Secondary patency rate					56	53	39	-	29
Overall limb salvage rate					85	80	74	-	69

CLI : critical limb ischemia

that a composite PTFE-vein graft seems to yield the same results as a PTFE with a distal Miller cuff. Regarding composite-sequential grafts in femorocrural bypasses, analysis of patency after below-knee and above-knee intermediate PTFE anastomoses has revealed a trend toward better patency with above-knee anastomoses, although this fact never has reached statistical significance [59].

ALLOGRAFTS

Cryopreserved allograft veins have been used with very poor results in patients with unsuitable autogenous vein (secondary patency of 36% at 1 year) [63]. In contrast, encouraging results have been reported with the use of cryopreserved arterial allografts (primary patency rate ranged from 60% to 76% at 1 year and from 43% to 46% at 2 years), although the number of grafts at risk available for late follow-up was quite small [61,62]. This may decrease the reliability of conclusions drawn regarding these groups.

■ LIMB SALVAGE RATES

Limb salvage rates theoretically should meet or exceed the patency rates of femorocrural bypass grafts when ischemia is the primary cause of limb threat. Because of this, lower extremity salvage rates were

Table III PATENCY AND LIMB SALVAGE RATES OF PTFE GRAFTS FOR FEMORODISTAL BYPASS PROCEDURES

1st author [ref.]	*Year*	*Graft N*	*PTFE %*	*CLI %*	*Crural bypass %*	*1 yr %*	*2 yrs %*	*3 yrs %*	*4 yrs %*	*5 yrs %*
Cranley [40]	1982	13	100							
Primary patency rate						35	35	35	35	-
Christenson [41]	1985	54	100							
Primary patency rate						54	52	49	49	49
Hobson [42]	1985	41	100							
Primary patency rate						21	12	12	12	12
Veith [1]	1986	98	100	76.1						
Primary patency rate						48	35	33	12	-
Limb salvage						-	-	-	57	-
Rafferty [43]	1987	76	100							
Primary patency rate						31	21	19	17	14
Rutherford [28]	1988	35	40	94	100					
PTFE		*14*								
Primary patency rate						25	25	17	-	-
*HUV**		*21*								
Primary patency rate						7	7	7	-	-
Flinn [44]	1988	75								
Primary patency rate						58	45	37	37	-
Tyrrell [45]	1989	30	100							
Primary patency rate						40	-	-	-	-
Whittemore [46]	1989	21	100							
Primary patency rate						25	12	12	12	12
Johnson [47]	1991	37	100							
Primary patency rate						37	26	21	7	-
Londrey [6]	1991	33	100	92	88					
Primary patency rate						38	26	7	7	7
Quiñones-Baldrich [48]	1992	28								
Primary patency rate						44	40	22	22	22
Limb salvage						-	-	37	-	-
Schweiger [2]	1993	211	100	100	100					
Primary patency rate						46	-	30.9	-	22.6
Secondary patency rate						60	-	38.1	-	25.4
Limb salvage						-	63	-	-	51
Fichelle [49]	1995	31								
Primary patency rate						61	48	43	42	31
Parsons [3]	1996	66								
Primary patency rate						71	47	39	36	28
Secondary patency rate						78	61	55	51	43
Limb salvage						84	71	71	66	66
Current series	1998	28								
Primary patency rate						45	35	21	10	-
Limb salvage						45	35	28	18	9

* HUV: human umbilical vein

Table IV PATENCY AND LIMB SALVAGE FOR PTFE GRAFTS AND ADJUNCTIVE PROCEDURES

1st author [ref.]	*Year*	*Graft N*	*Vein PTFE %*	*CLI %*	*Crural bypass %*	*1 yr %*	*2 yrs %*	*3 yrs %*	*4 yrs %*	*5 yrs %*
Jacobs [52]	1992	30	PTFE	100	93.4					
Primary patency rate			&			71	71	-	-	-
Limb salvage			AVF			-	-	83	-	-
Cheshire [17]	1992	41	100	100	100					
Primary patency rate						-	-	38	-	-
Secondary patency rate						-	-	48	-	-
Wijesinghe [53]	1998	51	100	100	100					
Primary patency rate						64	-	51	-	-
Secondary patency rate						66	-	52	-	-
Limb salvage						85-80	-	-	-	-
Wolfe [4]	1991	55								
Primary patency rate						-	-	52	-	-
Taylor [52]	1992	83			100					
Primary patency rate						74	64	58	-	54
Morris [23]	1993	201								
Primary patency rate						51	40	29	-	-
Limb salvage						74	-	64	-	-
Fichelle [49]	1995	65								
Primary patency rate						63	54	45	-	45
Current series	1998	42								
Primary patency rate						52	41	34	25	12
Limb salvage						67	63	58	-	-

AVF: arteriovenous fistula

better with autogenous vein grafts than with prosthetic or composite-vein grafts. Limb salvage rate for autogenous veins ranged from 65% to 94% after 3 years, and from 70% to 86% after 5 years of follow-up, whilst for prosthetic grafts the limb salvage rates in the best series [3] were 71% and 66%, respectively. Limb salvage rate for composite-vein grafts ranged from 59% to 84% after 2 years of follow-up. Although the use of prosthetic grafts results in significantly lower patency rates, it appears to be effective for limb salvage. On the other hand, there are some patients with limb loss despite a patent femorocrural bypass graft. The most common causes of failure of foot salvage despite a patent bypass have been large and deep infection or extensive gangrene in patients with end-stage renal disease or diabetes [20,63].

■ MORTALITY AND SURVIVAL RATES

After femorocrural bypass the 30-day mortality rate ranged from 2% [64] to 15% [23], with an average rate of 7%. At 5 years, survival rate ranged from 38.4% in Taylor's series [20] to 75% in Wijesinghe's series [53]. Many of the gross differences among reported results are due to the population under study. These differences reflect the severity of atherosclerosis, the advanced age of the patients or the inappropriate inclusion of patients with non-critical limb ischemia. After bypass surgery the relative risk

Table V PATENCY AND LIMB SALVAGE FOR PTFE COMPOSITE VEIN BYPASSES TO THE CRURAL LEVEL

1st author [ref.]	*Year*	*Graft N*	*PTFE %*	*CLI %*	*Crural bypass %*	*1 yr %*	*2 yrs %*	*3 yrs %*	*4 yrs %*	*5 yrs (%)*
Feinberg [7]	1990	108		96.6	100					
PTFE		*87*								
Primary patency rate						34.6	29.9	-	-	-
Limb salvage							62.3	-	-	-
*HUV**		*21*								
Primary patency rate						12.1	6	-	-	-
Londrey [6]	1991	45	100	92	88					
Primary patency rate						49	44	37	28	28
Limb salvage						-	-	-	-	82
McCarthy [59]	1992	67	100	100	100					
Primary patency rate						72	64	48	40	-
Limb salvage						-	84	-	70	-
Chang [32]	1995	103	100	100						
Composite - Grafts		*48*	*100*							
Primary patency rate						68	69	54	38	38
Secondary patency rate						82	84	67	58	58
Limb salvage						-	-	-	-	53 at 8 yrs
Composite - Sequential		*55*	*100*							
Primary patency rate						68	47	35	35	35
Secondary patency rate						82	69	62	62	52
Limb salvage						-	-	-	-	65 at 8 yrs
Fichelle [49]	1995	53		100						
Primary patency rate						58	44	36	33	29
Karacagil [58]	1996	94	100	85.3	100					
Primary patency rate						28	23	18	-	-
Limb salvage						63	59	55	-	-

* HUV: human umbilical vein

for mortality was significantly less for the patent and the non-amputated patients as compared to those amputated [65]. The results once again stress that amputation is the main predictor of mortality. The survival was also lowest in the gangrene group (31% vs. 72% at 5 years). In a multivariate analysis [33], presence of ischemic tissue lesions (ulcer or gangrene) increased the mortality risk by a factor of 2.7 (CI: 1.5-5.1).

■ FACTORS AFFECTING LONG-TERM GRAFT PATENCY

The influence of a number of factors on the patency of femoroperoneal and femorotibial bypasses has been established, while the role of other factors remains uncertain. Although factors such as smoking at the time of operation [26,28], ankle/brachial index [26], placement of the graft (anatomic or subcutaneously) [66], and short- or long-term medication (aspirin vs. warfarin) [3,44] have been shown to be important, most reports have failed to show statistical significance. Multivariate analysis in our series showed that the use of prosthetic graft material (regression coefficient 0.2580, $p < 0.01$) and the occlusion of pedal vessels (regression coefficient 0.1645, $p < 0.01$) were the only independent factors correlating with graft failure. Although limbs with good pedal circulation showed significantly better patency rates compared to limbs with poor pedal circulation, the significance of this variable was dependent on the conduit graft utilized, since the patency of autologous vein bypasses were less affected (Figures 1-3). The presence of gangrene, end-stage renal disease, and diabetes had a significant effect on mortality (Fig. 4).

Inflow State

The inflow site had no impact on the results in the majority of series published [6,20,27], which is in agreement with the fact that the grafts may originate from a distal site provided there is no proximal hemodynamically significant stenosis. The condition of the inflow vessels was exceptionally important in the report of Panayiotopoulos et al. [16] and Shortell et al. [67]. They found that even apparently modest disease of the inflow tract had an adverse effect on the patency rate (56% vs. 16% at 2 years).

Distal Runoff and Patency of Pedal Vessels

Utilizing completion arteriography, O'Mara et al. [68] demonstrated that patency of femorotibial and femoroperoneal grafts was greatly improved when at least one of the two pedal arches was patent. In Schweiger's series [2] an occluded plantar arch resulted in a primary patency of 21.6% at 1 year, compared with a 66.9% in patients with a complete arch. Panayiotopoulos et al. [16] reported a 58% graft patency at 3 years compared with almost complete failure at 1 year (6% patency) for grafts associated with diseased pedal vessels. Karacagil et al. [58], in a series of composite-grafts to femorocrural vessels, reported a primary patency of only 4% at 1 year in limbs with poor runoff. In contrast, the impact of arch integrity on the patency of femorocrural reconstruction has not been confirmed by other authors [27].

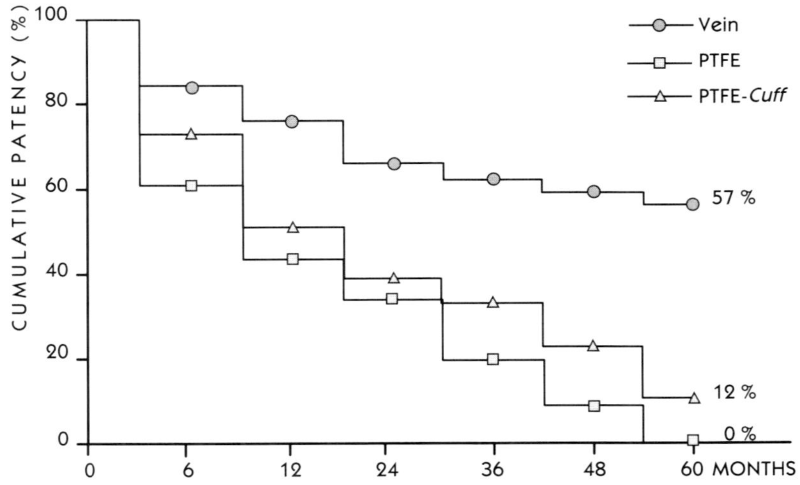

PATIENTS AT RISK

Vein grafts (n=165)	117	93	50	28	21	7
PTFE grafts (n=28)	15	10	6	2	1	0
PTFE-cuff grafts (n=42)	25	16	8	(4)	(2)	(1)

FIG. 1 Life-table with cumulative primary patency rates for all vein grafts, PTFE grafts, and PTFE with adjunctive vein cuff in our series. Number at risk in parentheses indicates a standard error exceeding 10%. Difference between patency of vein grafts and PTFE grafts is significant ($p < 0.001$).

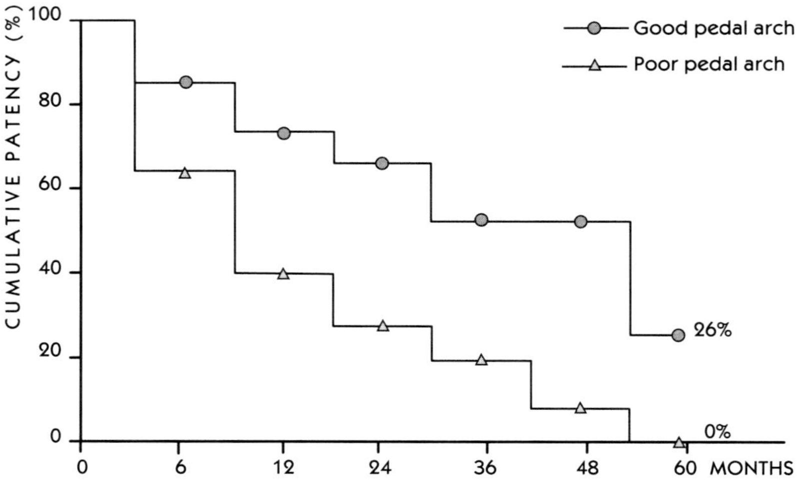

PATIENTS AT RISK

Prosthetic grafts with good pedal arch (n=21)	*15*	*(11)*	*(7)*	*(2)*	*(2)*	*1*
Prosthetic grafts with poor pedal arch (n=49)	*25*	*15*	*7*	*4*	*1*	*0*

FIG. 2 Life-table of primary patency rates for prosthetic grafts with good pedal arch compared with prosthetic grafts with poor pedal arch ($p < 0.001$). Number at risk in parentheses indicates a standard error exceeding 10%.

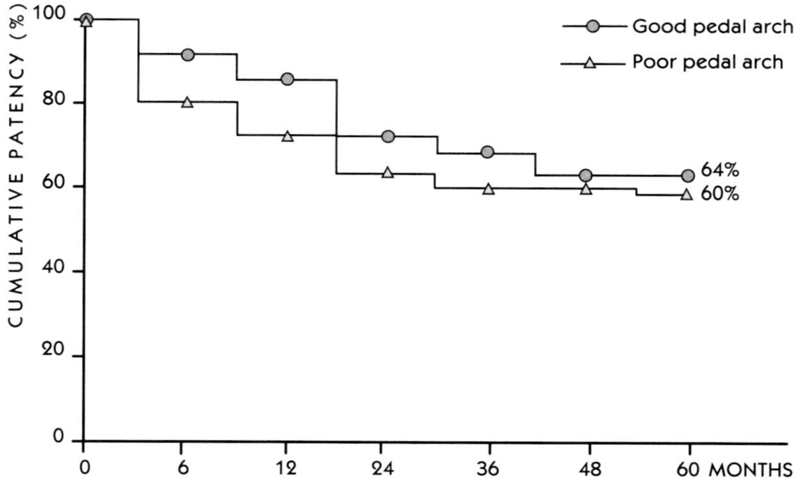

PATIENTS AT RISK

Vein grafts with good pedal arch (n=67)	*51*	*40*	*22*	*(13)*	*(11)*	*(6)*
Vein grafts with poor pedal arch (n=98)	*66*	*53*	*28*	*15*	*(10)*	*(1)*

FIG. 3 Life-table of primary patency rates for vein grafts with good pedal arch compared with vein grafts with poor pedal arch. Number at risk in parentheses indicates a standard error exceeding 10%.

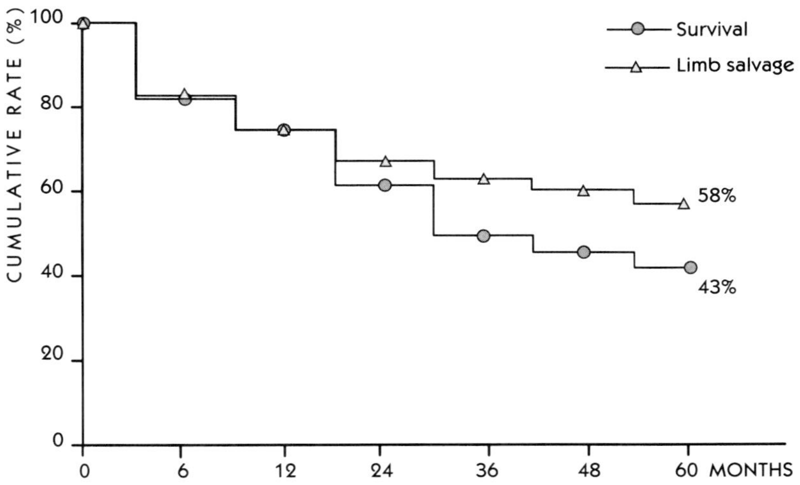

PATIENTS AT RISK						
Survival (n=195)	*160*	*142*	*102*	*77*	*70*	*47*
Limb salvage (n=235)	*164*	*130*	*80*	*51*	*42*	*27*

FIG. 4 Limb salvage and survival rate for femorocrural bypass operations in our series.

Primary and Secondary Procedures

Given that vein grafts, either in-situ or reversed, are mostly used in primary procedures, secondary prosthetic grafts markedly affect outcome, since they were performed in more complicated circumstances. In 1985 Ascer et al. [69] reported a 40% patency rate at 2 years in a series of 207 tibial PTFE grafts, and in 1987 the same group [70] found only a 17% patency rate for secondary grafts at 2 years. Veith et al. [1] showed a 18% patency rate at 3 years when PTFE graft was obligatory used because the vein was unavailable, in comparison with a 27% patency rate in a randomized group at 3 years follow-up. In Schweiger's series [2], primary patency rates for primary and secondary procedures was 52.4% and 21.6% after 2 years; and 42% and 13.8%, respectively, after 4 years of follow-up.

Site of Distal Anastomosis

The majority of the studies analyzed [6,16,20,26,27,67] showed that the site of distal anastomosis or the level at which the anastomosis was performed had no effect on graft patency or limb salvage, provided that the graft is placed beyond all areas of disease. Only Morris et al. [23] reported that patency was significantly higher when anastomosed to the tibioperoneal trunk and peroneal artery compared to the posterior tibial artery, although they did not use peripheral resistance measurements or arteriographic correlation with the status of the distal outflow vessels.

Indication for Surgery

Results from the *Swedish Vascular Registry* [65] showed that the patients with gangrene had significantly more amputations as compared to those with rest pain and ischemic ulcers both at 1 month and at 1 year (10.8% vs. 4.4%, and 24.9% vs. 12.1%, respectively, $p< 0.0001$). Luther and Lepäntalo [33] reported in a study with 187 autogenous vein reconstructions that the postoperative occlusion rate (30-days) for patients with gangrene was higher than for patients with rest pain or ischemic ulcer (23% vs. 6% and 8%, respectively). Thereafter the patency was not influenced by the severity of the preoperative ischemia. Similarly, the short-term amputation rate for patients with gangrene was higher than for patients with rest pain and ulcer (28% vs. 10% and 17%, respectively). Regarding the use of prosthetic grafts, Schweiger et al. [2] reported that PTFE grafts of patients operated upon for necrosis or gangrene showed worse results beyond the first year as compared with results in PTFE grafts operated on for rest pain. Therefore, differences in outcome for these various stages of critical limb ischemia are usually recorded, and it may be reasonable to describe the results of a reconstruction separately for these groups [65,71].

Age and Gender

Comparison of the older and younger groups has shown no statistically significant difference in rates of operative mortality, limb salvage or graft patency [26,27,30,72]. However, in the *Swedish Vascular Registry* [65] a higher mortality rate was found in the older group and the rate of patients *alive and improved* at 1 month and at 1 year was significantly lower.

The majority of series did not identify gender as an independent risk factor for either patency, limb salvage or survival [26,27,64,67]. Recently, Luther and Lepäntalo [33] have shown in a multivariate analysis that the combination of female sex and diabetes independently had a negative influence. For non-diabetic women, patency was 60% at three years, while it fell to 52% already at 18 months for diabetic women ($p < 0.05$). For non-diabetic women leg salvage was 90% at 3 years and for diabetic women 48% at 18 months.

Diabetes

In a large series of Shah et al. [9] comprising 681 femorotibial arterial reconstructions for limb salvage, the outcome in diabetic patients (n= 387) was essentially identical to that seen in non-diabetic patients (n= 294). Primary patency and limb salvage in diabetic patients was 90,8% and 96% after one year; and 74.1% and 86% after five years of follow-up. Primary patency and limb salvage in non-diabetic patients during the same period of time was 89.7% and 99%; and 76.4% and 94%, respectively. Other authors [6,8,20,26,27,64,67] have confirmed these results. The majority of series have reported that the operative mortality and long-term survival of non-diabetics is significantly better than for diabetics [33]. In the Panayiotopoulos' series [18], the survival rate was similar at 3 years, but in diabetic population it fell to 17% at 4 years compared with 60% in non-diabetic patients. In addition, more diabetic than non-diabetic patients were subjected to amputation. Tordoir et al. [27] reported a significant higher amputation rate of 25% compared to 11% in non-diabetic patients.

End-Stage Renal Disease

The presence of renal failure was associated in all series with a significant negative effect on the limb salvage rates, despite a patent vascular reconstruction [20]. Difficulties with healing and an increased rate of postoperative infections represent the main reasons for the higher amputation rate. In the series of Johnson et al. [73] the limb salvage rate at 1 year decreased from 74% to 51% in patients admitted with gangrene. The results seem to be different if we consider patients not requiring dialysis (moderate renal failure) and patients requiring dialysis (end-stage renal disease). Peltonen et al. [74] reported patency and leg salvage rates at one year of 81% and 79% in non-dialysis patients, and 47% and 37% in dialysis patients. The patient survival rate at 1 year was 68% in non-dialysis patients and 10% in dialysis patients. None of the patients in dialysis was alive with salvaged legs 4 months after revascularization.

■ REINTERVENTIONS

In a retrospective analysis of 129 reconstructions to crural vessels, Cheshire et al. [22] reported that secondary operations were required in 67% after 3 years of follow-up, with a mean of two operations per patient. Considering reintervention alone, 72% of operations were performed for recurrence of ischemia in the operated leg. Irreversible failure of femorocrural grafts resulted in amputation in 21% of cases, unsuccessful reintervention was reported in 45% of cases, and successful secondary intervention in 34% of cases. Failure to reintervene after femorocrural graft failure would have reduced overall limb salvage from 68 to 48% with loss of 22 salvaged limbs. Similarly, no significant differences between primary PTFE and subsequent PTFE reconstructions were reported by Morris et al. [23]. These authors believe that the success of reintervention shown in their studies justifies a policy of an aggressive approach to the management of recurrent ischemia after arterial grafting. On the contrary, Dennis et al. [75] reported that only one-third of their secondary PTFE bypass grafts to tibial vessels (8 of 24) remained patent for a mean interval of 14.8 months, and the 52% of patients were dead at a mean follow-up of 27.5 months. Consequently, they suggest that an amputation should be considered if prosthetic material is the only choice to treat patients requiring reintervention.

■ COST

Cheshire et al. [17] in a series of femorocrural reconstruction for critical leg ischemia, showed that the mean cost of revascularization including secondary operations was 44% cheaper than the cost of amputation when autologous vein was used and 6% cheaper than amputation when PTFE grafts were used. In the Panayiotopoulos series [16] the mean cost of successful revascularization was 4 400 dollars and the cost of a primary amputation was 12 730 dollars, while a failed graft leading to a major amputation accounted for 17 500 dollars. Luther [19] reported that an amputation of a previously mobile patient was about twice as expensive on a cost/year level as a reconstruction, including also those followed by an amputation. In contrast, for patients who were non-ambulatory or who lived in nursing homes an amputation was often an inevitable and the cheapest treatment (Fig. 5).

■ QUALITY OF LIFE

The few studies that have assessed functional outcomes in populations with limb-threatening ischemia have been limited by a too short follow-up period [12-15]. Predictably, and in keeping with the experience of others, mobility was better in patients after limb salvage than after primary amputation. Preoperative independence and ambulation have been reported as important predictors of

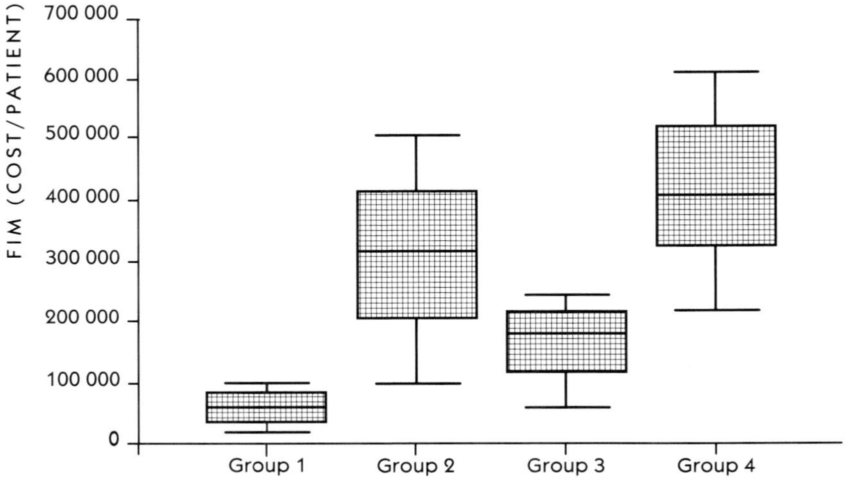

FIG. 5 Mean costs associated with institutionalized patients undergoing primary amputation (Group 1), independently living patients undergoing primary amputation (Group 2), patients with successful reconstruction (Group 3), and patients with failed reconstruction leading to amputation (Group 4). *(Reproduced with permission, from M. Luther. Surgical treatment for chronic critical leg ischaemia: a 5-year follow-up of socioeconomic outcome. Eur J Vasc Endovasc Surg 1997;13: 452-459).*

postoperative independence and ambulation [25]. However, only 68% of patients who were ambulatory and living independently before surgery maintained this status at the time of last follow-up. In addition, previous functional outcome assessments have not directly considered the adverse long-term effects of delayed wound healing, episodes of recurrent ischemia, and the need for repeat operations. In the Nicoloff series [24], only 14% of patients had an uncomplicated operation, relief of symptoms, uncomplicated wound healing, no need for repeat surgery, and maintenance of functional status. For 86% of patients a large portion of the final months of life was occupied with treatment for limb ischemia.

■ DISCUSSION

The major determinant impacting on patency of femorocrural bypass grafts is unquestionably the conduit utilized; the patency of autogenous saphenous vein grafts of any type is far superior to that of prosthetic grafts [1,6,16,27,28,67]. The superiority of the autogenous saphenous vein in the reversed or in-situ position as the conduit of choice for infrapopliteal vascular reconstruction is widely acknowledged [8,9,20]. This is particularly true for long bypasses to tibial vessels with compromised runoff. Because of this, every effort should be made to establish and preserve vein graft including use of alternative vein sources (arm and lesser saphenous vein), distal graft origins (from a patent distal superficial femoral artery or popliteal artery), and intensive screening programs to enhance graft longevity [6]. Only after all of the above options are exhausted, we consider a prosthetic bypass to a crural vessel. Prosthetic grafts to tibial arteries perform so badly that many surgeons believe these operations should not be performed in the absence of vein [42]. Several techniques using a segment of vein (composite-vein graft [7,53,59], the Miller cuff [50,51] and the Taylor patch [52]) have been credited with surprisingly good results. Although these techniques have improved the patency rate of PTFE grafts to crural vessels, our results remain 25 to 30% worse than those of autologous saphenous vein.

On the other hand, one of the most important factors affecting the results of femorocrural reconstructions is the status of the distal outflow vessels [2,16]. Similar to the popliteal trifurcation in femoropopliteal bypass, the status of small arteries of the foot may be the best indicator of runoff for bypass grafting to the crural vessels [68]. The integrity of the plantar arch has been suggested as a reliable predictor of tibial graft patency [29]. Indeed, the number of patent tibial arteries did not significantly influence patency when the angiography showed at least a single patent calf vessel [2,52].

High-quality arteriography, with particular emphasis on the foot, is absolutely essential for operative planning and appropriate decision making [76]. The inability of conventional preoperative arteriography to identify patent vessels below the knee in patients with critical limb ischemia is well known, and we prefer to use liberal intraoperative and selected repeat arteriography to identify distal vessels. Inadequate arteriography may fail to visualize anatomic features favorable for limb salvage and mistakenly deem the extremity as non-reconstructable [64]. If any concern remains regarding the adequacy of the arteriogram we prefer to do a surgical exploration of the vessels.

The poor long-term survival of patients with critical ischemia suggests that, for the majority, concern over long-term patency is irrelevant. However, the more distal the femorotibial reconstruction, the closer the patency rate and the limb salvage rate become. This may be different for tibial bypasses that originate in the popliteal artery where failure of the reconstruction may be tolerated by the extremity as a result of existing collateral flow [3]. Patency and limb salvage are evidently not always the same, patent reconstruction without limb salvage are reported, as well as limb salvage with occluded grafts [33]. There are some patients who need a graft for a limited time only until peripheral lesions have healed [1,3]. There are, however, patients who undergo successful revascularization but fail to heal and still progress to major lower extremity amputation [20]. This is especially true for diabetic patients with large soft tissue infection [18,27] and patients in dialysis with end-stage renal disease [20,73,74]. In these circumstances in which deep extensive infection or gangrene of the heel are present primary amputation should strongly be considered if we are to avoid the most costly form of amputation (i.e., bypass grafting followed by amputation) [63].

The main parameters affecting costs in femorocrural reconstruction are the length of hospital stay and the number of secondary procedures to maintain or restore graft patency [16,17]. Reconstruction to crural vessels is associated with a higher reintervention and secondary amputation rate than more proximal operations. The value of continuing attempts to prevent amputation once the graft has failed remains contentious. While some authors [22,23] recommend a policy of aggressive approach to the management of recurrent ischemia, others [61] suggest that if prosthetic material is the only choice to treat patients requiring secondary surgical procedures to the tibial arteries an amputation should be considered, because of the higher amputation rate and poorer life expectancy.

The issue of costs versus benefit of revascularization when compared with amputation has been addressed by several studies [20]. Successful revascularization is consistently less expensive than amputation, although failed revascularization followed by amputation is the most expensive alternative [16]. The significantly higher hospital cost of failed grafts resulting in amputation suggests that it is important to identify the patients who are not likely to benefit from revascularization, in whom primary amputation would be beneficial. Based in the socioeconomic findings reported by Luther [19], in which later care costs were calculated, there is no economic reason to perform an amputation in preference to a revascularization in independently living patients. In contrast, for institutionalized, immobile patients an amputation is often an inevitable and cheapest treatment.

Although published amputation series of lower-extremity ischemia cannot be correctly compared with revascularizations series because of the obvious differences in patient selection [20], revascularization apparently reduces mortality compared to amputation [65]. These data support the view that arterial reconstruction to crural vessels could be justified in all circumstances in preference to amputation. However, increasing emphasis has been placed on the importance of maintaining the functional status after treatment of leg threatening ischemia. In a recent study, Abou-Zamzam et al. [25] found that nearly all patients who were ambulatory and independent before bypass surgery for limb salvage maintained this function as long as their bypass graft remained patent. Not surprisingly, patients who were non-ambulatory or who lived in nursing homes before their limb threat retained this status postoperatively. Thus, preoperative independence and ambulation are important predictors of postoperative independence and ambulation. An aggressive policy of arterial reconstruction for limb salvage in ambulatory patients is commendable. However, arterial reconstruction of a threatened or gangrenous extremity in an immobile patient in a nursing home makes little sense. Although there is something to be said about the psychological and cosmetic benefit of leg salvage, the relative merit of leg preservation at all costs under these circumstances is highly debatable [77].

CONCLUSION

The decision to attempt a femorocrural bypass should be based on the individual lifestyle requirements of each patient in addition to factors relating to technical aspects of surgical approach to limb salvage [13], basically the availability of a venous conduit and the status of the distal outflow vessels. High-quality arteriography with particular emphasis on the foot is absolutely essential for operative planning and appropriate decision making.

REFERENCES

1 Veith FJ, Gupta SK, Ascer E et al. Six-year prospective multicenter randomized comparison of autologous saphenous vein and expanded polytetrafluoroethylene grafts in infrainguinal arterial reconstructions. *J Vasc Surg* 1986; 3: 104-114.

2 Schweiger H, Klein P, Lang W. Tibial bypass grafting for limb salvage with ringed polytetrafluoroethylene prostheses: results of primary and secondary procedures. *J Vasc Surg* 1993; 18: 867-874.

3 Parsons RE, Suggs WD, Veith FJ et al. Polytetrafluoroethylene bypasses to infrapopliteal arteries without cuffs or patches: a better option than amputation in patients without autologous vein. *J Vasc Surg* 1996; 23: 347-356.

4 Wolfe JHN, Tyrrell MR. Justifying arterial reconstruction to crural vessels - even with a prosthetic graft. *Br J Surg* 1991; 77: 897-899.

5 Londrey GL, Bosher LP, Brown PW et al. Infrainguinal reconstruction with arm vein, lesser saphenous vein, and remnants of greater saphenous vein: a report of 257 cases. *J Vasc Surg* 1994; 20: 451-457.

6 Londrey GL, Ramsey DE, Hodgson KJ et al. Infrapopliteal bypass for severe ischemia: comparison of autogenous vein, composite, and prosthetic grafts. *J Vasc Surg* 1991; 13: 631-636.

7 Feinberg RL, Winter RP, Wheeler JR et al. The use of composite grafts in femorocrural bypasses performed for limb salvage: a review of 108 consecutive cases and comparison with 57 in-situ saphenous vein bypasses. *J Vasc Surg* 1990; 12: 257-263.

8 Donaldson MC, Mannick JA, Whittemore A. Femoral-distal bypass with in-situ greater saphenous vein. Long-term results using the Mills valvulotome. *Ann Surg* 1991; 213: 457-465.

9 Shah DM, Chang BB, Fitzgerald KM et al. Durability of the tibial artery bypass in diabetic patients. *Am J Surg* 1988; 156: 133-135.

10 Sipponen J, Ala-Kulju K, Ketonen P et al. Femorotibial bypass grafting for lower limb ischemia. *Inter Angio* 1989; 8: 65-69.

11 Schneider JR, Walsh DB, McDaniel MD et al. Pedal bypass versus tibial bypass with autogenous vein: a comparison of outcome and hemodynamic results. *J Vasc Surg* 1993; 17: 1029-1040.

12 Thompson MM, Sayers RD, Reid A et al. Quality of life following infragenicular bypass and lower limb amputation. *Eur J Vasc Surg* 1995; 9: 310-313.

13 Johnson BF, Evans L, Drury R et al. Surgery for limb threatening ischaemia: a reappraisal of the costs and benefits. *Eur J Vasc Endovasc Surg* 1995; 9: 181-188.

14 Chetter IC, Spark JI, Scott DJA et al. Prospective analysis of quality of life in patients following infrainguinal reconstruction for chronic critical ischaemia. *Br J Surg* 1998; 85: 951-955.

15 Johnson BF, Singh S, Evans L et al. A prospective study of the effect of limb-threatening ischaemia and its surgical treatment on the quality of life. *Eur J Vasc Endovasc Surg* 1997; 13: 306-314.

16 Panayiotopoulos YP, Tyrrel MR, Owen SE et al. Outcome and cost analysis after femorocrural and femoropedal grafting for critical limb ischaemia. *Br J Surg* 1997; 84: 207-212.

17 Cheshire NJW, Wolfe JHN, Noone MA et al. The economics of femorocrural reconstruction for critical limb ischemia with and without autologous vein. *J Vasc Surg* 1992; 15: 167-175.

18 Panayiotopoulos YP, Tyrrell MR, Arnold FJL et al. Results and cost analysis of distal (crural/pedal) arterial revascularisation for limb salvage in diabetic and non-diabetic patients. *Diabetic Med* 1997; 14: 214-220.

19 Luther M. Surgical treatment for chronic critical leg ischaemia: a 5 year follow-up of socioeconomic outcome. *Eur J Vasc Endovasc Surg* 1997; 13: 452-459.

20 Taylor LM Jr, Hamre D, Dalman RL, Porter JM. Limb salvage vs amputation for critical ischemia. The role of vascular surgery. *Arch Surg* 1991; 126: 1251-1258.

21 Bergamini TM, George SM Jr, Massey HT et al. Pedal or peroneal bypass: which is better when both are patent? *J Vasc Surg* 1994; 20: 347-356.

22 Cheshire NJW, Noone MA, Wolfe JHN. Re-intervention after vascular surgery for critical leg ischaemia. *Eur J Vasc Surg* 1992; 6: 545-550.

23 Morris GE, Raptis S, Miller JH, Faris IB. Femorocrural grafting and regrafting: does polytetrafluoroethylene have a role? *Eur J Vasc Surg* 1993; 7: 329-334.

24 Nicoloff AD, Taylor LM Jr, McLafferty RB et al. Patient recovery after infrainguinal bypass grafting for limb salvage. *J Vasc Surg* 1998; 27: 256-266.

25 Abou-Zamzam AM Jr, Lee RW, Moneta GL et al. Functional outcome after infrainguinal bypass for limb salvage. *J Vasc Surg* 1997; 25: 287-297.

26 Sayers RD, Thompson MM, London NJM et al. Selection of patients with critical limb ischaemia for femorodistal vein bypass. *Eur J Vasc Surg* 1993; 7: 291-297.

27 Tordoir JHM, Van der Plas JPL, Jacobs MJHM, Kitslaar PJEHM. Factors determining the outcome of crural and pedal revascularisation for critical limb ischaemia. *Eur J Vasc Surg* 1993; 7: 82-86.

28 Rutherford RB, Jones DN, Bergentz SE et al. Factors affecting the patency of infrainguinal bypass. *J Vasc Surg* 1988; 8: 236-246.

29 Panayiotopoulos YP, Edmondson RA, Reidy JF, Taylor PR. A scoring system to predict the outcome of long femorodistal arterial bypass grafts to single calf or pedal vessels. *Eur J Vasc Endovasc Surg* 1998; 15: 380-386.

30 O'Mara CS, Kilgore TL Jr, McMullan MH et al. Distal bypass for limb salvage in very elderly patients. *Am Surg* 1987; 53: 66-70.

31 Bergamini TM, Towne JB, Bandyk DF et al. Experience with in-situ saphenous vein bypasses during 1981 to 1989: determinant factors of long-term patency. *J Vasc Surg* 1991; 13: 137-149.

32 Chang JB, Stein TA. The long-term value of composite grafts for limb salvage. *J Vasc Surg* 1995; 22: 25-31.

33 Luther M, Lepäntalo M. Femorotibial reconstruction for chronic critical leg ischaemia: influence on outcome by diabetes, gender and age. *Eur J Vasc Endovasc Surg* 1997; 13: 569-577.

34 Harward TRS, Coe D, Flynn TC, Seeger JM. The use of arm vein conduits during infrageniculate arterial bypass. *J Vasc Surg* 1992; 16: 420-427.

35 Hölzenbein TJ, Pomposelli FB Jr, Miller A et al. Results of a policy with arm veins used as the first alternative to an unavailable ipsilateral greater saphenous vein for infrainguinal bypass. *J Vasc Surg* 1996; 23: 130-140.

36 Sesto ME, Sullivan TM, Hertzer NR et al. Cephalic vein grafts for lower extremity revascularization. *J Vasc Surg* 1992; 15: 543-549.

37 Chalmers RTA, Hoballah JJ, Kresowik EF et al. The impact of color duplex surveillance on the outcome of lower limb bypass with segments of arm veins. *J Vasc Surg* 1994; 19: 279-288.

38 Ballard JL, Killeen D, Smith LL. Popliteal-tibial bypass grafts in the management of limb-threatening ischemia. *Arch Surg* 1993; 128: 976-981.

39 Marks J, King TA, Baele H et al. Popliteal-to-distal bypass for limb-threatening ischemia. *J Vasc Surg* 1992; 15: 755-760.

40 Cranley JJ, Hafner CD. Revascularization of the femoropopliteal arteries using saphenous vein, polytetrafluoroethylene, and umbilical vein grafts. Five- and six-year results. *Arch Surg* 1982; 117: 1543-1550.

41 Christenson JT, Broome A, Norgren L, Eklof B. Revascularization of popliteal and below-knee arteries with polytetrafluoroethylene. *Surgery* 1985; 97: 141-149.

42 Hobson RW II, Lynch TG, Jamil Z et al. Results of revascularization and amputation in severe lower extremity ischemia: a five-year clinical experience. *J Vasc Surg* 1985; 2: 174-185.

43 Rafferty TD, Avellone JC, Farrell CJ et al. A metropolitan experience with infrainguinal revascularization. Operative risk and late results in northeastern Ohio. *J Vasc Surg* 1987; 6: 365-371.

44 Flinn WR, Rohrer MJ, Yao JST et al. Improved long-term patency of infragenicular polytetrafluoroethylene grafts. *J Vasc Surg* 1988; 7: 685-690.

45 Tyrrell MR, Grigg MJ, Wolfe JH. Is arterial reconstruction to the ankle worthwhile in the absence of autologous vein? *Eur J Vasc Surg* 1989; 3: 429-434.

46 Whittemore AD, Kent KC, Donaldson MC et al. What is the proper role of polytetrafluoroethylene grafts in infrainguinal reconstruction? *J Vasc Surg* 1989; 10: 299-305.

47 Johnson WC, Squires JW. Axillo-femoral (PTFE) and infrainguinal revascularization (PTFE and umbilical vein). Vascular Registry of the New England Society for Vascular Surgery. *J Cardiovasc Surg* 1991; 32: 344-349.

48 Quiñones-Baldrich WJ, Prego AA, Ucelay-Gomez R et al. Long-term results of infrainguinal revascularization with polytetrafluoroethylene: a ten-year experience. *J Vasc Surg* 1992; 16: 209-217.

49 Fichelle J, Marzelle J, Colacchio G et al. Infrapopliteal polytetrafluoroethylene and composite bypass: factors influencing patency. *Ann Vasc Surg* 1995; 9: 187-196.

50 Miller JH, Foreman RK, Ferguson L, Faris I. Interposition vein cuff for anastomosis of prosthesis to small artey. *Aust N Z J Surg* 1984; 54: 283-285.

51 Raptis S, Miller JH. Influence of a vein cuff on polytetrafluoroethylene grafts for primary femoropopliteal bypass. *Br J Surg* 1995; 82: 487-491.

52 Taylor RS, Loh A, McFarland RJ et al. Improved technique for polytetrafluoroethylene bypass grafting: long-term results using anastomotic vein patches. *Br J Surg* 1992; 79: 348-354.

53 Wijesinghe LD, Beardsmore DM, Scott DJA. Polytetrafluoroethylene (PTFE) femorodistal grafts with a distal vein cuff for critical ischaemia. *Eur J Vasc Endovasc Surg* 1998; 15: 449-453.

54 Dardik H, Berry SM, Dardik A et al. Infrapopliteal prosthetic graft patency by use of the distal adjunctive arteriovenous fistula. *J Vasc Surg* 1991; 13: 685-691.

55 Paty PSK, Shah DM, Saifi J et al. Remote distal arteriovenous fistula to improve infrapopliteal bypass patency. *J Vasc Surg* 1990; 11: 117-118.

56 Jacobs MJHM, Reul GJ, Gregoric ID et al. Creation of a distal arteriovenous fistula improves microcirculatory hemodynamics of prosthetic graft bypass in secondary limb salvage procedures. *J Vasc Surg* 1993; 18: 1-9.

57 Jacobs MJHM, Gregoric ID, Reul GJ. Prosthetic graft placement and creation of a distal arteriovenous fistula for secondary vascular reconstruction in patients with severe limb ischemia. *J Vasc Surg* 1992; 15: 612-618.

58 Karacagil S, Holmberg A, Narbani A et al. Composite polytetrafluoroethylene/vein bypass grafts: conventional distal vein segment or vein cuff? *Eur J Vasc Endovasc Surg* 1996; 12: 337-341.

59 McCarthy W, Pearce W, Flinn WR et al. Long-term evaluation of composite sequential bypass for limb-threatening ischemia. *J Vasc Surg* 1992; 15: 761-770.

60 Harris RW, Schneider PA, Andros G et al. Allograft vein bypass: is it an acceptable alternative for infrapopliteal revascularization? *J Vasc Surg* 1993; 18: 553-560.

61 Simeón JM, Serrano M, Miralles A et al. Homoinjertos arteriales criopreservados: experiencia clínica durante tres años. *Angiología* 1998; 5: 215-222.

62 Gournier JP, Adham M, Favre JP et al. Cryopreserved arterial homografts: preliminary study. *Ann Vasc Surg* 1993; 7: 503-511.

63 Carsten CG III, Taylor SM, Langan EM III, Crane MM. Factors associated with limb loss despite a patent infrainguinal bypass graft. *Am Surg* 1998; 64: 33-37.

64 Elliot BM, Robison JG, Brothers TE, Cross MA. Limitations of peroneal artery bypass grafting for limb salvage. *J Vasc Surg* 1993; 18: 881-888.

65 Zdanowski Z, Troëng T, Norgren L. Outcome and influence of age after infrainguinal revascularisation in critical limb ischaemia. The Swedish Vacular Registry. *Eur J Vasc Endovasc Surg* 1998; 16: 137-141.

66 Blankensteijn JD, Gertler JP, Brewster DC et al. Intraoperative determinants of infrainguinal bypass graft patency: a prospective study. *Eur J Vasc Endovasc Surg* 1995; 9: 375-382.

67 Shortell CK, Ouriel K, DeWeese JA, Green RM. Peroneal artery bypass: a multifactorial analysis. *Ann Vasc Surg* 1992; 6: 15-19.

68 O'Mara CS, Flinn WR, Neiman HL et al. Correlation of foot arterial anatomy with early tibial bypass patency. *Surgery* 1981; 89: 743-752.

69 Ascer E, Veith FJ, Gupta SK et al. Six year experience with expanded polytetrafluoroethylene arterial grafts for limb salvage. *J Cardiovasc Surg* 1985; 26: 468-472.

70 Ascer E, Collier P, Gupta SK, Veith FJ. Reoperation for polytetrafluoroethylene bypass failure: the importance of distal outflow site and operative technique in determining outcome. *J Vasc Surg* 1987; 5: 298-310.

71 Wolfe JHN, Wyatt MG. Critical and subcritical ischaemia. *Eur J Vasc Endovasc Surg* 1997; 13: 578-582.

72 Humphreys WV, Evans F, Watkin G, Williams T. Critical limb ischaemia in patients over 80 years of age: options in a district general hospital. *Br J Surg* 1995; 82: 1361-1363.

73 Johnson BL, Glickman MH, Bandyk DF, Esses GE. Failure of foot salvage in patients with end-stage renal disease after surgical revascularization. *J Vasc Surg* 1995; 22: 280-286.

74 Peltonen S, Biancari F, Lindgren L et al. Outcome of infrainguinal bypass surgery for critical leg ischaemia in patients with chronic renal failure. *Eur J Vasc Endovasc Surg* 1998; 15: 122-127.

75 Dennis JW, Littooy FN, Greisler HP, Baker WH. Secondary vascular procedures with PTFE grafts for lower extremity ischemia in a male veteran population. *J Vasc Surg* 1988; 8: 137-142.

76 Jacobs MJHM, Ubbink DT, Ten Hoedt R, Biasi GM. Current reflections of the vascular surgeon on the assessment and treatment of critical limb ischaemia. *Eur J Vasc Endovasc Surg* 1995; 9: 473-478.

77 Skillman JJ. It can be fixed, but should it be? *Ann Surg* 1993; 218: 713-714 (editorial).

19

BYPASS TO THE ANKLE AND FOOT ARTERIES

MALCOLM H. SIMMS, ANDREW H. GARNHAM

The efficacy of bypass reconstruction for the relief of limb ischemia in patients with peripheral arterial occlusive disease was established over the past half century by pioneers such as Dos Santos, LeReiche, Kunlin, Courbier and Hall in Europe, and DeBakey and Linton in the USA. More recent developments in radiological imaging, optical magnification, and surgical sutures and instrumentation have enabled vascular surgeons to consider reconstruction of arteries of 2 mm diameter or less. Throughout the 1970s and 1980s, large series of femorotibial bypass operations were reported [1,2], demonstrating good results with the use of autogenous vein grafts [3,4]; the present decade has seen a surge of interest in bypass to arteries of the foot [5,6]. Debate continues over the usefulness and durability of bypass grafting to small distal arteries and what follows constitutes a personal perspective.

■ GENERAL PRINCIPLES

When considering arterial bypass reconstruction surgeons should remember that the size of the potential outflow bed is proportional to the size of the recipient artery. Small is not necessarily beautiful in this context and if correction of peripheral ischemia can be accomplished by a bypass to a large proximal artery, then this should be the preferred strategy. However, the objective of vascular reconstruction is to restore pulsatile arterial flow at systemic pressure throughout the extremity and in many patients the extent of occlusion necessitates bypass to arteries in the ankle and foot.

The principal determinant of symptomatology in peripheral vascular disease is the level and extent of atherosclerotic occlusion, which can be related to recognizable anatomical levels. In the case of the lower limb these levels comprise the aortic, iliac and femoral bifurcations, then the superficial femoral/popliteal transition at the adductor hiatus, followed by the popliteal trifurcation, the malleolar anastomosis and the pedal arch. Experience has shown that occlusion at a single level is likely to produce claudication in distal muscle beds. The development of critical ischemia usually requires occlusion at two successive levels. In practice, that means that correction of critical ischemia will require bypass from the level of the last palpable pulse to a patent artery two levels beyond. Thus, when critical ischemia is evident in a patient who has no palpable pulse below the groin, limb salvage can usually be effected by femorocrural bypass.

Similarly, if the popliteal pulse is the most distal palpable, bypass to arteries in the foot will be necessary. Extrapolation of this theory implies that if the occlusive process commences distal to the popliteal trifurcation then the disease must extend beyond the pedal arch, making successful reconstruction very unlikely. However, occasional successes are obtained with a tibiopedal bypass in diabetic patients [7]. This simplistic concept of arterial assessment is rarely contradicted in practice and exceptions can usually be attributed to confounding factors such as unsuspected stenosis of the profunda femoral artery or intercurrent disorders such as neuropathy, cardiac failure, venous hypertension, vasculitis or trauma.

■ CLINICAL PRESENTATIONS

Atherosclerosis is a common systemic disorder, the onset and progression of which is determined by well-recognized risk factors, including genetic susceptibility. Acquired factors such as smoking, obesity, hyperlipidemia and diabetes mellitus, which are amenable to modification, exert a variable influence over the distribution over the atherosclerotic occlusive process, producing distinct patterns of disease.

Cigarette smoking tends to affect large and medium sized arteries, producing the common presentations of aortoiliac and femoropopliteal occlusion. A small proportion of young male smokers develop thromboangiitis obliterans (Buerger's disease). In this condition the occlusive process spreads proximally from the digital arteries in the foot and bypass reconstruction is usually impossible.

When atherosclerotic occlusion preferentially affects arteries at or below the knee, certain etiological factors are prevalent, of which diabetes mellitus is the most important. In diabetes, not only does the atherosclerotic process focus on the popliteal trifurcation and beyond but also the associated sensory neuropathy impairs protective sensation in the foot. In addition, the premature onset of medial calcification makes arteries brittle, and therefore more prone to traumatic occlusion.

This extensive medial calcification affecting small caliber arteries, known as Monckeberg's sclerosis when it occurs in old age, is also seen prematurely in patients with chronic renal failure, often in association with secondary hyperparathyroidism. As well as hastening the occlusive process, medial calcification of arteries complicates the assessment of disease severity by elevating the cuff occlusion pressures obtained with doppler sphygmomanometry and also produces technical problems with hemostasis and vessel anastomosis during bypass reconstruction.

Other systemic disorders that appear to favor the development of infrapopliteal atherosclerosis include the connective tissue disorders, notably rheumatoid arthritis. Rarely, acute limb ischemia due to small vessel thrombosis may arise in patients with hyperviscosity or hypercoagulability but in such cases the role of surgery is limited and thrombolysis and anticoagulation offer the best hope of limb salvage.

Bypass to ankle and foot arteries is occasionally required in cases of lower limb trauma but here the outcome is more likely to depend on the extent of collateral damage to bones, soft tissues, veins and nerves than on the quality of the reconstruction.

The natural history of popliteal aneurysm involves the accumulation of thrombotic material in the expanding aneurysm wall. Displacement or embolization of this material during knee flexion risks the progressive obliteration of the ostia of the trifurcation arteries. When completion of this process produces aneurysmal thrombosis, the ensuing limb ischemia can be notoriously severe and limb salvage may involve bypass grafting to arteries of the ankle or foot.

■ HEMODYNAMIC ASSESSMENT

In infrapopliteal vascular occlusion, evaluation of the severity of ischemia depends more on clinical appearances than on doppler pressure testing, which cannot be relied upon because of the prevalence of vascular calcification in this group of patients. When ankle systolic pressures and pressure index calculations seem unfeasibly high, it is possible to detect occult low intraarterial pressure by the *Pole Test* [8]. This simply entails elevating the ischemic foot with the patient supine while insonating the dominant ankle artery with a doppler probe. If the doppler signal fades reproducibly at a given level above the heart, then this height in centimeters multiplied by a factor of 0.735, corresponds to ankle systolic pressure in mmHg.

The most useful application of doppler flowmetry in patients with infrapopliteal occlusion is in the evaluation of the pedal arch [9]. This is best undertaken with the patient seated upright on a couch with the leg hanging in dependency. Preliminary insonation of each calf artery in turn is useful to establish its location, patency and if possible, compressibility. The peroneal trunk is best detected in the distal calf by firm probe pressure on the posteromedial aspect of the fibula and its main divisions may be detected on the anteromedial and posteromedial aspects of the lateral malleolus. The deep plantar signal, representing the apex of the major pedal arch, is then sought on the dorsum of the foot at the base of the first metatarsal cleft. Once this signal is audible optimally, the three main calf arteries are occluded in turn by firm digital compression in the lower third of the calf. If compression of a particular digital artery obliterates the pedal arch signal reproducibly, this artery constitutes the dominant feed to the foot. If two arteries have to be compressed simultaneously to obliterate the foot signal, then the pedal arch is intact with bi-directional perfusion. In interpreting this test it must be remembered that the peroneal trunk can perfuse both anterior and posterior-tibial arteries through collateral channels at the malleolar level. When an indistinct pedal arch signal persists whichever of the tibial arteries are occluded manually, prospects for reconstruction are usually poor; either all crural arteries are heavily calcified or all are occluded and foot perfusion is maintained via unreconstructable non-axial collaterals.

■ VASCULAR IMAGING

Duplex scanning of the abdominal aorta and the arteries of both lower limbs is an important adjunct to assessment in infrapopliteal bypass. Proximal aneurysmal disease can be excluded and the patency of the femoropopliteal segment can be assessed for its suitability for transluminal angioplasty. The reliability of duplex ultrasound in assessing the status of the popliteal trifurcation and the condition of the crural and pedal vessels varies between centers and whenever there is doubt about the feasibility of percutaneous angioplasty, then digitally enhanced arteriography should be undertaken. If the duplex scan demonstrates extensive and complete occlusion of the popliteal trifurcation then it can be assumed with confidence that transluminal angioplasty will be ineffective. In such cases it is not essential to undertake preoperative femoral arteriography. At the time of operation the surgeon requires detailed images of the calf and foot arteries to demonstrate the optimal site for distal runoff and this can be provided reliably by an intraoperative arteriogram. The usual technique is to inject 50 ml of contrast rapidly through a 19 gauge needle below a femoral or popliteal artery clamp, waiting 5-10 seconds before obtaining the image [10].

The limitations of preoperative contrast imaging of the arteries of the calf and foot relate to the difficulties of opacifying small distal vessels in a slow-flowing vascular bed. Digital enhancement has brought great improvement in image quality, while other strategies such as proximal balloon occlusion, intra-arterial vasodilatation, and optimal timing of images have all improved the accuracy of distal arteriograms. However, in a minority of patients, doppler-detectable pedal arteries fail to opacify. Very occasionally, in some patients, surgical exploration may identify patent distal arteries that had previously escaped detection, either by dependent doppler insonation or by any form of arteriography. At operation, if external inspection of an artery is equivocal, its patency can be established without prejudice by stabbing it with the point of a micro-knife and observing the bleeding. This stab incision will seal within a few minutes without the need for continued clamping. If the arteriogram demonstrates collateral filling throughout the ankle and foot with no opacification of axial arteries, then it is safe to assume that the limb is not reconstructable. If, however, there is little sign of contrast reaching the foot then no assumptions about axial artery patency can be made. The patient may be counselled about the possibility of amputation, but if the foot is potentially useful, this should not proceed until blind intraoperative exploration of the foot arteries has been undertaken, starting with the dorsalis pedis and the anterior branch of the peroneal, and concluding with the plantar artery and its medial and lateral branches.

■ PLANNING RECONSTRUCTION

The planning of any vascular reconstruction involves a careful and detailed review of the patient, the limb and the proposed graft. The risk/benefit analysis for the patient must take into account their general health and lifestyle, the severity of the symptoms and the complexity and risk of the intervention. The assessment of the limb should include its overall usefulness to the patient, taking into account locomotor function as well as details of its vascular anatomy and hemodynamics and the extent of ischemic injury. Associated venous and dermatological disorders may be highly relevant.

In patients with symptoms and signs of infrapopliteal arterial occlusion, autogenous vein is the only graft material worthy of consideration [3,5]. Operative planning must therefore include an inventory of autogenous vein material suitable for grafting. In order of preference this should be: ipsilateral long saphenous vein, any large caliber superficial arm vein, including cephalic, ulnar or basilic [11], ipsilateral short saphenous vein and finally, contralateral long or short saphenous vein. If necessary, the status of these veins can be illustrated by duplex ultrasound scanning. When outflow sites below the groin, such as the superficial femoral or popliteal, are available, results are not prejudiced and the length requirement for veins is reduced. If it is essential to place the proximal anastomosis at common femoral level and insufficient vein is available, then composite sequential grafting can be considered, routing a prosthetic graft from the groin to the above-knee popliteal artery, then extending a vein graft from the same popliteal arteriotomy into the calf or foot. This approach offers graft patency rates similar to those obtained with vein-only grafts [12].

Regarding the distal runoff there is no intrinsic virtue in extending a vascular graft onto the foot and long grafts and extensive dissection should be avoided whenever possible. Many patients with extensive tibial artery occlusion retain a patent section of peroneal artery in the calf, which communicates via malleolar collaterals with the dorsalis pedis and plantar arteries. This is confirmed by the observation that a high proportion of patients in whom successful femoroperoneal bypass is undertaken will eventually develop a palpable foot pulse during follow-up.

■ GENERAL PREPARATION AND ANESTHESIA

Whenever possible at the time of surgery, the patient's condition should be optimized with respect to cardiovascular function, metabolism, nutrition and infection risk. In practice, when the condition of the ischemic leg demands urgent attention some compromise may be acceptable so long as the principle of *life before limb* is observed. In all cases it is essential to explain to the patient and his or her family the rationale, risks and complications of the procedure, based on local outcomes.

Dry gangrene of the ischemic extremity does not increase the risk of surgical wound sepsis and preoperative debridement is not required. Definitive amputation can be undertaken at the same time as revascularization or delayed as long as judged appropriate. Ulcerated areas should be swabbed preoperatively so that appropriate antiseptic and antibiotic cover can be provided at operation. When there is suspicion of deep-seated infection, preoperative drainage and debridement is essential, laying open tissue planes and avoiding primary suturing.

Systemic anticoagulation for peripheral vascular reconstruction is not necessary but subcutaneous heparin prophylaxis against deep venous thrombosis is desirable and anti-platelet medication such as aspirin should be continued. Additional antiplatelet cover during the perioperative period can be provided by intravenous infusion of Dextran 40, 500 ml over four hours, repeated on a daily basis.

Infrapopliteal vascular reconstruction can be performed using a variety of anesthetic techniques [13]. General anesthesia is convenient and acceptable to most patients but does have potential for cardiovascular and respiratory morbidity. Spinal or epidural anesthesia gives good results in patients with chronic lung disorders but in patients with high cardiovascular risk, local anesthesia may offer the safest option. A combination of percutaneous femoral nerve block and local skin infiltration will provide anesthesia from groin to knee. The sciatic nerve trunk can then be exposed by a standard above-knee popliteal exposure and infiltrated directly, achieving complete distal anesthesia for several hours. This method requires <100ml of 0.5% prilocaine solution and is safe and well tolerated. Adjunctive sedation is not required in most patients [14].

■ OPERATIVE TECHNIQUE

The surgical objective in distal artery bypass is to restore normal arterial perfusion with the minimum of tissue trauma, utilizing a graft combining excellent hemodynamics with durability, biocompatability, and low thrombogenicity. A significant proportion of patients that require bypass to ankle or foot arteries will exhibit a distal pattern of occlusion, permitting the use of the superficial femoral and popliteal arteries for the inflow anastomosis. In patients with an intact long saphenous vein, this means that only two-thirds of the full length of the vein is required, extending across the knee and ankle to the foot. It is therefore possible to harvest only the proximal two-thirds of the vein between groin and upper calf, sparing the vulnerable skin of the mid-calf and ankle from the trauma of vein harvesting. Two further advantages arise from utilizing fully-mobilized vein segments.

Firstly, there is no possibility of overlooking patent tributaries that could lead to residual arteriovenous fistula. Secondly, it is possible to route the graft through deep anatomical tunnels along its whole length, avoiding the risk of graft exposure in the event of superficial wound infection and dehiscence. With fully mobilized grafts the surgeon can orientate the vein in a reversed or non-reversed direction. The advantages of the non-reversed format are firstly, that the proximal anastomosis can be completed first, allowing the vein graft to be tunneled whilst pulsating. In this way, any flow impediment arising from kinks, twists or external compression can be detected and corrected before the distal anastomosis is completed.

The second advantage of the non-reversed format is that in the event of graft occlusion there is no impediment to the passage of balloon embolectomy catheters through the graft in any direction. Fears that venous valvulotomy might increase graft thrombogenicity or predispose to postoperative graft strictures due to myointimal hyperplasia, have not been borne out by experience. Modern disposable valvulotomes are effective and atraumatic when used with care and problems due to retained valve cusps should not arise. Occasionally the tapering configuration of the vein favors the reversed format. This is more frequent when segments of arm vein are employed. Indeed, when using the basilic/cephalic vein loop it is inevitable that one section of the graft will be reversed and the other non-reversed. In this case the non-reversed section should be sited proximally and valvulotomy is achieved through an antecubital side branch. This side branch can be left open to permit down-flushing of the distal, reversed section of graft while the distal anastomosis awaits completion. When using the reversed format for a free tunneled vein graft it is advisable to perform the distal anastomosis first in order to confirm unimpeded flow by down-flushing, before completing the proximal anastomosis.

Some further points regarding vein grafts are worth mentioning, some specific to arm veins. Any section of vein showing evidence of previous phlebitis should be discarded. Fibrous thickening of the vein wall may be apparent externally or may only be revealed by a lack of elastic distensibility on flushing with heparinized solution. The appearance of a white opacification on the adventitial surface of the vein on flushing indicates a splitting of the venous endothelium with dissection of the media, and this section of vein is unsuitable for grafting. Passage of the valvulotome may reveal further areas of intimal fibrosis and if a 2.5 mm valvulotome will not pass freely, the vein is best discarded.

Arm veins are thinner-walled and more distensible than leg veins and although they may appear too small when first dissected or when in spasm, it is very rare for any nondiseased segment of cephalic, ulnar, or basilic vein to fail to distend sufficiently to create an adequate graft. This property of distensibility necessitates increased caution in ligating tributaries in arm vein grafts. Ties on large antecubital tributaries are liable to become dislodged as the vein distends under arterial pressure. In order to avoid a troublesome postoperative hemorrhage it is best to double-tie or oversew any vulnerable points.

Morbidity in the arm due to vein harvesting is rare, with subcutaneous hematoma due to reactive hemorrhage being the most frequent complication. Swelling due to venous insufficiency should not occur but in this regard it is important when harvesting the basilic vein in the medial upper arm to stop at the point where the vein joins the brachial venae comitantes to form the axillary vein.

Graft Routing

A variety of instruments are available for the routing of vein grafts through deep anatomical tunnels. These include dedicated vascular tunnelers, long vascular clamps and tissue probes. In order to minimize soft tissue injury and bleeding, it is important to avoid force and over-dilatation.

Tunneling grafts through the popliteal fossa is a simple atraumatic maneuver requiring about 30°C of knee flexion. When the interosseous membrane has to be crossed to enter the anterior-tibial compartment, the membrane should be breached with a cruciform incision made under direct vision from a small, high anteriortibial incision. This avoids the risk of rupturing anteriortibial vessels by forcible blind tunneling.

The anteriortibial compartment is particularly narrow and a good method of threading a vein graft along this tunnel to the dorsalis pedis artery is by using the blunt end of a disposable valvulotome, passed from a high anteriortibial incision to the dorsum of the foot. This is used to engage and draw back a ligature, which can then be tied to the end of the vein graft to draw it in turn through the tunnel to the foot. When using this route for a vein graft, compression by the extensor retinaculum at the ankle joint should be assessed and if excessive, the retinaculum should be divided subcutaneously. Occasionally it is appropriate to tunnel a vein graft from the medial calf to the dorsalis pedis through a subcutaneous tunnel. This must be performed with precision. Too superficial a tunnel will cause skin necrosis, too deep may pass under an extensor tendon causing graft compression.

The Distal Anastomosis

Knowledge of basic vascular anatomy facilitates exposure of the foot arteries. The plantar artery and its medial and lateral divisions are easily found by dividing the flexor retinaculum along a parallel 2 cm posterior to the medial malleolus and extending into the foot. The anteriortibial artery becomes the dorsalis pedis at the lower border of the extensor retinaculum where it lies between and deep to the tendons of tibialis anterior and extensor digitorum longus. The extensor hallucis longus tendon is lateral proximally and crosses in front of the artery as it crosses the ankle joint, to lie medially in the foot. The anterior branch of the peroneal artery lies in the front of the lateral malleolus, just lateral to the extensor digitorum longus tendon and is occasionally the dominant artery of the dorsum of the foot. The dorsalis pedis artery is linked to the termination of the lateral plantar artery by the deep plantar artery, which passes through the base of the first metatarsal cleft to complete the primary pedal arch.

A precise anastomotic technique requires careful dissection of the artery from accompanying veins and soft tissue, with removal of loose adventitia and sometimes division of one or two small artery branches, to achieve adequate mobilization. Optical magnification using 2.5 or 3.5 x loupes is desirable for this task and is essential for the subsequent anastomosis. With adequate mobilization of the recipient artery it is a simple matter to secure hemostasis during arteriotomy by traction slings or microvascular clamps, even in the presence of significant medial calcification. On the rare occasions when calcification is too dense to permit external compression, hemostasis can be achieved by intraluminal flow arresters. The use of a proximal arterial tourniquet to produce a bloodless field introduces a number of potential complications and is not advised. An 8-10 mm long arteriotomy is usually adequate, preferably made with a fine microvascular blade, systemic heparinization is unnecessary but local vessel flushing with heparinized solution (10 units/ml) not only prevents local thrombosis but affords the opportunity to obtain a crude estimate of outflow resistance. Care must be taken to avoid traumatizing the arterial endothelium by careful technique and the use of smooth and loose-fitting cannulae. Occasionally runoff can be improved by gentle dilation with a smooth, tapered probe when it is not practical to extend the bypass to a more healthy distal section of artery.

The distal anastomosis should be performed with the finest practicable suture/needle combination, using monofilament polypropylene of 6, 7, or 8-0 and needles of 6-8mm length. The vein graft should be spatulated and left almost square rather than cut obliquely, in order to avoid stenosis at the heel or toe of the anastomosis and care should be taken to take precise perpendicular bites of the vessel walls, avoiding any adventitial prolapse or snagging. Endarterectomy of tibial and pedal vessels is best avoided and if necessary should not extend to the distal end of the arteriotomy. Loose intimal plaque must be carefully secured by inside to outside passages of the needle through the artery.

A continuous knot-free parachute suturing technique is simple and effective, commencing with a *double heel*, passing the two needles outside to inside at the 11 o'clock and 1 o'clock positions relative to the venotomy, and emerging inside to outside at 1 o'clock and 11 o'clock respectively on the artery. The suture slack can be taken up after extending 3-4 bites along each side of the anastomosis. The toe of the anastomosis can be completed with radially placed continuous suture bites without incurring any stenosis, and the suture line can be completed midway down the second side of the anastomosis. Before completing the anastomosis it is advisable to perform a final bleed-out and flush in order to wash out any retained debris or local thrombus.

Graft Monitoring

Although many methods of monitoring graft patency and flow are available, none are fully comprehensive and often the simplest observations will prove adequate. During distal revascularization procedures the foot should be covered in a transparent, sterile plastic bag or kept exposed so that changes in skin color and temperature can be observed clinically, and distal pulses can be palpated or insonated with doppler. The doppler probe can be covered with a sterile wrap, such as a surgical glove and applied directly to the graft and recipient artery. With a little experience the audible signal can be interpreted to evaluate flow velocity and distal resistance. Peripheral vasoconstriction can be minimized by maintaining adequate blood volume and body temperature. Vasodilator drugs such as papaverine and prostacyclin can be injected into the graft using a 25 gauge needle and can be applied topically around the distal anastomosis. In the case of prostacyclin the effect of intragraft injection in augmenting flow has been shown to persist up to seven days postoperatively [15].

Measurements of graft flow and peripheral resistance can be obtained in various ways. Electromagnetic flow meters and calibrated doppler flow meters measure volume flow in mls per minute, and if graft pressure is known it is possible to calculate peripheral resistance. Similarly computerized syringe drivers are now available that will measure graft outflow capacity at constant pressure, enabling both flow and resistance to be calculated. With any of these methods a graft flow of less than 99 ml per minute predicts a poor prognosis and more than 165 ml per minute, a good prognosis [16].

Completion arteriography is necessary only if simpler methods fail to demonstrate satisfactory graft flow and distal perfusion. Problems with inflow or along the course of the graft will produce a poor graft down-flow, which can be detected clinically and dealt with appropriately. The main value of completion arteriography is in demonstrating technical problems at the anastomosis, or the nature and extent of runoff obstructions. Anastomotic revision, distal embolectomy or graft extension to improve runoff are all options for consideration.

Sequential and Bifurcated Outflow

Lower limb revascularization differs from coronary artery bypass grafting in that there is fairly good intercommunication between the runoff beds of the crural arteries in contrast to those of the coronary arteries. In femorodistal bypass therefore the objective should be to restore pulsatile flow to the artery that provides best perfusion of the pedal arch. In some patients, however, two or more patent crural arteries will appear to communicate with the pedal arch. In such cases, if technically practicable, creation of a dual outflow may be achieved either by using a bifurcated graft (Fig. 1) or by carrying out sequential

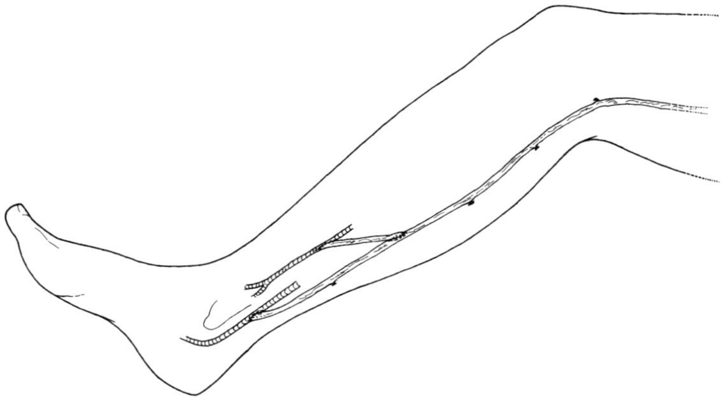

FIG. 1 Venous graft with creation of a bifurcation to revascularize two crural arteries.

anastomoses with a proximal side-to-side anastomosis and a distal end-to-side anastomosis from a single vein trunk. Intra-operative graft flow measurements show that this approach achieves a 30% (40% increase in flow down the graft trunk but there is no evidence that this improves late graft patency rates).

A comparable situation arises when there is extensive occlusive disease at the popliteal trifurcation, creating an isolated popliteal segment with a patent calf or foot artery distal. In such cases it is possible to perform an intermediate side-to-side anastomosis onto the patent popliteal segment before extending the distal trunk of the vein graft to form an end-to-side anastomosis on the recipient crural or pedal artery (Fig. 2). Unpublished flow studies in such grafts tend to suggest that there is a greater degree of separation of flow into the popliteal and distal beds with an increase in flow down the proximal graft trunk of 50% or greater.

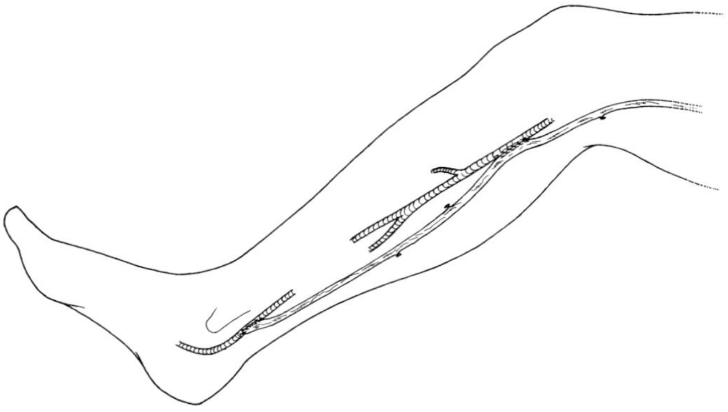

FIG. 2 Sequential venous graft side-to-side to patent popliteal segment, extending to a recipient crural or pedal artery.

CLINICAL EXPERIENCE AND OUTCOMES

Despite developments in percutaneous angioplasty and, in particular, subintimal angioplasty, the management of critical lower limb ischemia due to infrapopliteal atherosclerosis remains predominately surgical. The past decade has seen a continued increase in lower limb revascularization surgery with a shift towards a more distal pattern of reconstruction, reflecting increased confidence in the efficacy of bypass grafting to small distal vessels. This trend is demonstrated in my own practice, as illustrated in Fig. 3. Of the 130 grafts characterized by distal inflow or pedal outflow, 52 were constructed from the superficial femoral or popliteal artery to a tibial artery, 21 were femoropedal and 62 were popliteal to pedal.

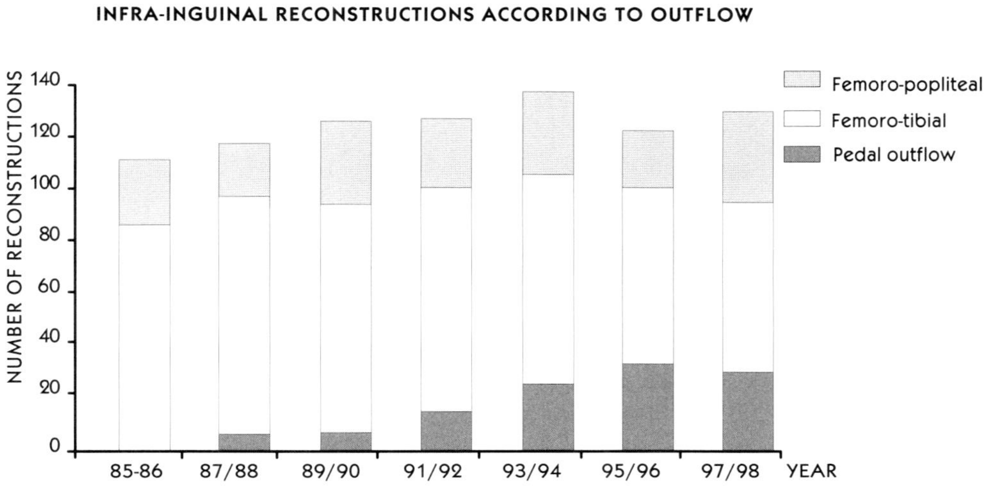

FIG. 3 Evolution of different bypass techniques in our own experience.

Comparison of the femorotibial population (A) to the distal/pedal population (B) shows a similar age and sex profile in both groups but a higher proportion of diabetics in Group B (47% compared to 25%). Group B patients were also considered less medically fit than Group A patients resulting in an increased utilization of local infiltration anesthesia (23% compared to 7.5%). With tibial artery outflow, the choice of runoff artery was anterior-tibial 43%, posterior-tibial 30% and peroneal 27%. Graft material was in-situ long saphenous vein in 59%, arm vein in 19% and composite sequential 22%. At 30 days postoperative, mortality was 3%, amputation 11% and graft patency 82%, falling to 73% at one year. Graft failure after six months was unlikely to lead to limb loss. After one year, graft failure was increasingly unlikely, whereas mortality remained constant at 5% per year. Of the 83 grafts to the foot, 79% were to the dorsalis pedis artery and 21% to the common plantar artery or its major branch. Arterial inflow sites for pedal bypass were: femoral (14.4%), superficial-femoral (3.6%), above knee popliteal (60.4%), below knee popliteal (19.2%) and anterior-tibial (2.4%). All pedal bypass grafts were constructed from autogenous vein. At 30 days, mortality was 6.1%, amputation 22% and graft patency 65.8%. Late follow-up is incomplete in this group but it is clear that late mortality rates exceed graft failure rates and that in survivors, some grafts remain patent for more than 4 years.

■ CONCLUSION

Our data and the results in the literature [6,17,18] show that bypass grafts to the arteries of the ankle and foot are effective and durable. Patients who develop critical leg ischemia associated with peripheral occlusive arterial disease have a high incidence of intercurrent disease and a diminished life expectancy. Numerous studies have pointed out the benefits of limb salvage surgery over amputation in terms of improved quality of life [19-21]. There is also evidence to show that distal reconstructive surgery in expert hands consumes less hospital and community resources than amputation [22]. Successful limb salvage surgery relieves suffering and helps patients to remain independent during the latter years of their lives. However, it is an exacting discipline in which errors of judgement or technique seldom escape retribution, particularly when bypassing to ankle or foot arteries.

Deciding when not to attempt a difficult infra-inguinal reconstruction in preference to amputation or palliation remains one of the most difficult issues for specialists in this field. Improving the efficacy of surgical intervention requires a better understanding of the patho-physiology of limb ischemia. This will derive not only from advances in basic science but from scrupulous data recording and audit by vascular surgeons.

REFERENCES

1 Leather RP, Shan DM, Karmody AM. Infrapopliteal arterial bypass for limb salvage: increased patency and utilization of the saphenous vein used "in-situ". *Surgery* 1981; 190: 1000-1008.
2 Graham JW, Hanel KC. Vein grafts to the peroneal artery. *Surgery* 1981; 89: 264-267.
3 Bergan JJ, Veith FJ, Bernhard VM et al. Randomization of autogenous vein and polytetrafluorethylene grafts in femoro-distal reconstruction. *Surgery* 1982; 92: 921-930.
4 Hickey NC, Thomson IA, Shearman CP, Simms MH. Aggressive arterial reconstruction for critical lower limb ischaemia. *Br J Surg* 1991; 78: 1476-1478.
5 Andros G. Bypass grafts to the ankle and foot. A personal perspective. *Surg Clin N Am* 1995; 75: 715-729.
6 Pomposelli FB Jr., Marcaccio EJ, Gibbons GW et al. Dorsalis pedis arterial bypass: durable limb salvage for foot ischemia in patients with diabetes mellitus. *J Vasc Surg* 1995; 21: 375-384.
7 Woelfle KD, Lange G, Mayer H et al. Distal vein graft reconstruction for isolated tibioperoneal vessel occlusive disease in diabetics with critical foot ischaemia - does it work? *Eur J Vasc Surg* 1993; 7: 409-413.
8 Smith FC, Shearman CP, Simms MH, Gwynn BR. Falsely elevated ankle pressures in severe leg ischaemia: the pole test - an alternative approach. *Eur J Vasc Surg* 1994; 8: 408-412.
9 Roedersheimer LR, Feins R, Green RM. Doppler evaluation of the pedal arch. *Am J Surg* 1981; 142: 601-604.
10 Ricco JB, Pearce WH, Yao JS et al. The use of operative prebypass arteriography and doppler ultrasound recordings to select patients for extended femoro-distal bypass. *Ann Surg* 1983; 198: 646-653.
11 Holzenbein TJ, Pomposelli FB Jr, Miller A et al. Results of a policy with arm vein used as the first alternative to an unavailable greater saphenous vein for infra-inguinal bypass. *J Vasc Surg* 1996; 23: 130-140.
12 McCarthy WJ, Pearce WH, Flinn WR et al. Long-term evaluation of composite sequential bypass for limb threatening ischemia. *J Vasc Surg* 1992; 15: 761-770.
13 Pierce ET, Pomposelli FB Jr, Stanley GD et al. Anesthesia type does not influence early graft patency or limb salvage rates of lower extremity arterial bypass. *J Vasc Surg* 1997; 25: 226-233.
14 Mackay CA, Razik W, Simms MH. Local anaesthetic for lower-limb revascularisation in high-risk patients. *Br J Surg* 1997; 84: 1096-1098.
15 Smith FC, Tsang GM, Watson HR, Shearman CP. Iloprost reduces peripheral resistance during femoro-distal reconstruction. *Eur J Vasc Surg* 1992; 6: 194-198.
16 Davies AH, Magee TR, Baird RN, Horrocks M. Intra-operative measurement of vascular graft resistance as a predictor of early outcome. *Br J Surg* 1993; 80: 854-857.
17 Davidson JT III, Callis JT. Arterial reconstruction of vessels in the foot and ankle. *Ann Surg* 1993; 217: 699-710.
18 Schneider JR, Walsh DB, McDaniel et al. Pedal bypass versus tibial bypass with autogenous vein: a comparison of outcome and hemodynamic results. *J Vasc Surg* 1993; 17: 1029-1040.
19 Albers M, Fratezi AC, De Luccia N. Assessment of quality of life of patients with severe ischemia as a result of infrainguinal arterial occlusive disease. *J Vasc Surg* 1992; 16: 54-59.
20 Pell JP, Donnan PT, Fowkes FG et al. Quality of life following lower limb amputation for peripheral arterial disease. *Eur J Vasc Surg* 1993; 7: 448-451.
21 Thompson MM, Sayers RD, Reid A et al. Quality of life following infragenicular bypass and lower limb amputation. *Eur J Vasc Endovasc Surg* 1995; 9: 310-313.
22 Cheshire NJ, Wolfe JH, Noone MA et al. The economics of femoro-crural reconstruction for critical leg ischemia with and without autologous vein. *J Vasc Surg* 1992; 15: 167-174.

20

TECHNICAL IMPROVEMENT OF IN-SITU VENOUS GRAFTS

H. LOEPRECHT

The in-situ *bypass procedures for femoropopliteal occlusions were introduced by Hall in 1962 [1]. The main principle is to render the venous valve incompetent and occlude the side branches in order to avoid any arteriovenous fistula, which at least could lead to increased cardiac preload or cutaneous hyperemia with skin necrosis. This technique in particular evolved for tibial- and peroneal artery reconstruction, since the diameter discrepancy could be overcome with this special technique, and kinking of the bypass is negligible compared with a below knee poplitea in-situ reconstruction. However, most surgeons applied a continuous skin incision for side-branch occlusion resulting in 30% wound complications like skin and fat necrosis [2]. Valve destruction was mostly done in a blind manner with Hall-type valvulotomes or hooks. Reports of this early period provided an incidence of 10%–20% of remnant valves and incomplete transsected cusps, initiating intima hyperplasia and consecutive stenosis. In order to minimize the trauma to the skin, completion angiography and ultrasound were used to identify side branches and arteriovenous fistula.*

■ ENDOSCOPIC EQUIPMENT

In 1969 Vollmar et al. [3] introduced angioscopy for the intraoperative control after thromboendarterectomy with a rigid scope and later on with large diameter scopes. Only with the advent of small diameter scopes (1-2.1 mm diameter) was it also applicable for femorodistal and tibial reconstructions. The main prerequisites are a closed irrigation system and an introducer sheath. Nowadays, a video camera is mostly used to simplify the procedure, to share the findings with the whole team, and to get a sufficient documentation. Since the venous graft is localized subcutaneously, stearability of the scope is no more a fundamental necessity. The tip of the scope can be directed by manual pressure from outside. The volume of ringer solution needed for a visualization seldomly exceeds 300 ml. Since the scopes are sterilized with ethylene oxide the possible danger of infection can be negated. Each advancement of an angioscope in the endovascular route leads to endothelial damage. However, Hashizume et al. [3] could show experimentally that after four weeks all lesions had been covered with a new endothelial layer. Cautious manœuvering of the scope should not be hazardous to the graft and hence should not jeopardize early graft patency.

■ RENDERING VALVES INCOMPETENT

The first step is dissection of the greater saphenous vein at the level of the proximal and distal anastomosis. While harvesting the greater saphenous vein in the groin, a large diameter segment will mostly be found in the confluence part of the greater saphenous vein. Therefore the first valve should be transected in this position with scissors, since the valvulotome mostly will not accommodate the diameter of the vein completely. This is not necessary, however, if a more distal central anastomosis has been selected. The scope is introduced from the proximal orifice of the graft together with an irrigation catheter and the valvulotome entering from below. The valve-cutter is advanced towards the scope and then retracted in two directions. With a Hall-type valvulotome direct view during transsection is not possible. On the contrary, with a Mills hook this is feasible under direct vision. The great advantage of intravascular visualization is the completeness of valve-ablation and identification of anatomical variations like a doubling of the greater saphenous vein or postphlebitic webs (Table) [4].

Table PREDICTION OF ABNORMALITIES WITHIN THE GRAFT

	Angioscopy %	*Angiography* %
Sensitivity	91	41
Specificity	100	100
Positive predictive value	100	100
Negative predictive value	98	92

n = 100 bypass procedures (Wölfle, Loprecht).

■ IDENTIFICATION OF SIDE BRANCHES

After rendering the valvecusps incompetent, the irrigation fluid displaces the blood in the total length of the bypass. The orifices of side branches can be seen directly and interruption of the irrigation shows a reentering of blood from the side branches. By translumination the position of the scope and the level of the venous orifice can be localized on the skin. With a small skin incision the side branch can be identified and ligated or occluded with an external clip. By displacing the blood and interruption of the irrigation all side branches can be recognized and excluded. It is questionable whether all side branches have to be interrupted and it has been recommended to only ligate efferent branches to the deep venous system. These branches can be identified by the competence of the valves of the perforating veins preventing reflux of blood.

In our experience completion angiography after performance of the proximal and distal anastomosis did show only singular small side branches with the additional need for ligation or fistulas in a second procedure. The time needed for this complementary intraoperative investigation has a range between 15 and 30 minutes depending on the experience of the investigator and the presence of anatomical variations.

Endovascular Occlusion of Side Branches

Rosenthal et al. [5] developed a highly steerable nitinol device (8F nickel-titanium catheter) to enlarge the endoscopic and endovascular possibilities in in-situ bypass procedures. They demonstrated the feasibility of engagement of side branches and deployment of occluding coils via the steerable catheter. The occlusion of side branches was complete in 82% and partially successful in 18%. The partially occluded side branches were completely occluded within a few days after the operation. However, in 11% of the deployed coils they observed a recoiling into the saphenous vein. The coils have been retrievable with an embolectomy catheter of a flushing maneuver. Rosenthal et al. also occluded side branches with a smaller 6.5F catheter under fluoroscopic control only. The incidence of total and partial occlusion was comparable to the initial experience (84% vs. 15%). Nevertheless, an additional 19% of patients suffered from arteriovenous fistula, which required surgical ligation. The cumulative patency meets the standard of conventional procedures in durability, cost and enhanced patient comfort.

Subcutaneous Endoscopic Side Branch Ligation

Cardiac surgeons specially developed the technique of harvesting the greater saphenous vein for coronary artery bypass procedures [6-9]. It was shown in the recent years that wound complications were lower, but the duration of the procedure longer. It was therefore quite logical that this technique would also be applied for side branch ligation during in-situ procedures. Rosenthal et al. [10] reported one case with combined endoscopic transluminal identification of side branches and subcutaneous endoscopic ligation of side branches using a rigid laparoscope and balloon distention as well as CO_2 insufflation. There are no larger series available to document the usefulness of this technique with respect to efficacy and cost-effectiveness.

■ GENERAL CONSIDERATIONS

Angioscopy represents a fascinating possibility to control endovascular maneuvers, especially in demanding operations like tibial and peroneal reconstructions using the in-situ bypass technique. The efficacy of valve ablation can be judged and identification of side branches can be performed safely [4]. The main problem imposed in the in-situ technique are wound complications, related to a continuous skin incision. This problem, affecting one-third of the patients, can be markedly reduced to an incidence of 4-5% [5]. Non-transsected valve cusps can be completely avoided. Angioscopy can also assess the quality of the conduit, which is important because abnormal veins are prone to early occlusion [11]. The identification of side branches is easily achieved and therefore only small skin incisions for the side branch ligations are necessary. The incidence of undetected branches in angioscopically controlled series is low. The endoluminal side branch occlusion with coils is feasible; however, it requires experience and the technique significantly extends operating time. Furthermore, it is mostly done under fluoroscopic control [12]. The endovascular side branch occlusion is at present time an approved technique in single centers. The limitation of angioscopy in in-situ veins are related to the diameter of the scopes. Below a diameter of 3 mm there is a tendency toward increased endothelial damage, which could endanger the function of the arterial graft. The endothelial lesion will be covered with a new layer within four weeks [3,12]. The danger of infection or fluid overload by angioscopy is minimal, using the currently available instruments and closed irrigation systems.

With relation to primary patency and cost, there is no significant advantage of angioscopy in in-situ procedures. In conclusion, the efficacy of angioscopy to render venous valves incompetent is improved in comparison to conventional methods and wound complications are reduced. However, improved patency rates of in-situ bypass procedures using angioscopic assisted techniques cannot be expected.

REFERENCES

1 Hall VK. The great saphenous vein used in-situ as an arterial shunt after extirpation of the vein valves. *Surgery* 1962; 51: 492-495.
2 Schwartz ME, Harrington EB, Schanzer H. Wound complications after in-situ bypass. *J Vasc Surg* 1988; 7: 802-807.
3 Hashizume M, Yang Y, Galt S et al. Intimal response of saphenous vein to intraluminal trauma by stimulated angioscopic insertion. *J Vasc Surg* 1987; 5: 862-868.
4 Wölfle KD, Bruijnen H, Zügel N et al. In-situ saphenous vein grafts for femoro-distal bypass: evaluation of endoscopically monitored vavulotomy of graft patency. *Cardiovasc Surg* 1993; 1: 600-694.
5 Rosenthal D, Herring MB, O'Donnovan TG et al. Endovascular infrainguinal in-situ saphenous vein bypass: a multicenter preliminary report. *J Vasc Surg* 1992; 16: 453-458.
6 Allen KB, Griffith GL, Heimansohn DA et al. Endoscopic versus traditional saphenous vein harvesting: a prospective randomized trial. *Ann Thor Surg* 1998; 66: 26-31.
7 Folliguet TA, Le Bret E, Moneta A et al. Endoscopic saphenous vein harvesting versus "open" technique. A prospective study. *Eur J Cardio Thorac Surg* 1998; 13: 662-666.
8 Jordan WD Jr, Voellinger DC, Schroeder PT et al. Video-assisted saphenous vein harvest: the evolution of a new technique. *J Vasc Surg* 1997; 26: 405-414.
9 Cable DG, Dearani JA, Pfeifer EA et al. Minimally invasive saphenous vein harvesting: endothelial integrity and early clinical results. *Ann Thor Surg* 1998; 66: 139-143.
10 Rosenthal D, Tucker JG, Atkins CP et al. Extraluminal endoscopic-assisted ligation of venous tributaries for infraguinal in situ saphenous vein bypass: a preliminary report. *Cardiovasc Surg* 1996; 4: 512-514.
11 Sales CM, Goldsmith J, Veith FJ. Prospective study of the value of prebypass saphenous vein angioscopy. *Am J Surg* 1995; 170: 106-108.
12 Piano G, Schwartz LB, Foster L et al. Assessing outcomes, costs, and benefits of emerging technology for minimally invasive saphenous vein in-situ distal arterial bypasses. *Arch Surg* 1998; 133: 613-618.
13 Wilson YG, Davies AH, Southgate K et al. Influence of angioscopic vein graft preparation on development of neointimal hyperplasia in an organ culture model of human saphenous vein. *J Endovasc Surg* 1996; 3: 436-444.

21

PROSTHETIC FEMOROCRURAL BYPASS AND ADJUNCTIVE PROCEDURES

JEAN-MICHEL JAUSSERAN, JOHN BUDD, SERGE ROUCHET
JEAN-FRANÇOIS GARBE, MICHEL FERDANI, ANTHONY LAMBERT
GIANFRANCO FADDA, LORENZO DI GIULIO

For more than a decade it has been shown that autogenous veins are the material of choice for infrainguinal and especially femorocrural bypasses [1,2].

In the absence of saphenous vein due to prior use, too small diameter [3], or diffuse varicosity, a prosthetic graft may be used. Polytetrafluoroethylene (PTFE) and polyester (Dacron) grafts have been applied for limb salvage at the level of the crural arteries with or without adjunctive procedures in order to improve graft patency. The main purpose of adjunctive procedures with prosthetic grafts is to diminish the main cause of graft failure: myointimal hyperplasia [4]. In this chapter a technical description of the different procedures is followed by a report of the results of a multicenter study performed between 1994 and 1998 using Dacron grafts.

■ CAUSES OF PROSTHETIC GRAFT FAILURES

Myointimal hyperplasia (MIH) is one of the major causes of graft failure, especially in cases involving prosthetic material. It develops preferentially at the level of the distal anastomosis both in the graft and on the floor of the recipient artery. Arteries with a diameter of 1 or 2 mm are prone to early occlusion due to the process of MIH. Hyperplasia is a non-specific reaction which rapidly develops after surgery and can occur following all kinds of arterial intimal damage. The intensity of MIH is far more apparent in association with direct contact between a prosthesis (PTFE or Dacron) and a recipient artery as compared to venous-arterial interface, probably due to interaction between the graft and blood elements, platelets in particular.

Another reason for the development of MIH is the limited compliance of the graft which is not in harmony with the compliance of the recipient artery. Sauvage et al. [5] have shown that prosthetic grafts are prone to occlusion if the blood flow reaches the critical thrombotic threshold level, due to the higher thrombogenecity as compared to native blood vessels. If grafts are connected to vessels with a high flow resistance, the risk of occlusion is obviously increased.

The attempts to impregnate prosthetic grafts with heparine focus on minimizing these complications and aim at reaching the same wall characteristics as autogenous veins, which accept considerably lower ranges of blood flow without thrombosing.

Kinking of prosthetic grafts, especially at the level of the knee, constitutes another cause of graft failure [6]. Composite grafts (prosthesis-vein) and external support of synthetic grafts decrease the occurrence of these failures.

■ ADJUNCTIVE PROCEDURES WITH PROSTHETIC GRAFTS

The Linton patch [7], conceived before the process of MIH was known, consists of a venous patch on the recipient artery. The patch has to be large enough to allow the prosthetic graft to be anastomosed to the patch, thereby avoiding direct contact between the prosthetics and the artery. Because of their small diameter it is often technically difficult to perform a Linton patch on crural arteries.

A composite or sequential prosthetic-vein graft is possible to construct if a venous segment of sufficient length can be harvested. With such a graft, crossing the knee joint with the prosthetic part can be avoided and the distal anastomosis is performed between the venous segment and the artery. However, MIH can develop at the prosthesis-venous anastomosis.

The principal author performs another procedure half-way down the sequential graft and the venous cuff. It requires only the preparation of a venous segment of 3 cm length and 3 to 4 mm diameter. The vein is anastomosed end-to-side to the recipient artery (Fig. 1). The proximal part of the venous segment is tied and sutured to an adjacent muscle to stabilize the vein. Following a venotomy of 10 to 12 mm the prosthetic graft is anastomosed end-to-side to the venous segment. This technique avoids longitudinal end-to-end anastomoses with discrepancies in diameters, is easy to perform, and seems to offer the same advantages as other *cuffs*.

The technique of the Taylor patch [8] combines a long distal anastomosis with an interposition of a lozenge-shaped venous patch onto the toe of the anastomosis (Fig. 2). This technique is rather difficult to perform on an artery of 1 to 2 mm and does not prevent the prosthesis-artery contact.

The first important series with the Miller cuff was published in 1984 [9], however, it was initially described by Siegman [10]. It consists of a venous cuff which is circumferentially sutured to the arteriotomy with subsequent end-to-end anastomosis with the prosthetic graft (Fig. 3). The Miller cuff has a poor anatomic profile and the bulbous vein cuff can cause severe turbulence due to the distensibility of the vein. At Saint Mary's Hospital a technique has been developed which aims to incorporate the advantages of the Taylor patch and Miller cuff and still retains the superior long-term patency rates that they maintain over the use of PTFE alone. The technique constructs a circumferential venous cuff at the arteriotomy whereafter the cuff is obliquely transected, creating an optimal anatomic and hemodynamic profile (Fig. 4). The technique has a reputation of being rather difficult.

The creation of an arteriovenous fistula at the level of the distal anastomosis is a procedure to prevent prosthetic graft thrombosis by lowering the peripheral resistance. The aim of a fistula is to increase the flow in the graft beyond the thrombotic threshold. The technique of Dardik [11], which is the most commonly applied method, consists of creating a common posterior wall by anastomosing the adjacent walls of the recipient artery and a comitant vein, thus forming a common ostium to which the prosthetic graft is anastomosed (Fig. 5). It is also possible to combine an arteriovenous fistula and a Miller cuff [12], carrying the advantage to decrease the MIH and to diminish the peripheral resistance.

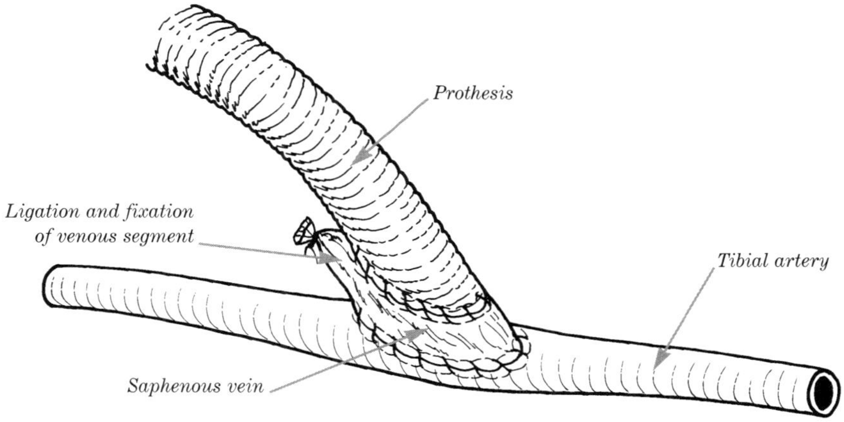

FIG. 1 Anastomosis technique of one of the authors (JMJ).

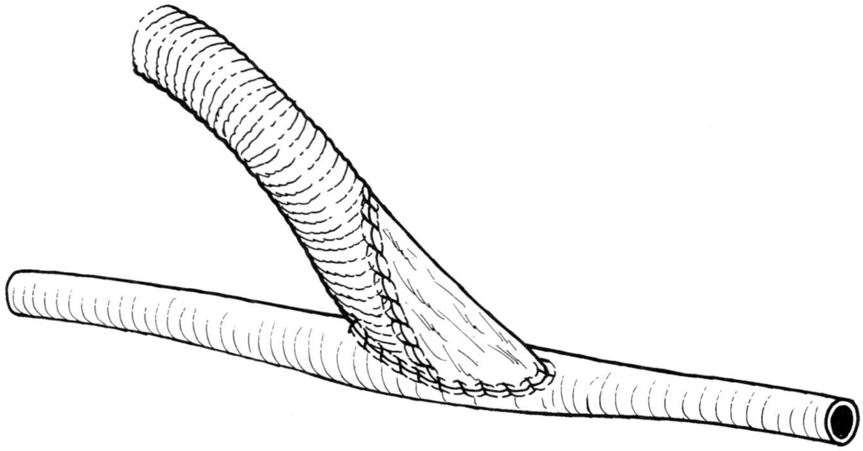

FIG. 2 Taylor patch.

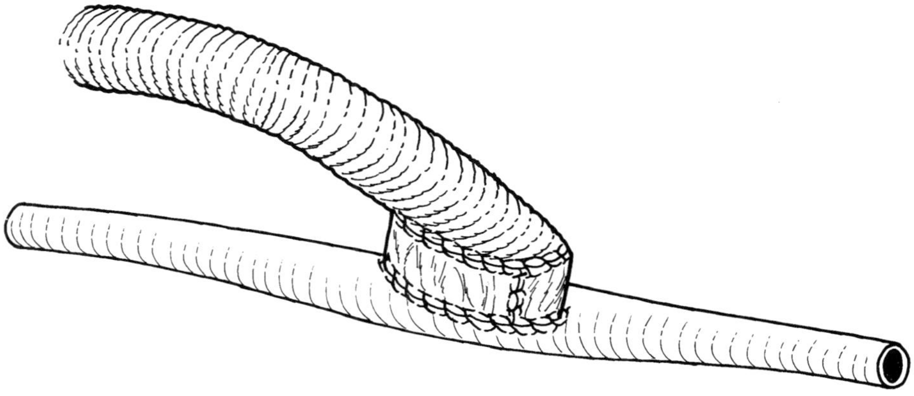

FIG. 3 Miller *cuff*.

■ MULTICENTER STUDY

The authors have performed a prospective, multicenter study comprising 45 patients in order to evaluate the results of small caliber prosthetic grafts to infragenicular arteries. Thin walled Dacron grafts (Intervascular, la Ciotat, France) were either heparin coated or not. All patients received a Dacron graft to a crural artery and were operated between January 1994 and August 1998.

Twenty eight males (62%) and 17 females (38%) had a mean age of 75 years (range 42-91). At the time of surgery the clinical stage comprised 21 patients (47%) with rest pain, gangrene in 18 (40%), acute ischemia in 3 (6.5%), and severe claudication with ankle pressure lower than 50 mmHg in another 3 patients (6.5%). Risk factors included smoking in 20 patients (45%), hypertension in 18 (40%), myocardial ischemia in 17 (38%), hyperlipidemia in 10 (22%), diabetes in 4 (8%) and renal failure in 2 patients (4%). Twenty-seven patients (60%) already underwent previous vascular reconstructions in the target limb; half of them more than one procedure. In all patients the saphenous vein was not available.

Thin walled Dacron grafts were reinforced with rings in 35 cases and heparin coated in 33 cases. In 39 patients (86%) the graft diameter was 6 mm and in 6 patients it was 5 mm. The proximal anastomosis was performed at the common femoral artery in 31 cases (68%), at the superficial femoral or supragenicular popliteal artery in 8, at a previously implanted graft in 5, and at the profunda femoral artery in 1 patient. The distal anastomosis was chosen at the anterior tibial artery in 25 cases (55%), posterior tibial artery in 11 (24%), and the peroneal artery in 9 patients (20%). The distal anastomosis was directly

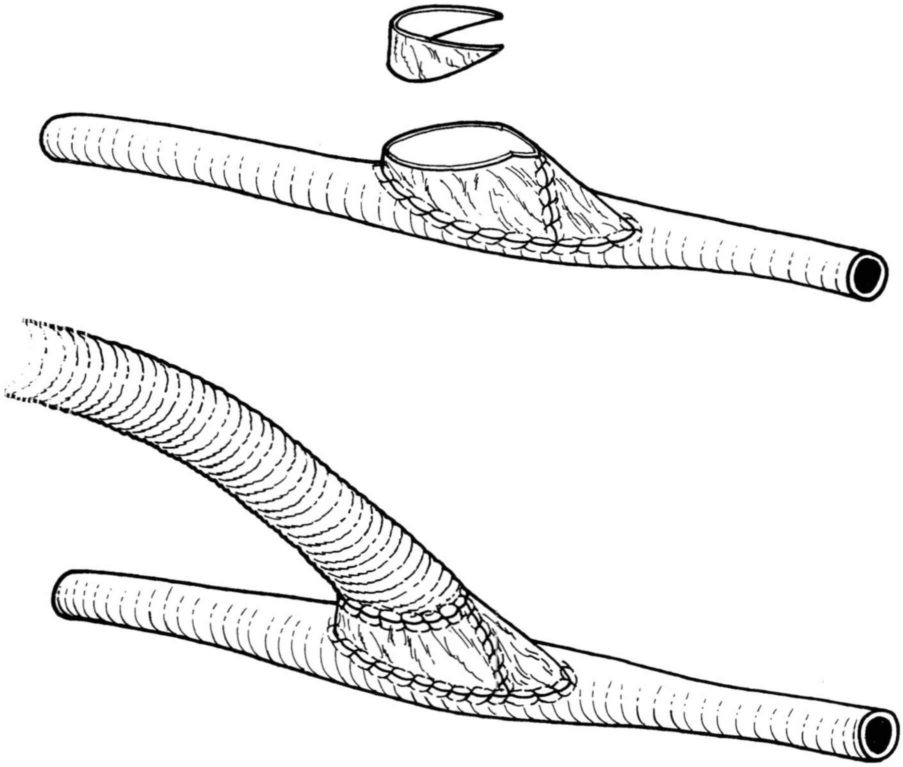

FIG. 4 Saint Mary's adjunctive procedure.

between the graft and the artery in 15 cases (33%) and in 21 a venous cuff and in 9 patients a venous patch was interpositioned.

In total, 67% of the patients received an adjunctive procedure. All procedures were performed under heparinization which was continued for 48 hours. Following, most patients received antiplatelet drugs.

■ RESULTS

Within the first month, one patient died from myocardial infarction. The mean follow-up (interval between intervention and last visit) was 39 months. The actuarial method was used to calculate the primary and secondary patency, limb salvage and patient survival.

Primary graft patency at 1 and 2 years were 45.2% and 41.1%, respectively with a confidence interval of 14.6%. The secondary patency rate at 1 year did not differ due to failing reinterventions (Fig. 6). Limb salvage rate among patients with critical ischemia was 75%, suggesting a salvage gain of 34% when compared with graft patency during the same period. Ten patients underwent a major amputation

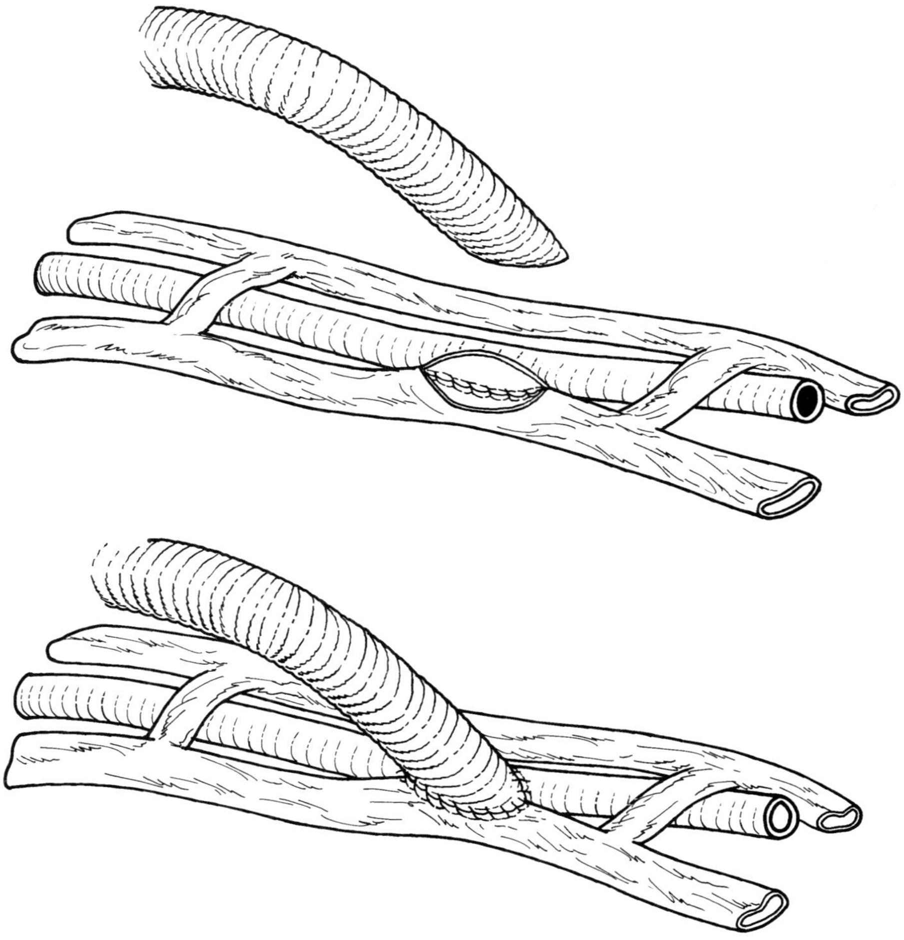

FIG. 5 Creation of arteriovenous fistula.

during follow-up, six of these before the third postoperative month. Cumulative survival rate was 90% at 1 year with a confidence interval of 10% and 83% at 2 years (Fig. 7). During 3 years follow-up, eight patients died, mainly from myocardial infarction.

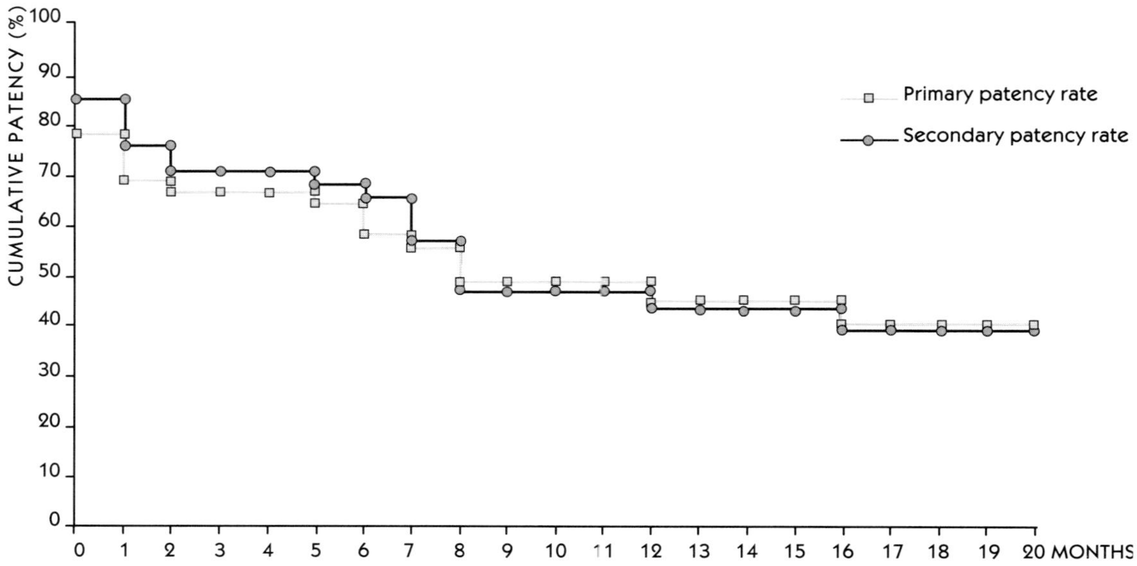

FIG. 6 Primary and secondary patency rate curves of 45 interventions in this series.

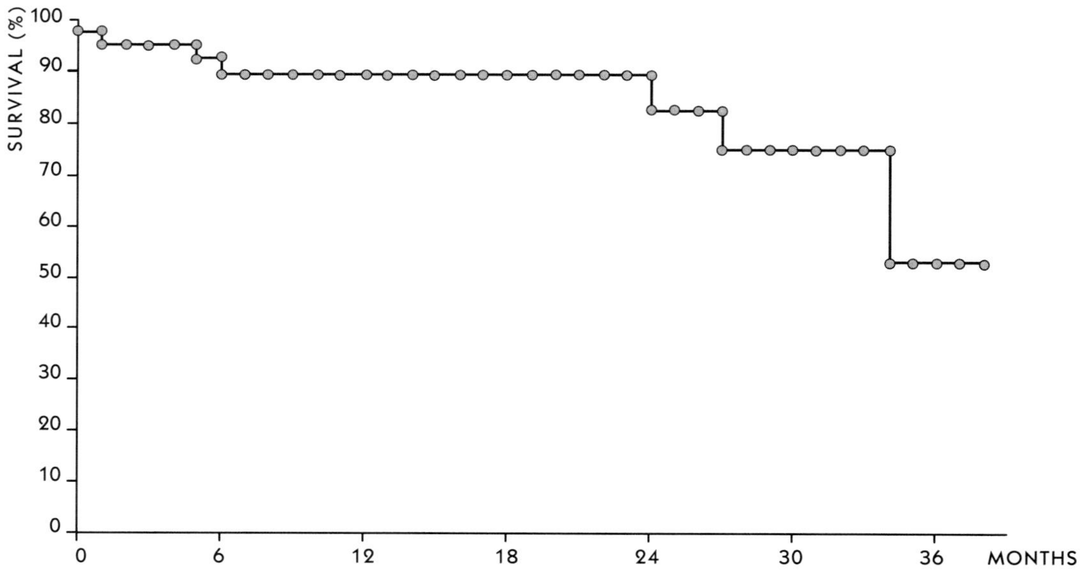

FIG. 7 Cumulative survival rate of 45 operated patients.

■ DISCUSSION

In the absence of a suitable venous conduit, femorocrural bypass surgery can be performed with arm veins, polyester Dacron grafts or arterial and venous allografts. Analyzing the results in the literature it is not always straightforward because few studies report on crural artery reconstructions only. The majority of statistical analyses regroup all infragenicular grafts, i.e. distal popliteal and crural bypasses. Conte et al. [13] reported a primary patency rate of 62% at 2 years in a series of 332 patients who received a saphenous vein bypass to a crural artery. After 5 years the primary and secondary patency rate were 51% and 61%, respectively. These kind of results obtained with saphenous vein, either reversed or in-situ, should be used as a reference when discussing and comparing results of femorocrural grafts. If the greater saphenous vein is not available, arm veins or lesser saphenous vein can be used. In a recent study, Londrey et al. [14] described 257 alternative autogenous vein grafts anastomosed to an infrapopliteal artery in 90% of the cases. Secondary patency rate at 3 and 5 years was 52% and 43%, respectively. This technique demands meticulous handling and precise dissection of the veins and, if necessary, liberal use of venectomy and venovenostomies to achieve high quality conduits.

PTFE grafts are extensively used in lower limb revascularizations. Parson et al. [15] performed 63 femorocrural PTFE grafts without any adjunctive procedure and described a primary patency rate of 39% and 28% at 3 and 5 years, respectively. Secondary patency rate at 3 and 5 years was 55% and 43% respectively, and limb salvage was 71% and 66%, respectively. They discussed that these results justify this strategy, in fact being a better option than amputation in patients without autologous vein. Polyester grafts, however, have not been used on a large scale for crural reconstructions and few series are published. In our experience the primary patency rate at 2 years was 41.1% ± 14.6% and limb salvage rate was 74.9% ± 10%. These results are more or less similar as compared to PTFE grafts.

In our series we did not correlate the results of the heparin coated grafts (33/45) because of the limited number of cases.

Recently, heparin coated polyester grafts were subject to a prospective randomized study with 189 above knee femoropopliteal grafts [16]. Heparin coating did not improve mid-term patency significantly. Nevertheless, patency rates were comparable to those obtained with PTFE grafts. Thirty five grafts in our series were reinforced with rings. At present, no comparative study on polyester reinforcement has been described.

Parson et al. [15] emphasized postoperative anticoagulant therapy and duplex surveillance to improve patency by early correction of restenosis. In our series, as in the experience of Schweiger et al. [17], patients were treated with aspirin.

During recent years, arterial allografts have regained interest for the treatment of critical ischemia and limb salvage. Magne et al. [18] have performed 74 grafts, 50 of which at the level of the crural arteries. Their primary patency rate at 1 year and 30 months was 72% and 30% respectively, and secondary patency was 79% and 50% at the same intervals, respectively. These encouraging early results are important, despite the difficulties with harvesting and preservation, the issue of immunology, and late graft degradation. Cryopreserved venous allografts have also been applied, however, clinical results are disappointing. Martin et al. [19] published a series of 115 cases with cryopreserved saphenous veins and reported a secondary patency rate at 2 years of 25%.

Important technical components of a femorocrural bypass comprise the level of the proximal anastomosis, the route of the graft and the different fashions of the distal anastomosis, with and without adjunctive procedures. The proximal anastomosis should be performed at a normal caliber and more or less undiseased artery, being the common femoral in the majority of cases. In patients who already underwent several groin dissections, the profunda femoral artery might serve as an inflow source. A venous or prosthetic supragenicular graft can also be used for the proximal anastomosis, as was the case in 5 of our 45 patients.

The routing of the prosthesis can either be anatomic or extra-anatomic, however, anatomic tunneling is often hazardous in patients who were operated before. Kinking at the level of the knee and compression in superficial routes can be avoided by reinforced prosthetic grafts. Several adjunctive procedures at the level of the distal anastomosis have been performed to alleviate the discrepancy of graft- (5 to 6 mm) and artery (2 mm) diameters. Various studies have tried to ameliorate the results of prosthetic grafts using additional procedures. Composite grafts offer an interesting solution if a venous segment with sufficient length can be harvested: the prosthesis-vein anastomosis is located above the knee and the venous segment crosses the joint and is sutured to the recipient artery. Recently, Favre et al. [6] reported on 77 femorocrural PTFE-vein composite grafts and described a primary patency rate of 53% and 32% at 6 months and 3 years, respectively. Secondary patency rates at the same time intervals were 60% and 39%, respectively. These results are comparable to those obtained with PTFE only.

Among the adjunctive procedures, the Taylor patch showed a patency for prosthetic grafts to the tibioperoneal trunc of 58% and 54% at 3 and 5 years, respectively.

Stonebridge et al. [20] performed a randomized multicenter study comparing prosthetic grafts with and without a Miller cuff. One and two year patency rates for grafts with a venous cuff were 78% and 49%, respectively as compared to 62% and 19%, respectively for grafts without the cuff. The St. Mary's booth [21] has been performed with PTFE femorocrural grafts showing long-term patency rates of 20% to 25% inferior to those obtained with in-situ venous grafts (Table). In the hands of the authors, additional venous cuffs seem to offer excellent results. However, other recent studies illustrate no beneficial effects [22]. Nevelsteen et al. [23] currently announced the results of a randomized multicenter study *(Belgian Miller Cuff Study Group)*. Femorocrural grafts with a venous cuff carried a patency rate at two years of 37% as compared to 53% for grafts without a cuff (77 cases). Creation of an adjunctive arteriovenous fistula aims for improved hemodynamic profiles at the distal anastomosis and reduced MIH with consequent improved graft patency. Dardik et al. [11] has described a series of 200 umbilical vein grafts and an arteriovenous fistula. Jacobs et al. [24] performed a study comparing 14 crural prosthetic grafts with and 12 grafts without an arteriovenous fistula. One year graft patency with fistula was 64% as compared to 21% without fistula. Harris et al. [12] published a study on infrapopliteal venous grafts, PTFE conduits with arteriovenous fistula and PTFE grafts with combined fistula and venous cuff. Cumulative patency rate at two years was 68% for venous grafts, 61% for PTFE grafts associated with combined fistula and cuff and 28% for PTFE conduits with fistula only.

Table PRIMARY PATENCY OF FEMOROCRURAL GRAFTS USING DIFFERENT TECHNIQUES

Bypass technique	*1st author [ref.]*	*Year*	*N*	*Primary patency at short follow-up (months)*	*%*	*Primary patency at longer follow-up (months)*	*%*
Saphenous vein	Conte [13]	1995	332	24	62	60	51
Other veins	Londrey [14]	1994	257	36	52	60	43
Arterial allografts	Magne [18]	1997	50	12	72	30	50
PTFE	Parson [15]	1996	63	36	39	60	28
Composite PTFE + vein	Favre [6]	1998	77	6	53	36	32
PTFE + patch	Taylor [8]	1992	45	36	54	60	54
PTFE + *cuff*	Stonebridge [20]	1995	96* + 50**	12	65	24	59

* Popliteal artery ** Crural arteries

The European Consensus on critical limb ischemia describes in recommendation 25 that a reconstructive procedure should be attempted if there is a 25% change of saving a useful limb for the patient for at least one year. In our series, patient survival rate was 83% at two years. Graft patency and limb salvage rate at two years was 41% and 75%, respectively, suggesting that our indications corresponded to the standards of the consensus document.

Graft patency and limb salvage do not entirely reflect the success of the surgical procedure. Robison et al. [26] emphasized the significance of jeopardized healing of distal skin incisions, especially at the level of the foot. Careful tunneling, discreet wound closure, and avoiding swelling of the foot are prerequisites in preventing wound complications. Nicoloff et al. [27] followed a group of 77 patients for four years. In addition to graft patency and limb salvage they scored patient's autonomy, delayed wound healing, and number of reinterventions. Delayed healing was observed in 23% of patients and, despite a graft patency of 77% at five years, 22% of patients never healed their surgical wounds or preexisting skin ulcers. Sixty one percent of the patients underwent a reintervention of which the result was optimal in 14% only. This realistic study probably requires reconsideration of the indications for surgery in patients with critical limb ischemia. Comparisons of cost-benefit analyses of femorocrural bypasses and major amputations in patients with critical ischemia are in favor of distal revascularization [28]. The cost for a patient with a major amputation continue during time whereas the cost after an *ideal* revascularization are limited to the procedure and hospital stay.

■ CONCLUSION

The results of PTFE and polyester femorocrural grafts in the treatment of critical ischemia justify their application in the absence of autogenous vein and should be preferred over primary major amputation. Besides the arterial anatomy, the indication for this surgical strategy should depend on patient's life expectancy and degree of autonomy.

REFERENCES

1 Veith FJ, Gupta SK, Ascer F et al. Six year prospective multicenter randomized comparison of autologous saphenous vein and expanded polytetrafluoroethylene grafts in infrainguinal reconstructions. *J Vasc Surg* 1986; 3: 104-114.

2 Bergan JJ, Veith FJ, Bernhard VM et al. Randomization of autogenous vein and polytetrafluoroethylene grafts in femoro-distal reconstruction. *Surgery* 1982; 92: 921-930.

3 Harris PL, Veith FJ, Shanik GD et al. Prospective randomized comparison of in-situ and reversed infrapoplitealveingrafts. *Br J Surg* 1993; 80: 173-176.

4 Sottiurai VS, Yao JTS, Flinn WR, Batson RC. Intimal hyperplasia and neointima: an ultrastructural analysis of thrombosed grafts in humans. *Surgery* 1983; 93: 809-817.

5 Sauvage LR, Walker MW, Berger K et al. Current arterial prosthesis: experimental evaluation by implantation in the carotid and circumflex coronary arteries of the dog. *Arch Surg* 1979; 114: 687-691.

6 Favre J. P, Bourrelle S, Raisky D et al. Revascularization of crural arteries by composite grafts. *Ann Vasc Surg*, in press.

7 Linton RR, Wilde WL. Modifications in the technique for femoro-popliteal saphenous vein bypass autografts. *Surgery* 1970; 67: 234-248.

8 Taylor RS, Loh A, MC Farland RJ et al. Improved technique for polytetrafluoroethylene bypass grafting: long-term results using anastomotic vein patches. *Br J Surg* 1992; 79: 348-354.

9 Miller JH, Foreman RK, Ferguson L, Faris I. Interposition vein cuff for anastomosis of prosthesis to small artery. *Aust NZ J Surg* 1984; 54: 283-285.

10 Siegman PA. Use of the venous cuff for graft anastomosis. *Surg Gynecol Obstet* 1979; 148: 930-934.

11 Dardik H, Berry SM, Dardik A et al. Infrapopliteal prosthetic graft patency by use of the distal adjunctive arteriovenous fistula. *J. Vasc Surg* 1991; 13: 685-691.

12 Harris PL, Bakran A, Enabi L, Nott DM. PTFE grafts for femoro-crural bypass. Improved results with combined adjuvant venous cuff and arteriovenous fistula ? *Eur J Vasc Surg* 1993; 7: 528-533.

13 Conte MS, Belkin M, Donaldson M. L, et al. Femorotibial bypass for claudication: do results justify an aggressive approach ? *J Vasc Surg* 1995; 21: 873-881.

14 Londrey GL, Bosher LP, Brown PW et al. Infrainguinal reconstruction with arm vein, lesser saphenous vein, and remnants of greater saphenous vein: a report of 257 cases. *J Vasc Surg* 1994; 20: 451-457.

15 Parsons RE, Suggs WD, Veith FJ et al. Polytetrafluoroethylene bypasses to infrapopliteal arteries without cuff or patches: a better option than amputation in patients without autologous vein. *J Vasc Surg* 1996; 23: 347-356.

16 Clement C, Plagnol P, Patra P et al. Etude prospective randomisée de 189 revascularisations fémoro-poplitées sus-articulaires par prothèse en polyester tricotée, paroi mince, enduite de collagène avec et sans fixation complémentaire d'héparine. *Ann Vasc Surg*, in press.

17 Schweigger H, Klein P, Lang W. Tibial bypass grafting for limb salvage with ringed polytetrafluoroethylene prostheses: results of primary and secondary procedures. *J Vasc Surg* 1993; 18: 867-874.

18 Magne JL, Fatard Ph, Farah I et al. Arterial allografts in limb salvage procedures. In: A. Branchereau, M. Jacobs (eds). *Long-term results of arterial interventions.* Armonk, Futura Media Service, 1997, pp 321-328.

19 Martin RS, Edwards WH, Mulherin JL et al. Cryopreserved saphenous veins allografts for below knee lower extremity revascularisation. *Ann Surg* 1994; 219: 664-672.

20 Stonebridge PA, Prescott RJ, Howlett J, Ruckley LV. Randomised trial comparing infraguinal polyhtetrafluoroethylene bypass grafting with and without vein interposition cuff at the distal anastomosis. *J Vasc Surg* 1997; 26: 543-50.

21 Williams I, Wolfe J. Long-term results of PTFE grafts for femoropopliteal and crural disease. In: A. Branchereau, M. Jacobs (eds.). *Long-term results of arterial interventions.* Armonk, Futura Media Service, 1997, pp 279-288.

22 Fichelle J, Marzelle J, Colacchio G et al. Infrapopliteal polytetrafluoroethylene and composite bypass; factors influencing patency. *Ann Vasc Surg* 1995; 9: 187-196.

23 Nevelsteen A. Vein interposition cuff at the distal anastomosis: a valuable adjunctive procedure in infrainguinal synthetic arterial reconstruction ? *Ann Vasc Surg*, in press.

24 Jacobs MJHM, Reul GJ, Gregoric ID et al. Creation of a distal arteriovenous fistula improves microcirculatory hemo dynamics of prosthetic graft bypass in secondary limb salvage procedures. *J Vasc Surg* 1993; 18: 1-9.

25 Anonymous. Second European Consensus Document on Chronic Critical Leg Ischemia. *Circulation* 1991; 84 (suppl IV): 1-26.

26 Robison JG, Ross JP, Brothers TE, Elliott BM. Complications de la cicatrisation des incisions distales des pontages au pied: Analyse des facteurs de risque. *Ann Chir Vasc* 1995; 9: 53-59.

27 Nicoloff AD, Taylor LM, Mc Lafferty RB et al. Patient recovery after infrainguinal bypass grafting for limb salvage. *J Vasc Surg* 1998; 27: 256-266.

28 Cheshire MJW, Wolfe JHN, Moore MA et al. Comparative cost between bypass to infrapopliteal arteries and amputation. *J Vasc Surg* 1992; 15: 167-175.

22

INFRAINGUINAL BYPASS IN THE ABSENCE OF SAPHENOUS VEIN: ALTERNATIVE VENOUS CONDUITS

JACQUES WATELET, DIDIER PLISSONNIER, JACQUES TESTART

The greater saphenous vein (GSV) is the material of choice for infrainguinal bypass procedures, but is not available in a percentage of cases varying from 10% to 30%, and even 40% in some studies in which the number of redo procedures is high. The poor results of prosthetic infrainguinal bypasses justify a preferably complete exploration of the autologous venous arsenal available in the absence of an ipsilateral GSV. In these circumstances, the contralateral GSV remains the first choice for the majority of the centers, except when the contralateral limb is also critically ischemic, requiring revascularization on a more or less short notice.

The veins of the upper extremity and lesser saphenous veins provide a source of autologous venous material, which is scarcely used, but often available. The venous arsenal comprises also the superficial femoral and popliteal veins, which nowadays are used anecdotally.

UPPER EXTREMITY VEINS

Arm veins offer a frequently available venous arsenal. In 1969, Kakkar et al. showed that the cephalic vein had a length sufficient to be implanted between the common femoral and the distal popliteal artery and could resist pressures up to 400 mmHg. Excellent results were obtained using this vein [1]. The cephalic vein may be harvested in its totality, from the infraclavicular fossa to the anatomical snuffbox, by means of a continuous incision: the length of the vein thus obtained allows the creation of a femoro-infragenicular bypass. The diameter of the conduit after implantation is about 5 to 7 mm; 1 to 2 mm more than that of the GSV. The almost uniform caliber allows a reversed usage. One might as well use the basilic vein or even its brachial segment together with the antebrachial cephalic vein (Fig. 1). These two combinations yield a sufficient length so as to perform a femoropopliteal bypass. When the veins of the forearm cannot be used, one could harvest the basilic, median cubital, and cephalic veins in one piece (Fig. 2), as was shown by LoGerfo et al. [2]. These veins are larger than the veins of the forearm and offer in total a sufficient length to perform an infragenicular bypass (Fig. 3). Harvesting in one piece is possible only if the median cubital vein is patent: this can routinely be checked by physical examination using a cuff. However, duplex scanning is most frequently required to complete the data from clinical examination; it specifies the localization of collaterals and possible vessel wall lesions caused by prior punctures. The proximal and distal vein diameter are determined by means of a cuff and the investigation ends by marking the vein and its side branches, which facilitates the harvesting. As soon as the use of upper extremity veins is considered, one should prohibit any venous infusion or puncture.

When the caliber of the venous network of the lower arm as determined by duplex scanning using a cuff appears to be of too small a diameter to perform a bypass (i.e. 2.5 mm), some have proposed performing a temporary arterialization of the antebrachial venous network in order to increase the diameter up to 3 or 4 mm. An arteriovenous fistula may be created at the wrist between the cephalic vein and the radial artery through a lateral anastomosis. Other anastomotic sites are possible depending on

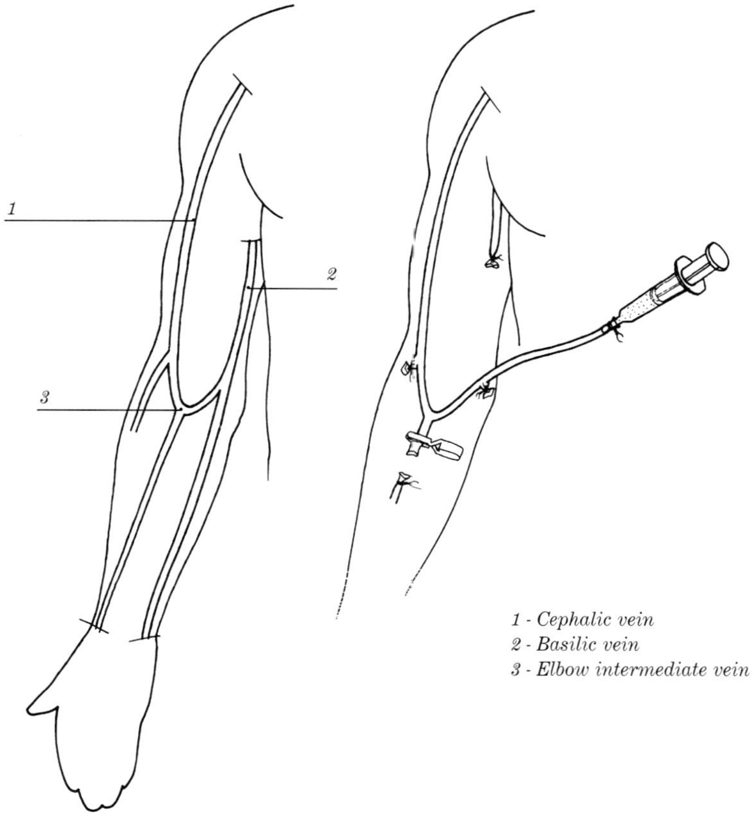

FIG. 1 Different possibilities with arm veins.

the anatomy of the venous network: an anastomosis between the basilic vein at the wrist and the ulnar artery or a lateral anastomosis in the elbow crook between the median cubital vein and brachial artery. A delay of 10 to 15 days would be sufficient to obtain a substantial dilatation of the venous network.

TECHNIQUE

General anesthesia is imperative. Harvesting and arterial access surgery performed simultaneously by two teams reduces the operation time. The harvesting of veins of the upper limb by means of continuous incisions should be performed meticulously, because of the thin wall of the vein. The side branches should be skeletized before ligation to avoid strictures. When the vein is exposed the surroundings of the vein are infiltrated with heparinized saline containing papaverine to prevent spasm. Despite these precautions, spasm frequently complicates the harvesting. This can be alleviated by injecting heparinized saline or blood in the distal part of the vein. The risk of excessive pressure is avoided if one takes care not to clamp the distal end of the vein during injection.

If the basilic and cephalic veins are harvested in one piece, the basilic vein is used non-reversed, because of its larger caliber, and the cephalic vein reversed. Thus, the congruence of the anastomoses is

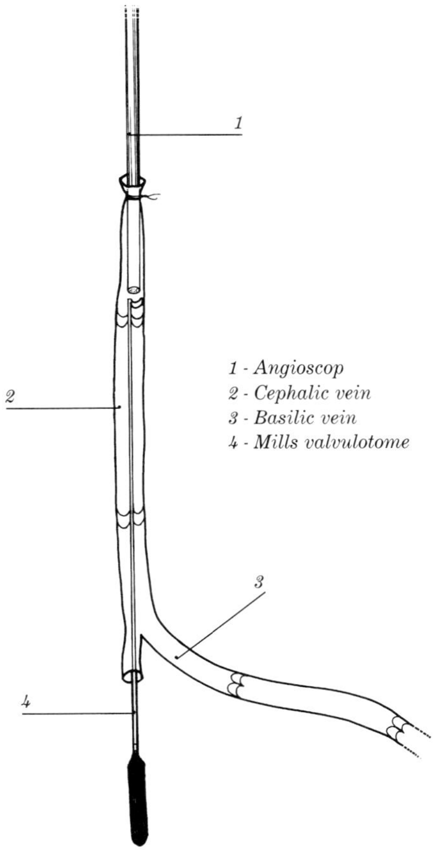

FIG. 2 Use of intermediate elbow vein, as described by LoGerfo [2].

Fig. 3 Completion angiography of fedo below-knee femoropopliteal bypass with reversed cephalic vein.

better, especially between the distal part of the cephalic vein and an infragenicular popliteal or crural artery. After completing the proximal anastomosis, the destruction of the venous valves of the basilic vein, when used non-reversed, is performed by means of a valvulotome, introduced through a remainder of the antebrachial cephalic vein kept for this purpose. After checking the flow, this remainder is closed and the distal anastomosis can be made.

The implantation should be made preferably through a subcutaneous tunnel to facilitate the following surveillance, since the veins of the upper extremity show a tendency to dilate in the long run, justifying clinical and ultrasonographic surveillance. The subcutaneous tunnel also allows autosurveillance by the patient and a greater accessibility for endovascular or surgical treatment of complications.

Results

The actuarial patency rate of infrainguinal revascularizations by means of upper extremity veins are very divergent, ranging from 43% to 85% after one year in the different series [5]. According to Stonebridge et al., the strongly varying proportion of diabetic patients among the series (27%-80%) may explain the differences in results [5]. In 102 infrainguinal bypass procedures (of which 52% were infrapopliteal) using upper extremity veins, Harris et al. [6] obtained a patency of 82% after one year and 60% after five years.

The series of Balshi et al. [1] comprised 36 bypass procedures of which 58% were infrapopliteal; the bypasses using upper extremity veins in one piece had primary and secondary patency rates after one year of 73% and 82%, respectively.

The bypass procedures performed by means of anastomosing several upper extremity veins had primary and secondary patency rates of 53% and 66%, respectively. In the same series infrapopliteal bypasses had a secondary patency of 79% after one year. In the series of Stonebridge et al. [5], a statistically significant reduction of postoperative thrombosis was found due to intraoperative angioscopic control.

Gentile et al. [7] performed 187 femoropopliteal (n=54) and femorocrural (n=133) bypass procedures using venous material other than the GSV. Upper extremity veins were used for 144 bypasses (77%); only venous in 97 cases and partially venous in 47 cases. The lesser saphenous vein was used in 11 cases and the superficial femoral vein in 5 cases. One hundred fourteen bypasses were constructed from several venous segments of different origin, with a single intermediate anastomosis in 87 cases (76%) and two or more anastomoses in 27 cases. The primary assisted patency rate was 63% after five years for femoropopliteal bypasses and 72% for femorocrural bypasses. These results were comparable with those obtained by the author using the GSV as bypass material.

The largest experience with upper extremity venous material is reported by Hölzenbein et al. [8]. The author's first choice was upper extremity veins as bypass material in the absence of ipsilateral GSV. Vein selection was ameliorated by the systematic application of angioscopy, which allowed guided corrections to improve the quality of the conduit in 52% of the cases. Thus, 199 bypasses were performed with a single vein, while 51 were performed with composite veins; 114 bypasses were femorotibial, 41 were femoropopliteal and 33 were popliteodistal. The cephalic vein was used in 86% of the bypasses and was the only bypass material in 50% of the cases. The basilic and cephalic veins used in one piece, according to the technique of LoGerfo, consisted of 35.6% of the bypass procedures. The overall primary assisted patency rate after one year was 70.6% and the secondary patency 77%. After three years, the primary patency was 52% for primary bypass procedures, 42% for redo's and 57% for bypass procedures using upper extremity veins. Other authors have presented more modest results. Calligaro et al. [9] found primary assisted and secondary patency rates after two years of 46% and 55%, respectively. In their series, the best results were obtained with bypass procedures using a single vein, yielding a primary assisted and secondary patency rate after two years of 55% and 64%, respectively.

In a study of Sesto et al. [4] comprising 34 bypasses, of which 25 were infragenicular, the primary and secondary patency rates after three years were 40% and 46%, respectively.

■ LESSER SAPHENOUS VEIN

Long ignored because of harvesting problems, the lesser saphenous vein (LSV) forms another alternative in the absence of the GSV. According to Weaver et al [10] the LSV could be used in 85% of the cases to perform a bypass procedure. In a study by Ouriel [11] the LSV was available in 72% of the patients requiring an infrapopliteal revascularization. As clinical assessment of the LSV is often difficult, it may be necessary to order a duplex scan to study the anatomy of the vein, the proximal and distal caliber, the location of its main side branches, and its junction with the popliteal vein.

TECHNIQUE

Harvesting is performed with the patient in prone position, through a continuous incision along the ultrasound markings. The two combined LSV's enable the creation of a femorocrural bypass. When a bypass is indicated from the distal popliteal artery to a calf artery, the LSV may be used *in-situ* using the technique described by Shandall et al. [12]. While the patient is in the supine position, the popliteal artery is approached through a conventional medial retrotibial incision, widely extended distally to be able to approach the terminal part of the LSV, releasing it from below the aponeurosis. After transection of its distal part the vein is mobilized sufficiently to pull it up to the popliteal artery. The marked side branches distal to the transection are ligated. The lower part of the vein is approached separately and can be anastomosed, either with the posterior tibial artery requiring a short subcutaneous tunnel, or with the anterior tibial artery after creation of a tunnel across the muscles of the deep posterior compartment of the calf and the interosseous membrane. After finishing the proximal anastomosis the valvulotomy can be performed, either according to Leather's technique, with a Mills valvulotome (Fig. 2), or the standard way using a conventional valvulotome.

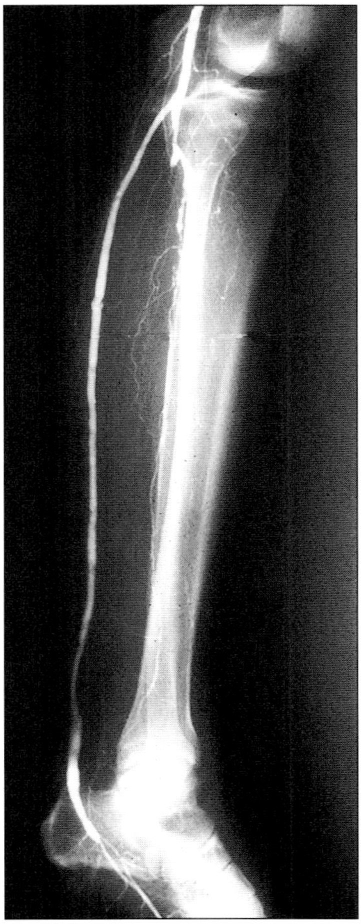

Fig. 4 Popliteal-plantar bypass with in-situ lesser saphenous vein (posterior approach).

A modification of this in-situ LSV technique is performing the whole or part of the bypass procedure on the patient in the prone position. The junction of the LSV is dissected through a bayonet incision in the popliteal fossa, or through a large transverse incision opposite or somewhat above the popliteal fold, depending on the data from the preoperative ultrasound localization. The LSV transected close to its junction with the popliteal vein can easily be anastomosed with the adjacent popliteal artery. Valvulotomy is performed after transection of the LSV at the ankle.

Subsequently the bypass can be anastomosed to the posterior tibial artery, either in its lower third or more distally in its retro or inframalleolar part down to its bifurcation (Fig. 4). To perform such an anastomosis, the ankle and the dorsum of the foot are supported on a block and the foot is kept in valgus. The anastomosis may also run towards the fibular artery, of which the access to the distal part is simple using the posterior route down to its bifurcation [13]. The anastomosis can even be made distally at the posteromedial branch of the fibular artery when it is hypertrophied and supplies the malleolar arch. Finally, the distal anastomosis may run towards the lower anterior tibial artery or the dorsal pedal artery after closing the incisions and reinstalling the patient in supine position. The bypass is then tunneled either through a subcutaneous tunnel anterior to the medial malleolus or across the interosseous membrane.

Results

In the study of Weaver et al. [10] the LSV was utilized in 29 cases in order to perform a revascularization of the infragenicular popliteal artery in four cases and of the crural arteries in 25 cases, with an actuarial patency rate of 60% after three years. In 14 cases the LSV was used together with autologous material with a patency rate of 47% after one year and 38% after 18 months.

Ouriel [11] performed a posterior popliteocrural revascularization using the LSV in 21 patients of which 81% were diabetics. The proximal anastomosis of the bypass was placed on the supragenicular popliteal artery in four cases and infragenicular in 17 cases; the site of the distal anastomosis was the tibiofibular trunk in six cases, the fibular artery in nine cases, the posterior tibial artery in four cases, and the anterior tibial artery in two cases. The primary patency rate was 83% after 18 months.

For Calligaro et al. [9] the results of infrapopliteal bypass procedures using LSV were disappointing, showing a primary assisted patency rate of 23.5% after two years, comparable to the results of infrapopliteal bypasses using PTFE (26%).

The results obtained by means of revascularizations with or without composite LSV are therefore in disagreement. This material appears useful in particular for popliteodistal revascularizations through a posterior approach, using the LSV in-situ or reversed.

■ SUPERFICIAL FEMORAL AND POPLITEAL VEINS

The application of the superficial femoral vein for infrainguinal bypass procedures has been proposed by Schulman et al. [14]. The superficial femoral and popliteal veins are harvested through a single or double incision creating a cutaneous bridge. Sidebranches are ligated with clips. The common and profunda femoral veins are preserved so as to maintain venous drainage of the lower extremity. The harvesting requires transsection of the tendon of the third adductor muscle and, when necessary, those of the internal muscles and tendons of the knee. Given the fact that the junction of the superficial and common femoral veins is located 8 cm below the arterial bifurcation, the loss of length often forces the creation of short bypasses, starting from the superficial femoral of profunda femoral artery. Valvulotomy is performed using the standard technique. The graft is applied in-situ and tunneled deeply.

The anastomoses are made over a length of at least 3 cm, allowing a gradual reduction of the caliber of the graft, in order to improve the hemodynamic conditions and to reduce the incidence of intimal hyperplasia. Veins with a caliber exceeding 12 mm, inducing a high failure rate according to Schulman, should be rejected. The author observed no venous trophical disturbances and postoperative edema was gradually resolved in a few months. The results as obtained by Schulman et al. at the femoropopliteal level are equal to those of a comparative series of bypasses using reversed saphenous vein, and are comparable with those of the better series of saphenous vein bypasses (primary and secondary patency rates after three years of 82% and 89%, respectively) [14]. However, no series in the literature so far has confirmed these results.

■ COMPOSITE VEINS

In case of insufficient length of the main venous segment, the various venous materials may be sutured together in multiple combinations in order to create an infrapopliteal bypass. Special care should be taken while creating venovenous anastomoses, of which some authors underline the adverse effect on the long-term patency, requiring a more intense ultrasound surveillance. The anastomoses are made end-to-end using a rod for support, if necessary. In small-caliber veins, opposite longitudinal incisions allow the creation of a long oblique anastomosis. Too great a mismatch may advocate the creation of a lateral anastomosis. The anastomoses are created under optical magnification using monofilament 6x0 or 7x0 stitches. When an anastomosis is made on small-caliber veins less than or equal to 4 mm, separate stitches may be made on one hemicircumference.

In the study of Harris et al. [6] 41 venous bypasses of this type were created with autologous venous material derived from upper and lower extremities and anastomosed in 80% of the cases with infrapopliteal vessels, with a actuarial patency rate of 73% after one year.

The study of Chang et al. [15] comprising 184 bypasses created with veins of various origin (upper extremity: 11%, LSV: 28%, ipsi or contralateral GSV: 61%), of which 84% were infrapopliteal, showed primary and secondary patency rates after four years of 45% and 61%, respectively. These patency rates were much poorer than those obtained by the author with in-situ, non-composite bypasses. Belkin [16] has stressed the importance of ultrasound surveillance, showing that 29% of the bypasses of this type required surgical or endoluminal interventions because of stenoses at the venovenous anastomosis sites.

■ CONCLUSION

In the absence of ipsilateral greater saphenous vein, one may settle for an infrainguinal revascularization using prosthetic material at the supragenicular level with satisfactory results. On the other hand, at the infragenicular level the superiority of the venous material is such that its use should be preferred. The use of the contralateral greater saphenous vein is the first choice in the absence of critical ischemia and arterial lesions that may require a revascularization of that extremity on short or intermediate term.

The data from the literature appear to prefer as second choice the use of upper extremity veins, of which the results may be optimized by means of intraoperative angioscopy.

The use of the lesser saphenous vein has the drawback to require harvesting in the prone position. The creation of an infragenicular bypass requires bilateral harvesting. As a matter of fact, the lesser saphenous vein is especially suited for infragenicular popliteodistal revascularizations of the posterior tibial artery and its branches, as well as the distal fibular artery. The use of the deeper superficial femoral and popliteal veins should be further evaluated.

REFERENCES

1 Balshi JD, Cantelmo NL, Menzoian JO, LoGerfo FW. The use of arm veins for infrainguinal bypass in end-stage peripheral vascular disease. *Arch Surg* 1989; 124: 1078-1081.

2 LoGerfo FW, Paniszyn CW, Menzoian J. A new arm vein graft for distal bypass. *J Vasc Surg* 1987; 5: 889-891.

3 Beals RL. Surgically created arteriovenous fistula to augment the cephalic vein: use as an arterial bypass graft. *N Engl J Med* 1971; 285: 29-30.

4 Sesto ME, Sullivan TM, Hertzer NR et al. Cephalic vein grafts for lower extremity revascularization. *J Vasc Surg* 1992; 15: 543-549.

5 Stonebridge PA, Miller A, Tsoukas A et al. Angioscopy of arm vein infrainguinal bypass grafts. *Ann Vasc Surg* 1991; 5: 170-175.

6 Harris RW, Andros G, Salles-Cunha SX et al. Alternative autogenous vein grafts to the inadequate saphenous vein. *Surgery* 1986; 100: 822-827.

7 Gentile AT, Lee RW, Moneta GL et al. Results of bypass to the popliteal and tibial arteries with alternative sources of autogenous vein. *J Vasc Surg* 1996; 23: 272-280.

8 Hölzenbein TJ, Pomposelli FB, Miller A et al. Results of a policy with arm veins used as the first alternative to an unavailable ipsilateral greater saphenous vein for infrainguinal bypass. *J Vasc Surg* 1996; 23: 130-140.

9 Calligaro KD, Syrek JR, Dougherty MJ et al. Use of arm and lesser saphenous vein compared with prosthetic grafts for infrapopliteal arterial bypass: are they worth the effort ? *J Vasc Surg* 1997; 26: 919-927.

10 Weaver FA, Barlow CR, Edwards WH et al. The lesser saphenous vein: autogenous tissue for lower extremity revascularization. *J Vasc Surg* 1987; 5: 687-692.

11 Ouriel K. The posterior approach to popliteal-crural bypass. *J Vasc Surg* 1994; 19: 74-80.

12 Shandall AA, Leather RP, Corson JD et al. Use of the short saphenous vein "in-situ" for popliteal to-distal artery bypass. *Am J Surg* 1987; 154: 240-244.

13 Watelet J. Revascularisation des artères de jambe par voie postérieure. *Encycl Méd Chir* (Paris-France). Techniques chirurgicales. Chirurgie Vasculaire, F.a. 43-029 H, 1995, 3 p.

14 Schulman ML, Badhey MR, Yatco R. Superficial femoral-popliteal veins and reversed saphenous veins as primary femoro-popliteal bypass grafts: a randomized comparative study. *J Vasc Surg* 1987; 6: 1-10.

15 Chang BB, Darling RC 3rd, Bock DE et al. The use of spliced vein bypasses for infrainguinal arterial reconstruction. *J Vasc Surg* 1995; 21: 403-412.

16 Belkin. In discussion ref. 15.

23

INFRAINGUINAL ARTERIAL ALLOGRAFTS FOR LIMB SALVAGE

JEAN-NOËL ALBERTINI, XAVIER BARRAL, ALAIN BRANCHEREAU
JEAN-PIERRE FAVRE, HENRI GUIDICELLI
PIERRE-EDOUARD MAGNAN, JEAN-LUC MAGNE

The arterial allograft (AA) has been widely used for arterial reconstruction at the aortoiliac and infrainguinal level since the beginning of arterial reconstructive surgery in the 1950s. However, it was soon abandoned because of the high incidence of mid- and long-term degradation [1].

At the beginning of the 1990s, the AA was reintroduced for the management of aortic graft infections and critical limb ischemia [2,3]. The disappointing results provided by prosthetic grafts for these clinical indications can explain this regained interest. Furthermore, improvements in organ harvesting and preservation techniques, as well as the positive results achieved with heart valve allografts [4] contributed to the current interest in this technique.

■ HARVESTING TECHNIQUE

The allografts are harvested from brain dead donors during multiorgan donor procedures. The legal regulations for allograft harvesting are similar to those for multiorgan donation, in particular the requirements for virological screening. Donor age ranges usually from 18 to 40 years. The harvested arteries must be of an adequate diameter and free of atherosclerotic lesions. They are collected after completion of organ harvesting: heart and lungs, liver, pancreas and kidneys. The ascending aorta and the arch are taken together with the heart and most of the infrarenal aorta is removed during harvesting of liver and pancreas. The vascular surgeon collects the descending thoracic aorta, the distal aorta including the aortic bifurcation and iliac arteries, and the femoral and popliteal arteries down to the crural arteries. The collection must be performed under strict sterile conditions. Excessive stretching of the vessels should be avoided to prevent lesions of the intima and/or the media. Perfusion of the arteries with preservation fluid may ease harvesting (through cannulation of the internal iliac artery or the infrarenal aorta). Side branches should be transected far enough from their aortic origin to allow subsequent adequate ligation. The harvested arteries are rinsed and placed in a preservation solution with antibiotics (Table I). They are transported in an isothermic container filled with ice. Donor blood samples are included to allow virological screening and blood type assessment.

Table I COMPOSITION OF PRESERVATION SOLUTION FOR FRESH ALLOGRAFTS

Center A		*Center B*	
Belzer [5]		RPMI 1640 [6]	
Gentamycine	240 mg/l	Gentamycine	20mg/l
Colimycine	10^6 U/l	Metronidazole	500 mg/l
Lincomycine	400 mg/l	Curoxime	750 mg/l
		Fungizone	50 mg/l
		Hepes	510^{-4} mol/l
		Heparine	7500 U/l
		Human albumin	50mg/l

Center A: CHU Grenoble. Center B: CHU Timone, Marseille.

■ PRESERVATION OF FRESH ALLOGRAFTS

The first used allografts were normally stored in departments of vascular surgery, in a preservation solution with antibiotics at 4°C. Table I presents the composition of both solutions we used in our experience. In a series of 45 fresh allografts, the average preservation period was 14 days, ranging between 2 and 40 days. Fresh allografts can be stored for a period up to one month, after which major structural alterations will appear. This limitation impairs the organization of large artery banks for AA storage and reduces the possibility to respect immunological compatibility. With this preservation technique, virological contamination remains possible if the donor was in the seronegative phase of the disease (for hepatitis and AIDS in particular). The main advantage of the fresh AA is its immediate availability, allowing its use in emergency procedures. Furthermore, fresh AA might have a higher resistance to infection than cryopreserved AA. However, these two advantages are actually of limited interest in the management of critical limb ischemia.

■ CRYOPRESERVATION

Cryopreservation of AA includes a specific programmed freezing process and storage in liquid nitrogen (-196°C) or in gas nitrogen (-140°C). A cryoprotector is added to the initial preservation solution with a maximum delay of 48 hours. The most commonly used cryoprotector is dimethyl sulfoxyde (DMSO) at concentration rates ranging from 10% to 15% according to the centers. Freezing takes place in a freezer equipped with a temperature probe, according to a progressive and preprogrammed process. Monitoring the rate of temperature decrease allows control on water migration from the cell and thus the intracellular hydratation. Too high deshydratation will lead to a high intracellular osmolality and cell death. In the case of insufficient dehydratation, a large number of ice crystals will appear. The volume of these crystals increases during the freezing process, causing damages to the cellular structures. Cryoprotectors decrease the speed of crystal formation, decrease their growth rate, and change their shape. They also have direct effects on the cell that are not well known. Cryopreserved arteries have a long storage period, which allows the constitution of artery banks and solves availability issues. The second advantage is total virological safety with regard to hepatitis and AIDS. Serological tests are performed on the patients who have received organs from the donor (kidney, heart, liver). When these tests remain negative beyond a three-month period, there will be no risk of contamination by the tissues coming from the same donor. Cryopreserved allografts will thus be released after a delay of three months. The better results obtained in some studies with cryopreserved venous allografts as compared to fresh ones may speak in favor of a lesser immunogenicity, but this has not been proven. The main drawback of cryopreservation is the high cost of the technique, which is caused by the need for a complex infrastructure of equipment as well as skilled personnel. This should be taken into consideration when discussing the different types of grafts in the management of critical limb ischemia.

■ BIOMECHANICAL PROPERTIES

In theory, the main advantage of AA when compared to other grafts is that their biomechanical properties are similar to normal arteries. For fresh allografts, no study has ever analyzed the evolution of these properties according to the storage period at 4°C.

It has been suggested that cryopreservation would impair mechanical properties, especially viscoelasticity, which would carry a disadvantage as compared to fresh arteries. Our team has performed a comparative study measuring the viscoelastic properties of fresh and cryopreserved allografts of harvested human arteries. Only slight changes have been observed after cryopreservation [7], especially for arteries of elastic type (common carotid artery). The arteries which are commonly used as AA in clinical practice (for the iliac-femoropopliteal level in particular) did not show any significant changes. Evolution of the biomechanical properties of AA after implantation remains unknown. In our experience, ultrasound investigation during patient follow-up showed that the implanted AA remains compliant and that compliance rates seem comparable to healthy arteries. These comments are nevertheless based on solitary and subjective observations. Such data must be confirmed by a prospective study including compliance measurements.

■ IMMUNOGENICITY

AA immunogenicity remains a major problem, and the knowledge on the subject is limited. AA rejection has been described in immunologically incompatible models in the rat [8,9]. An injection containing the equivalent of lymphocytes CD4 and CD8 in the rat leads to endothelium destruction, followed by destruction of smooth muscle cells in the media. After destruction of the smooth muscle cells, the injected substances disappear, leaving a fibrous and thin media. Duration of the whole process is approximately three months. These histological changes are associated with a dilatation of the AA, mainly due to the destruction and reorganization of the elastic fibers and the collagen of the media, which occurs together with the destruction of the smooth muscle cells.

In our experience, the histological analysis of explanted grafts showed similar characteristics: absence of endothelium and smooth muscle cells, medial fibrosis and mononucleus cell infiltration (Figure 1). At least three hypotheses can be proposed to explain the disappearance of endothelial cells and smooth muscle cells: destruction by the immunocompetent cells of the receiver, spontaneous cell death when the allograft is not viable, and cytotoxicity of the preservation solution and of antibiotics in particular. While the existence of this rejection phenomenon can reasonably be accepted, the importance, the frequency, and above all the clinical significance (AA occlusion and/or degradation) are unknown.

■ SURGICAL TECHNIQUE

AAs are supplied by the tissue bank on request by written medical prescription. The surgeon must specify the type of bypass (femoropopliteal or femorocrural) and the number of AAs, as well as the blood

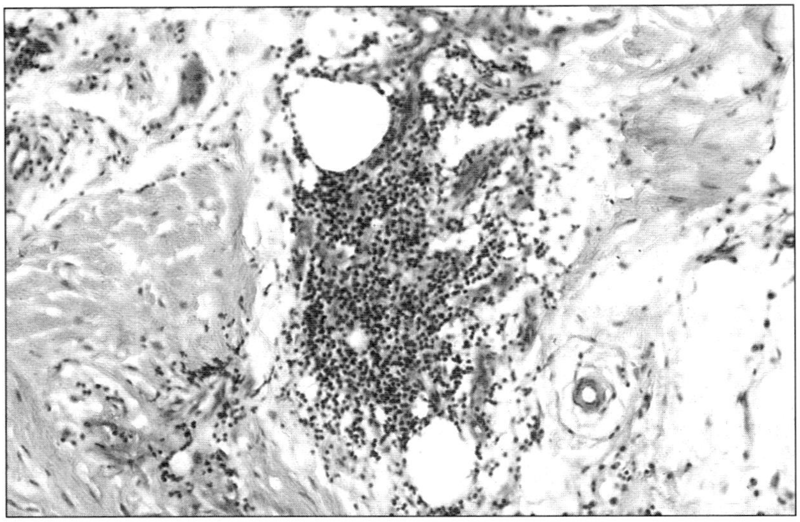

FIG. 1 Microscopic analysis of explanted AA showing chronic rejection characterized by lymphocyte infiltration.

type if he or she wishes to respect blood type compatibility. Thawing is generally performed in the operating theater, after the surgeon decides to use the AA.

The pouch containing the AA is placed in a water bath at a temperature of 37°C, allowing fast thawing. In order to prevent transverse breakages in the vessel wall, handling of the AA should be absolutely avoided until the last piece of ice has melted. The pouch is then opened and the DMSO/preservation solution is discarded. The AA is rinsed with heparinized serum and the side branches are ligated using nonabsorbable sutures (Figure 2). The AA should be handled with care during all these steps. While the AA is thawed and prepared, a second team takes care of vascular access.

The surgical technique is not different from the usual technique used for limb salvage procedures. The diameter of the distal part of the AA should correspond to the diameter of the receiving artery. One of the advantages of this technique is the excellent congruence of the anastomoses when the proximal part of the AA is a femoral artery and the distal part a crural artery. Because of the natural elasticity of the AA, length requirements are less strict than when a prosthesis or a saphenous vein is used. AAs have a significant longitudinal elasticity, providing an increase in the length which is much higher than with prosthetic or even autologous vein grafts. The following technique can be used when the surgeon prefers to use the distal part of the AA, consisting of a 4 to 5 cm segment of crural artery, to perform the distal anastomosis: after a rough estimation of the necessary graft length, both anastomoses are made. Clamps are removed and the graft is put under arterial pressure. The excess length of the graft is assessed and subsequently shortened in the area of the largest diameter. The use of metallic clamps should be avoided. For this purpose we use atraumatic bulldog clamps covered with latex sheaths (microclamp Fogarty, Baxter®). In our experience we always perform an intraoperative completion angiography (Figure 3) to verify the anatomic perfection of the bypass and the anastomoses, and to assess distal runoff as well as foot arteries patency (Figure 4).

■ THE FRENCH EXPERIENCE

A retrospective study on the results of infragenicular AA for limb salvage has been performed in three academic departments of vascular surgery in France: CHU Nord - Saint-Etienne, CHU Nord - Grenoble, CHU Timone - Marseille).

PATIENTS

Between 1991 and 1997, 165 infragenicular bypasses have been performed with AAs in 148 patients (90 men and 58 women). Mean age was 70 years (range 20 to 90 years). Surgical indication was rest pain in 54 patients and skin lesions in the other cases. The mean ankle pressure was 53 mmHg in nondiabetic patients (range 0 to 140 mmHg), and the mean transcutaneous oxygen pressure in diabetic patients was 10 mmHg (range 0 to 30 mmHg). Most of the patients had a poor runoff, with a mean runoff score, calculated according to *Society of Vascular Surgery* guidelines, of 5.5 (range 1 to 10) [10]. More than one quarter of the patients (n=45) had a score higher or equal to 8. One hundred and twenty-three patients (75%) had already undergone an infragenicular revascularization procedure prior to AA implantation.

SURGICAL TECHNIQUE

The recipient artery was the infragenicular popliteal artery in 34 patients, a crural artery in 114 patients, and a pedal artery in 17 patients. Thirty-one bypasses were composite grafts: a PTFE was used as an interposition at the proximal part when the AA was too short allowing matching of diameters of the distal donor and recipient artery. When the diameter of the AA did not correspond to the diameter of the recipient artery (mostly in the case of a leg artery of small diameter), a short segment of autologous saphenous vein was interpositioned. Seventeen bypasses were sequential; the patent intermediate segment was the popliteal artery in eight cases, the profunda femoral artery in two cases, the superficial femoral artery in two cases, and a crural artery in five cases.

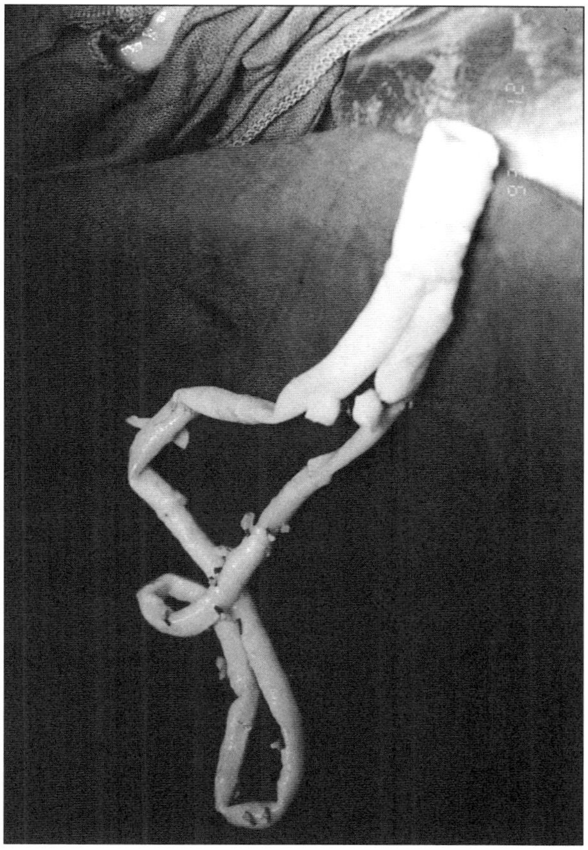

FIG. 2 Aspect of AA following thawing. The AA will be prepared in the OR just prior to implantation.

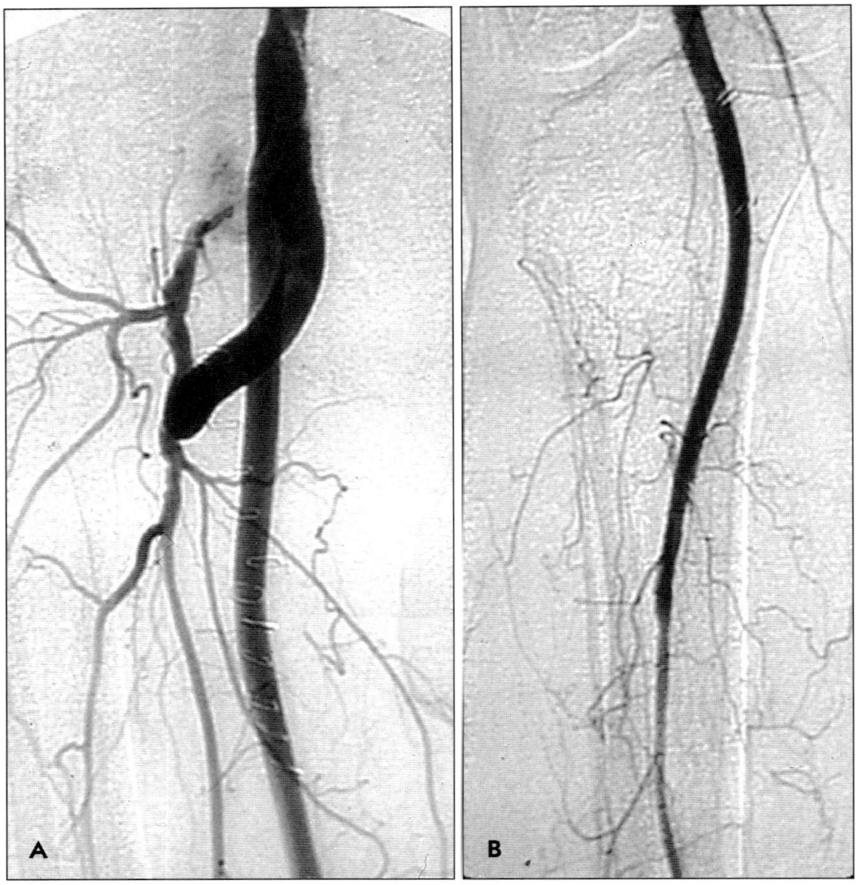

FIG. 3 Patient with CLI and previous failed revascularizations. Occlusive disease of common femoral and femoropopliteal arteries; absence of saphenous vein. Completion angiography showing a AA-sequential graft revascularizing the profunda femoral artery (A) and tibioperoneal trunk (B).

Early Results

Five patients (3.4%) died in the postoperative phase. There were 30 postoperative failures of which 28 were bypass occlusions. In one case the skin lesions deteriorated in spite of a patent bypass, and in one case a thrombus was detected on the control ultrasound and required surgical revision. The reasons for occlusion were: bad runoff (n=16), technical mistake (n=3), stenosis of the graft (n=1), poor inflow (n=1), sudden termination of anticoagulant medication (n=1), unknown (n=6). Fourteen patients underwent a surgical revision and one patient received a new AA. Five patients underwent amputation. The primary and secondary patency rates and the limb salvage rate at one month were 83%, 90%, and 98%, respectively.

Late Results

The mean follow-up was 31 months (range 1 to 81 months). During the follow-up period, 61 patients died, which accounts for an actuarial survival rate at five years of less than 50%. Sixty-five primary failures were registered: 56 occlusions, 5 stenoses of the distal anastomosis, 1 stenosis of the graft, 1 rupture and 2 aneurysms of the graft. The suspected causes of occlusion were progressive disease of distal

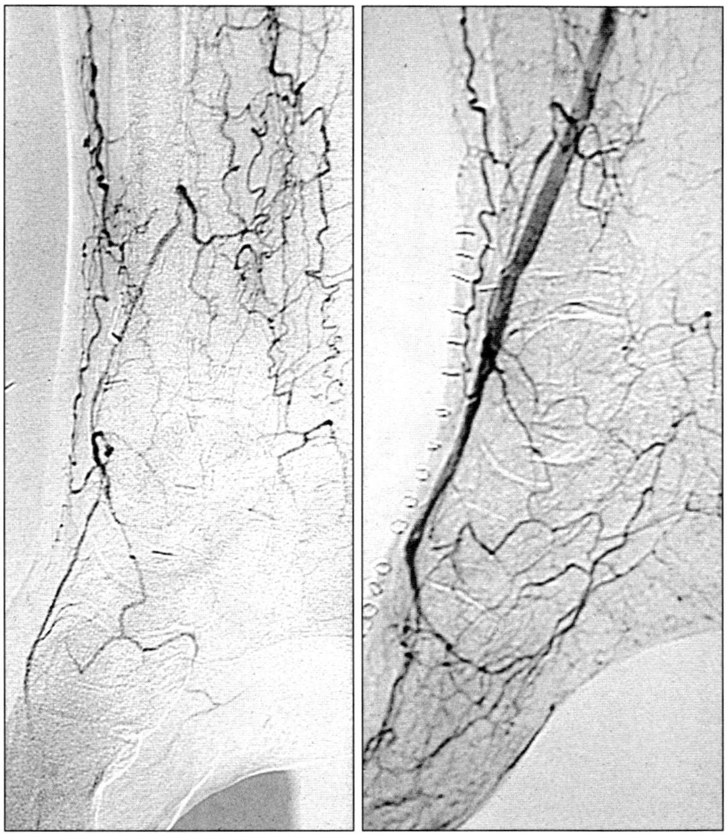

FIG. 4 Completion angiography of AA pedal bypass. Note the improved perfusion of foot arteries and acceptable runoff.

Table II CAUSES OF LATE PRIMARY OCCLUSIONS: DEGRADATIONS AND MISCELLANEOUS CAUSES

Degradations	Aneurysm	2
	Stenosis	2
	Rupture	1
	Dissection	1
Miscellaneous causes	Termination anti-coagulent medication	3
	Poor in-flow	3
	External compression	1
	Increase of proximal lesion	1
	Open ankle fracture	1

runoff vessels (n=15), myointimal hyperplasia (n=11), degradation of the graft (n=6), miscellaneous causes (n=8) (Table II) and unknown causes (n=16). Four patients received a new AA, 35 patients underwent a surgical revision, and thrombolysis with subsequent percutaneous transluminal angioplasty was performed in two patients. In 25 patients a major amputation was necessary. Primary and secondary patency rates for the overall series and for infrapopliteal bypasses are reported in Table III. The limb salvage rates at one year, three years, and five years were 83.8±6%, 76.4±7%, and 74.2±8%, respectively. Patency rate comparison according to preservation technique, ABO compatibility, site of distal anastomosis, distal runoff score, and number of previous revascularizations did not show any significant differences.

Table III PATENCY OF INFRAGENICULAR ARTERIAL ALLOGRAFTS FOR LIMB SALVAGE

Follow-up	*Infragenicular and infrapopliteal grafts*			*Infrapopliteal grafts*		
	1 year	*3 years*	*5 years*	*1 year*	*3 years*	*5 years*
Primary patency (%)	48.7 ± 8	34.9 ± 9	16.1 ± 11	43.3 ± 9	31.4 ± 9	14.1 ± 11
Secondary patency (%)	59.8 ± 8	42.1 ± 9	25.9 ± 11	56.5 ± 9	39.1 ± 10	22.1 ± 11

■ GENERAL CONSIDERATIONS

The main indication for the use of AA in critical lower limb ischemia is distal popliteal or infrapopliteal revascularization when the saphenous vein is unavailable or of poor quality. Alternatives include prosthetic bypasses (with or without adjunctive procedures), grafts made of arm veins or deep lower limb veins, and cryopreserved venous allografts. The results achieved with these different techniques are reported in Table IV-VI and are very disparate. Randomized studies would be the only way to find out which technique offers the best results. We have no experience with bypasses using arm veins or deep veins. The results reported with cryopreserved venous allografts are disappointing [26,27]. In our practice, the main alternative to AA is the use of a prosthetic graft with an adjunctive procedure; we believe that AA should be compared to this technique in a randomized study.

There is a discrepancy between the encouraging early results, which seem better than those obtained with prosthetic grafts, and the late results that show a high rate of occlusions in contradiction with a high limb salvage rate. This can be explained by a lower incidence of immediate technical problems due to the quality of the graft, while late failures are related to intimal hyperplasia and progression of the disease. It also suggests that the arterial demand of a threatened limb decreases once the acute phase has been traversed, as evidenced by healed skin lesions, culminating in salvage of a viable but chronically ischemic leg. These clinical findings and this hypothesis meet the criticism and shortcomings of the definition of critical ischemia and the need to identify patient subgroups according to the risk of limb loss.

AA immunology and its clinical consequences are not well understood. Nevertheless, because of their side effects and high cost, the currently available immunosuppressive treatments are not really usable in

Table IV PRIMARY PATENCY RATE OF INFRAPOPLITEAL PTFE GRAFTS WITHOUT ADJUNCTIVE PROCEDURE

Author [ref.]	*Year*	*Primary patency (%) - Follow-up*				
		1 year	*2 years*	*3 years*	*4 years*	*5 years*
Fichelle [11]	1995	61	48	43	42	-
Parsons [12]	1996	71	47	39	36	28
Schweiger [13]	1993	51	46	37	31	-
Davies [14]	1991	37	33	31	25	22
Veith [15]	1986	46	34	30	12	-
Rafferty [16]	1987	31	21	19	17	14

Table V PRIMARY PATENCY OF INFRAPOPLITEAL BYPASSES USING PTFE WITHOUT ADJUNCTIVE PROCEDURES

Type of adjunctive procedure	*Author [ref.]*	*Year*	*Primary patency (%) - Follow-up* 1 year	2 years	3 years	4 years	5 years
Cuff ± FAV	Wijesinghe [17]	1998	64	51	-	-	-
	Morris [18] ●	1993	29	29	27	23	-
	Morris [18] ◆	1993	25	20	17	14	-
	Harris [19]	1993	61	61	-	-	-
AVF	Ascer [20] ■	1996	78	70	61	-	-
Patch	Taylor [21]	1992	74	-	58	-	54
	Fichelle [11]	1995	-	-	45	-	-

● Primary procedure (no previous revascularization)
◆ Secondary procedure (at least one previous revascularization)
■ Assisted primary patency rate
AVF: arteriovenous fistula

Table VI PRIMARY PATENCY RATE OF INFRAPOPLITEAL BYPASSES WITH ARM VEINS AND CRYOPRESERVED ALLOGRAFTS

Technique	*Author [ref.]*	*Year*	*Number of patients*	*Primary patency (%) - Follow-up* 1 year	2 years	3 years	4 years	5 years
Arm veins	Andros [22]	1986	43	74	62	57	-	51
	Sesto [23]	1992	35	49	49	40	-	-
	Harward [24]	1992	43	67	58	49	-	-
Cryopreserved venous allografts	Walker [25]	1993	39	28	-	-	-	-
	Martin [26]	1994	115	37	19	-	3	-

the management of lower limb critical ischemia among elderly patients with multiple co-morbidity and limited life expectancy. Since there is no serious data showing the superiority of fresh AA, cryopreserved AA appears at this moment to be the best solution. Cryopreserved AA is managed by tissue banks according to strict protocols, thereby offering the best storage facilities and the guarantee for optimal preservation and virological and bacteriological protection. However, it is not yet possible to assess the risk of prion transmission, and what could be the preventive measures, as there is no screening test available and very little known on the subject.

■ CONCLUSION

Cryopreserved AA is a good alternative for infragenicular bypasses in patients with critical limb ischemia when there is no venous material available. The technique is feasible and technically attractive. Early results are excellent but late results are poor and comparable to the results obtained with other techniques in similar clinical situations. It would be recommended to conduct a prospective randomized trial comparing this technique to prosthetic bypasses with adjunctive procedures, which seems to be the most common alternative in these patients.

REFERENCES

1 Szilagyi DE, McDonald RT, Smith RF, Whitcomb JG. Biologic fate of human arterial homografts. *Arch Surg* 1957; 75: 506-529.

2 Bahnini A, Ruotolo C, Koskas F, Kieffer E. In situ fresh allograft replacement of an infected aortic prosthetic graft: eighteen months follow-up. *J Vasc Surg* 1991; 14: 98-102.

3 Gournier JP, Favre JP, Gay JL, Barral X. Cryopreserved arterial allografts for limb salvage in the absence of suitable saphenous vein: two-year results in 20 cases. *Ann Vasc Surg* 1995; 9 suppl: S7-14.

4 Barratt-Boyes BG, Roche AH, Subramanyan R et al. Long-term follow-up of patients with the antibiotic-sterilized aortic homograft valve inserted freehand in the aortic position. *Circulation* 1987; 75: 768-777.

5 Belzer OF, Southard JH. Principles of solid-organ preservation by cold storage. *Transplantation* 1988; 45: 673-676.

6 Moore GE, Gerner RE, Franklin HA. Culture of normal human leukocytes. *JAMA* 1967; 199: 519-524.

7 Rosset E, Friggi A, Novakowitch G et al. Effets de la cryopréservation sur les propriétés visco-élastiques des artères humaines. *Ann Chir Vasc* 1996; 10: 262-272.

8 Williams GM, ter Haar A, Krajewski C et al. Rejection and repair of endothelium in major vessel transplants. *Surgery* 1975; 78: 694-706.

9 Schmitz-Rixen T, Megerman J, Colvin RB et al. Immunosuppressive treatment of aortic allografts. *J Vasc Surg* 1988; 7: 82-92.

10 Rutherford RB, Baker D, Ernst C et al. Recommended standards for reports dealing with lower extremity ischemia: revised version. *J Vasc Surg* 1997; 26: 517-538.

11 Fichelle JM, Marzelle J, Colacchio G et al. Infrapopliteal polytetrafluoroethylene and composite bypass: factors influencing patency. *Ann Vasc Surg* 1995; 9: 187-196.

12 Parsons RE, Suggs WD, Veith FJ et al. Polytetrafluoroethylene bypasses to infrapopliteal arteries without cuffs or patches: a better option than amputation in patients without autologous vein. *J Vasc Surg* 1996; 23: 347-356.

13 Schweiger H, Klein P, Lang W. Tibial bypass grafting for limb salvage with ringed polytetrafluoroethylene prostheses: results of primary and secondary procedures. *J Vasc Surg* 1993; 18: 867-874

14 Davies MG, Feeley TM, O'Malley MK et al. Pontages sous-inguinaux en PTFE: sauvetage de membre ou temps perdu? Une expérience de 10 ans. *Ann Chir Vasc* 1991; 5: 519-524.

15 Veith FJ, Gupta SK, Ascer E et al. Six-year prospective multicenter randomized comparison of autologous saphenous vein and expanded polytetrafluoroethylene grafts in infrainguinal reconstructions. *J Vasc Surg* 1986; 3: 104-114.

16 Rafferty TD, Avellone JC, Farrell CJ et al. A metropolitan experience with infrainguinal revascularization. Operative risk and late results in northeastern Ohio. *J Vasc Surg* 1987; 6: 365-371.

17 Wijesinghe LD, Beardsmore DM, Scott DJA. Polytetrafluoroethylene (PTFE) femorodistal grafts with a distal vein cuff for critical ischaemia. *Eur J Vasc Endovasc Surg* 1998; 15: 449-453.

18 Morris GE, Raptis S, Miller JH, Faris IB. Femorocrural grafting and regrafting: does polytetrafluoroethylene have a role ? *Eur J Vasc Surg*; 7: 329-334.

19 Harris PL, Bakran A, Enabi L, Nott DM. ePTFE grafts for femoro-crural bypass: improved results with combined adjuvant venous cuff and arteriovenous fistula ? *Eur J Vasc Surg* 1993; 7: 528-533.

20 Ascer E, Gennaro M, Pollina RM et al. Complementary distal arteriovenous fistula and deep vein interposition: a five-year experience with a new technique to improve infrapopliteal prosthetic bypass patency. *J Vasc Surg* 1996; 24: 134-143.

21 Taylor RS, Loh A, McFarland RJ et al. Improved technique for polytetrafluoroethylene bypass grafting: long-term results using anastomotic vein patches. *Br J Surg* 1992; 79: 348-354.

22 Andros G, Harris RW, Salles-Cunha SX et al. Arm veins for arterial revascularization of the leg: arteriographic and clinical observations. *J Vasc Surg* 1986; 4: 416-427.

23 Sesto ME, Sullivan TM, Hertzer NR et al. Cephalic vein grafts for lower extremity revascularization. *J Vasc Surg* 1992; 15: 543-549.

24 Harward TRS, Coe D, Flynn TC, Seeger JM. The use of arm vein conduits during infrageniculate arterial bypass. *J Vasc Surg* 1992; 16: 420-427.

25 Walker PJ, Mitchell RS, McFadden M et al. Early experience with cryopreserved saphenous vein allografts as a conduit for complex limb salvage procedures. *J Vasc Surg* 1993; 18: 561-569.

26 Martin RS, Edwards WH, Mulherin JL et al. Cryopreserved saphenous vein allografts for below-knee lower extremity revascularization. *Ann Surg* 1994; 219: 664-672.

24

DISTAL ARTERIAL RECONSTRUCTION COMBINED WITH FREE MUSCLE FLAPS

MAURI LEPÄNTALO, FAUSTO BIANCARI, ERKKI TUKIAINEN

The incidence of critical leg ischemia is rapidly increasing with aging of the European population. Critical leg ischemia threatens the limb and often life of the patient, it causes suffering and reduces quality of life. Major amputation decreases the motility of this old patient population, rather few of whom are able to learn to walk with a prosthesis.

There is evidence that aggressive lower limb revascularization is clinically worthwhile and can decrease amputation rate by up to 60% [1]. However, infrainguinal reconstruction may be insufficient treatment in case of major ischemic tissue loss. Large, infected tissue defects with exposed tendons and bones usually do not heal despite successful bypass grafting and local wound care including several aggressive debridements and skin grafting attempts [2]. Local flap coverage is mostly inadequate for wound coverage, as there is only little viable tissue available due to advanced ischemia, infection and fibrosis.

Deep wound infection, which causes exposure of infrainguinal autogenous vein grafts, is another extremely serious condition, the treatment of which is under debate. In the infrapopliteal site, local wound care cannot prevent the problems related to inadequate graft coverage and it is often associated with a risk of delayed rupture of infected, functioning autogenous vein graft [3]. Local muscle flap coverage may have arterial inflow problems due to ischemia and limits related to the infectious status and small size of the transferred tissues.

Therefore, a number of patients with ischemic tissue loss are still at risk for major amputation despite a successful infrainguinal bypass graft. The experience with free tissue transfers in the management of traumatic soft tissue loss of the lower limb [4] suggested that this method could be used even for the management of major tissue loss secondary to atherosclerotic disease with or without diabetes. During the last decade, anedoctal observations provoked the gradual spread of this method into institutions devoted to leg salvage. Several series reported impressive results [2, 5-14]. Our preliminary results showed that early coverage with free muscle flaps may achieve leg salvage even in case of infected, functioning autogenous vein graft [14]. The aim of this chapter is to describe the technical aspects of combined lower limb revascularization and free flap transfer (LLR-FFT) and to present a critical review of the reported results.

■ TECHNIQUE

If no systemic disease and local condition contraindicate such a major leg salvage procedure, selective lower extremity angiography is performed to evaluate the extent of arterial disease and to choose appropriate inflow and outflow arteries for infrainguinal bypass grafting. In a few cases, surgical and/or endovascular treatment of aortoiliac stenotic lesions and endovascular treatment of diseased arteries of the lower extremity are performed as inflow procedures and the free flap pedicle is anastomosed to the native vessels. Most often a femorocrural or femoropedal bypass graft is needed for proper revascularization of the ischemic foot.

When simultaneous arterial reconstruction and microvascular free flap transfer is planned, a two-team approach is suggested to perform the bypass graft, to dissect the flap and to debride the ulcer (Fig. 1). The latissimus dorsi flap (Fig. 2) should be preferred over the rectus abdominis flap, as its long pedicle is usually not affected by atherosclerosis. However, the rectus abdominis muscle flap is preferred whenever the patient is at high risk and calls on for regional anesthesia and supine position. The use of the fasciocutaneous forearm flap, the tensor fasciae latae muscle flap, the gracilis muscle flap, the scapular flap, the temporalis fascia flap, the serratus muscle flap, and omentum has been described. The free fasciocutaneous or muscle flap transfer is then completed with skin graft coverage. Muscle flaps offer the advantage to cover large tissue loss with tendon and bone exposure, whereas fasciocutaneous flaps can provide superior contour and are preferred in smaller wounds of the foot and ankle. Serletti et al. [12] favor the use of the radial forearm flap when the contour is a potential problem. The scapular fasciocutaneous flap is preferred by these authors for the management of large wounds on the weight-bearing plantar surface because of increased skin thickness of this flap. In our experience, free muscle flap with skin graft was a valuable means for coverage of large, infected wounds of the ankle and foot. Although free muscle flaps tend to remain bulky, these contour problems subside with muscle atrophy, which occurs in a few months [15].

End-to-side arterial anastomosis and end-to-end anastomosis to the deep system vein are made with interrupted 8-0 polypropylene sutures with the aid of 4.5-fold magnifying loupes. End-to-end arterial anastomosis is made when an arterial graft to free flap is performed in patients with unreconstructable outflow vessels. When the free flap is used to cover exposed arterial graft an end-to-side anastomosis is made directly on the arterial bypass graft. Antibiotics are given for 3-4 weeks according to culture results, wound healing and kidney function. Continuous epidural anesthetic block is used during the first postoperative days. Bed rest with the slight limb elevation is indicated for about 7 days. Ambulation is started very gradually, with exercises in standing position for 10 minutes at the time to start with, but weight bearing not allowed until after 4-6 weeks.

Although Shestak et al. [13], Serletti et al. [12] and our experience [10] showed that indirect revascularisation of the leg by infrainguinal bypass to free muscle flaps in patients with unreconstructable arteries may succeed, this extremely aggressive procedure is seldom indicated and all efforts should be done to find an outflow artery for the distal bypass [16]. However, in such cases a neovascular network between the flap and the neighboring tissue may develop, and signs and symptoms of ischemia may subside [9,10,12,13].

Graft patency after LLR-FFT is assessed by direct pulse examination and doppler ankle/brachial index measurement, and the viability of the free flap is assessed clinically. Duplex scan examination and angiography are performed whenever signs and symptoms of leg and/or free flap ischemia occur. Early reintervention is mandatory in any case of early obstruction, the problem most often situating at the site of venous anastomosis. A variety of clinical problems resolved with this combined approach are presented in Figures 1-9.

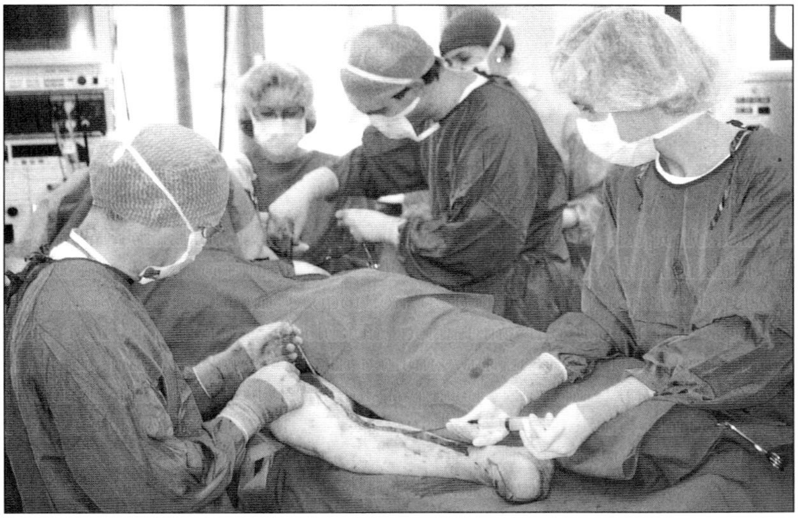

FIG. 1 Two-team approach: plastic surgeons raising the latissimus dorsi free flap and vascular surgeons performing infrainguinal bypass grafting.

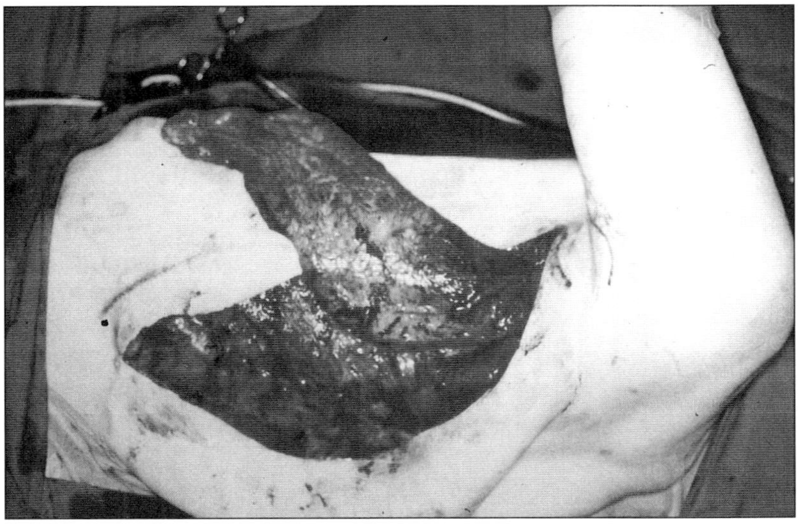

FIG. 2 Detail of raising the latissimus dorsi free flap.

■ RESULTS

Since 1989, 27 LLR-FFTs have been performed at our institution in 26 patients with chronic critical leg ischemia and major tissue loss. The free flap artery was anastomosed to the native arteries in three patients, one of whom had aortobifemoral reconstruction, whereas two had percutaneous transluminal angioplasty of the superficial femoral artery. In 23 patients, the free flap artery was anastomosed to the bypass conduit, in the majority of cases (59%) to a pedal artery. The revascularization procedure was performed simultaneously in 21 instances. In one patient without any suitable outflow arteries, a popliteal to free muscle flap bypass was carried out. At 2-year follow up, the bypass graft patency was 90%, secondary free flap viability was 78%, leg salvage rate was 80%, and 76% of patients were alive with legs. Three patients achieved long-term primary patency and leg salvage despite early free flap failure. The

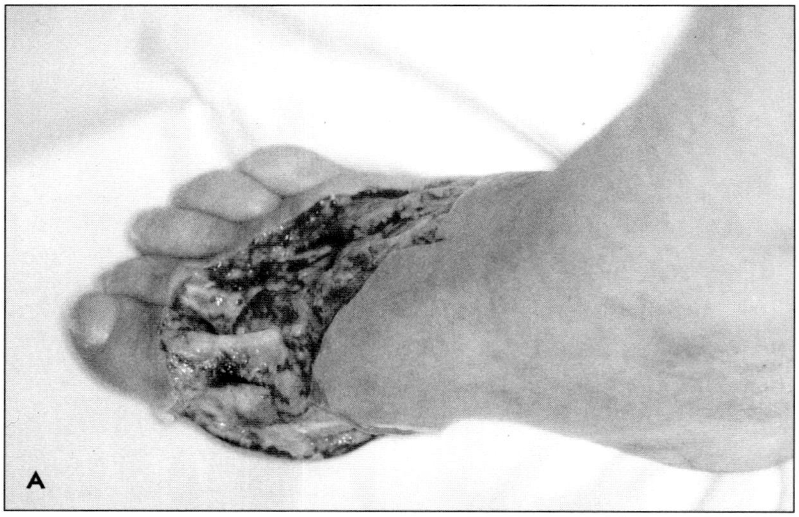

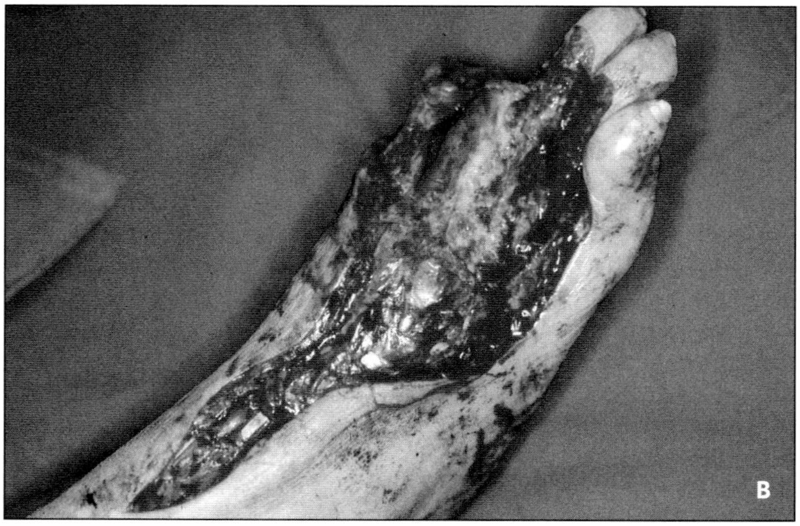

FIG. 3 Large ulcer of the dorsum of the foot. A - B Debridement of the ulcer and amputation.

autogenous vein graft to the free muscle flap occluded in one patient 6 months after operation, but the free flap survived 6 months despite inflow occlusion. Below-knee amputation was then required because of infection.

Cronenwett et al. [6] reported, in 1989, one of the first large series on this method. Fifteen revascularization procedures were followed by free tissue transfer to achieve leg salvage in 14 patients. Ninety-three percent of patients achieved leg salvage with preserved function after a mean follow-up of 24 months. Serletti et al. [12] reported 30 patients who had lower limb revascularization and free muscle, fasciocutaneous, and omentum flaps. In five patients with inadequate outflow arteries, bypass grafting to the free flap was carried out. After a mean follow-up of 22 months, 76% of patients had free

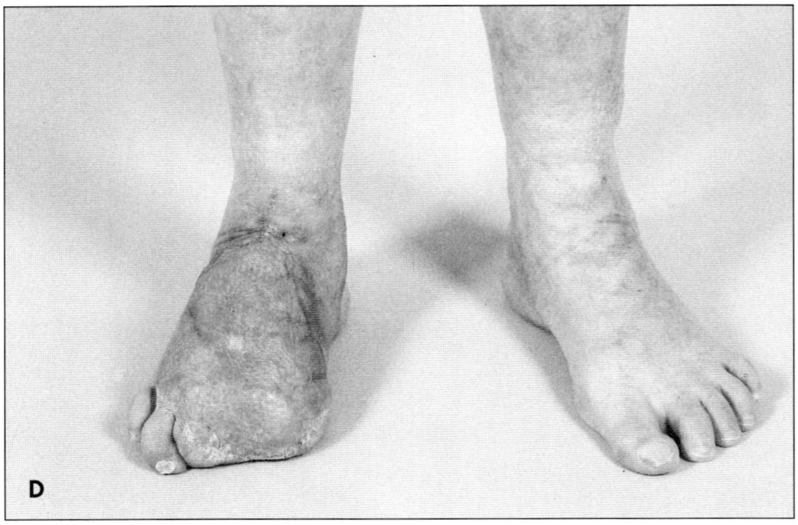

FIG. 3 C - Outcome at 18 days after combined LLR-FFT. D - Outcome at 6-month follow-up after combined LLR-FFT.

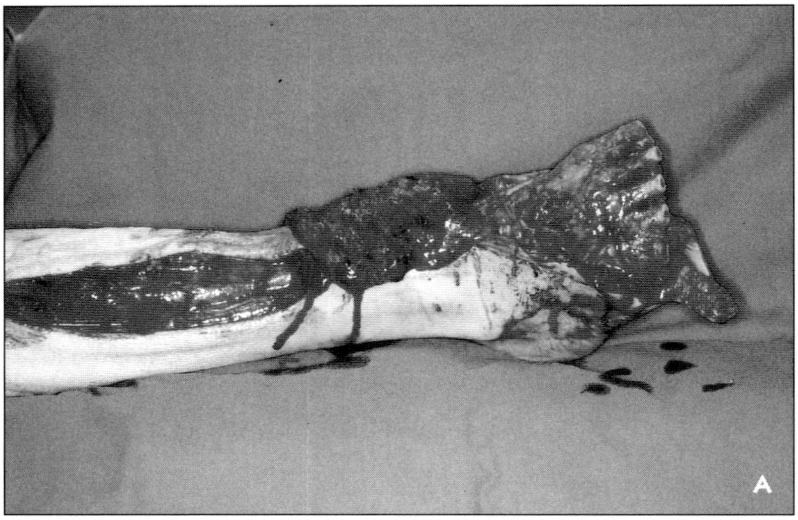

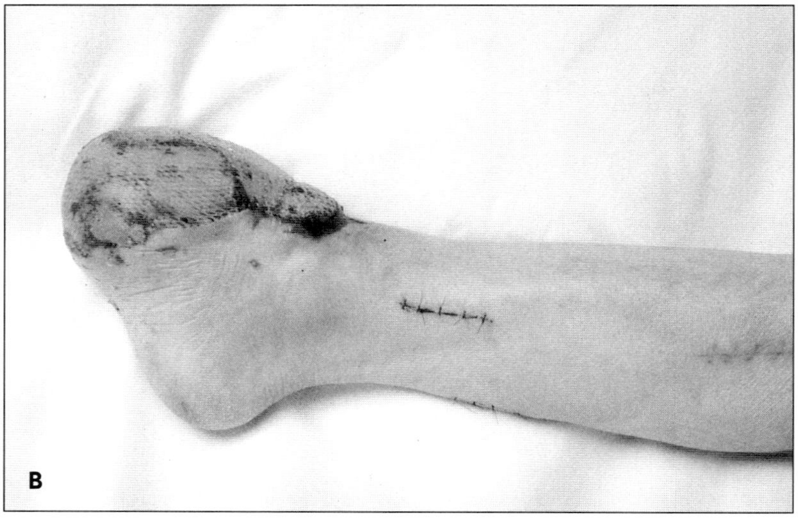

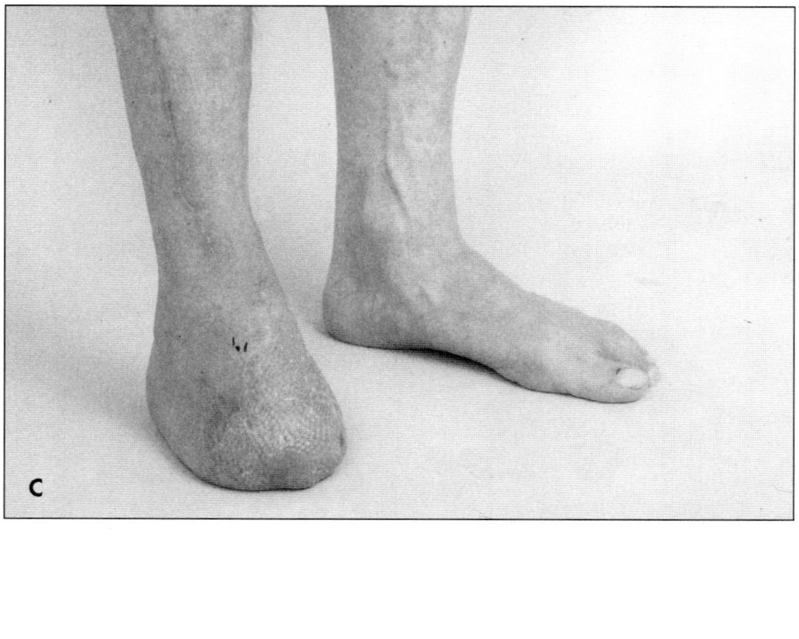

FIG. 4 A - Bypass and free muscle flap with vessel anastomoses done for the management of complications of a failed femorocrural bypass. B - Outcome after 9 days. C - Outcome at 1-year follow-up.

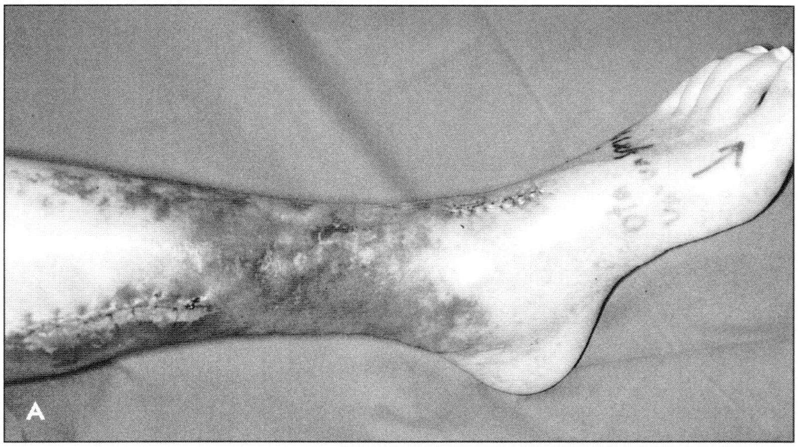

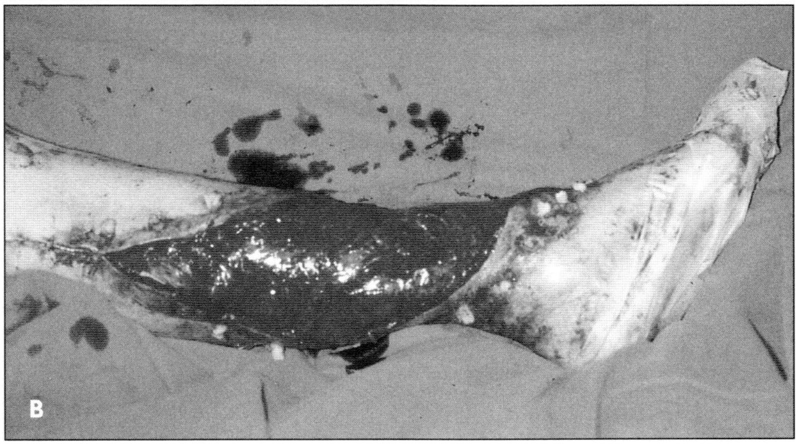

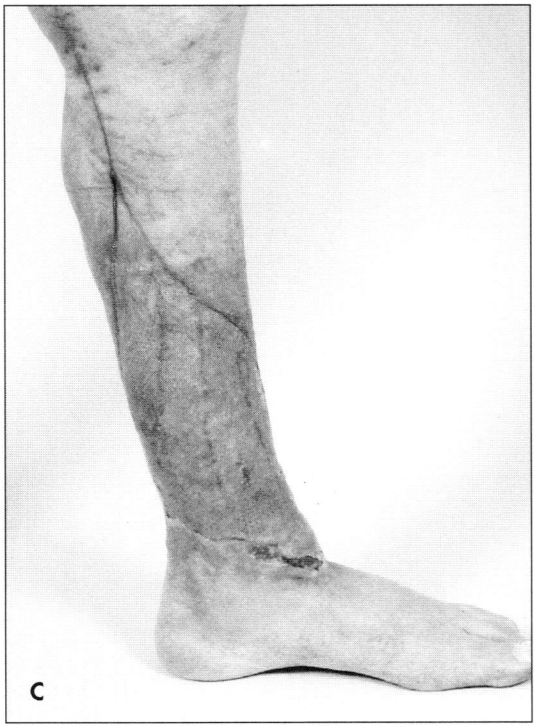

FIG. 5 A - Deep wound infection involving an autogenous saphenous vein graft. B - Debridement and muscle flap transfer. C - Final result.

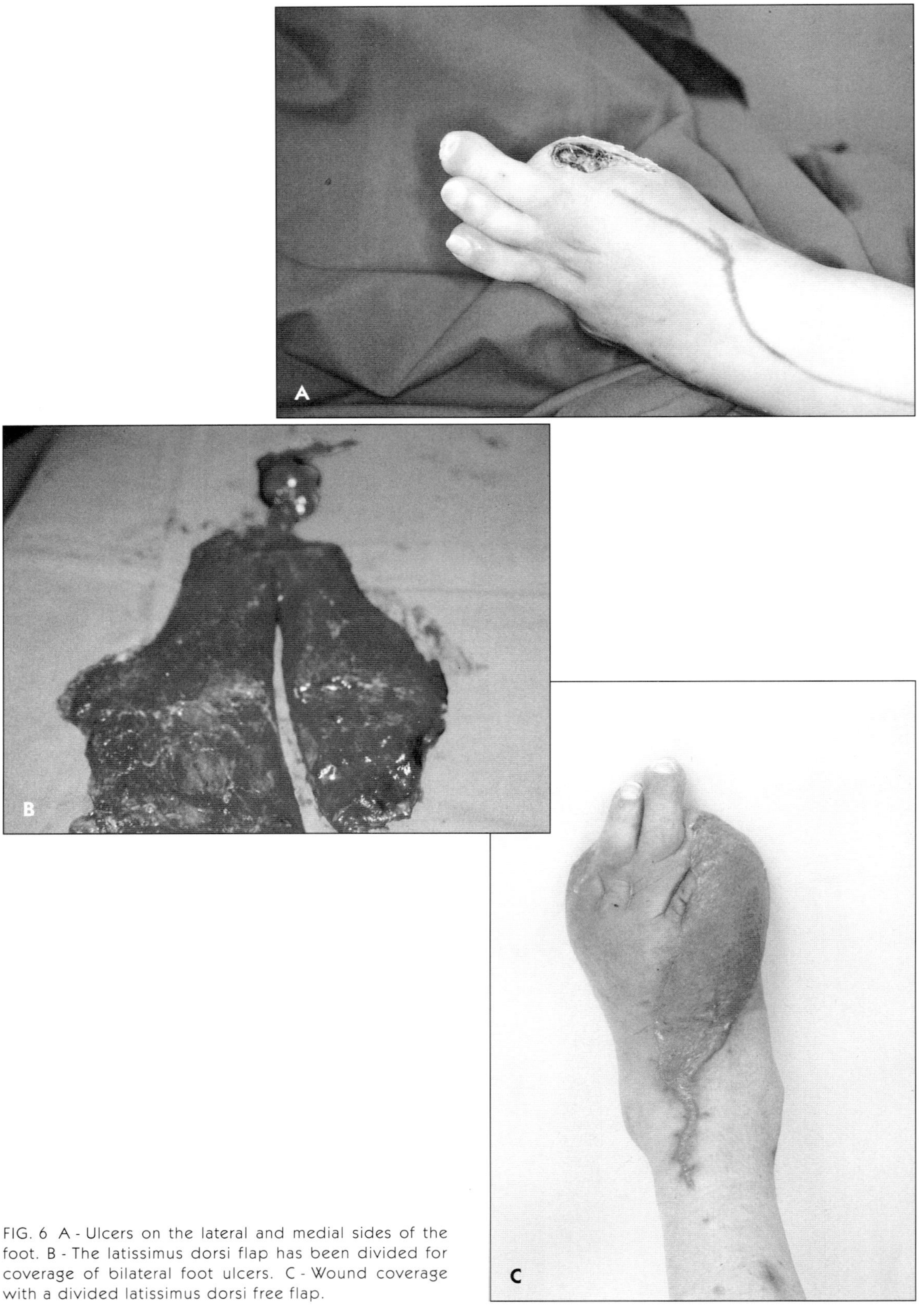

FIG. 6 A - Ulcers on the lateral and medial sides of the foot. B - The latissimus dorsi flap has been divided for coverage of bilateral foot ulcers. C - Wound coverage with a divided latissimus dorsi free flap.

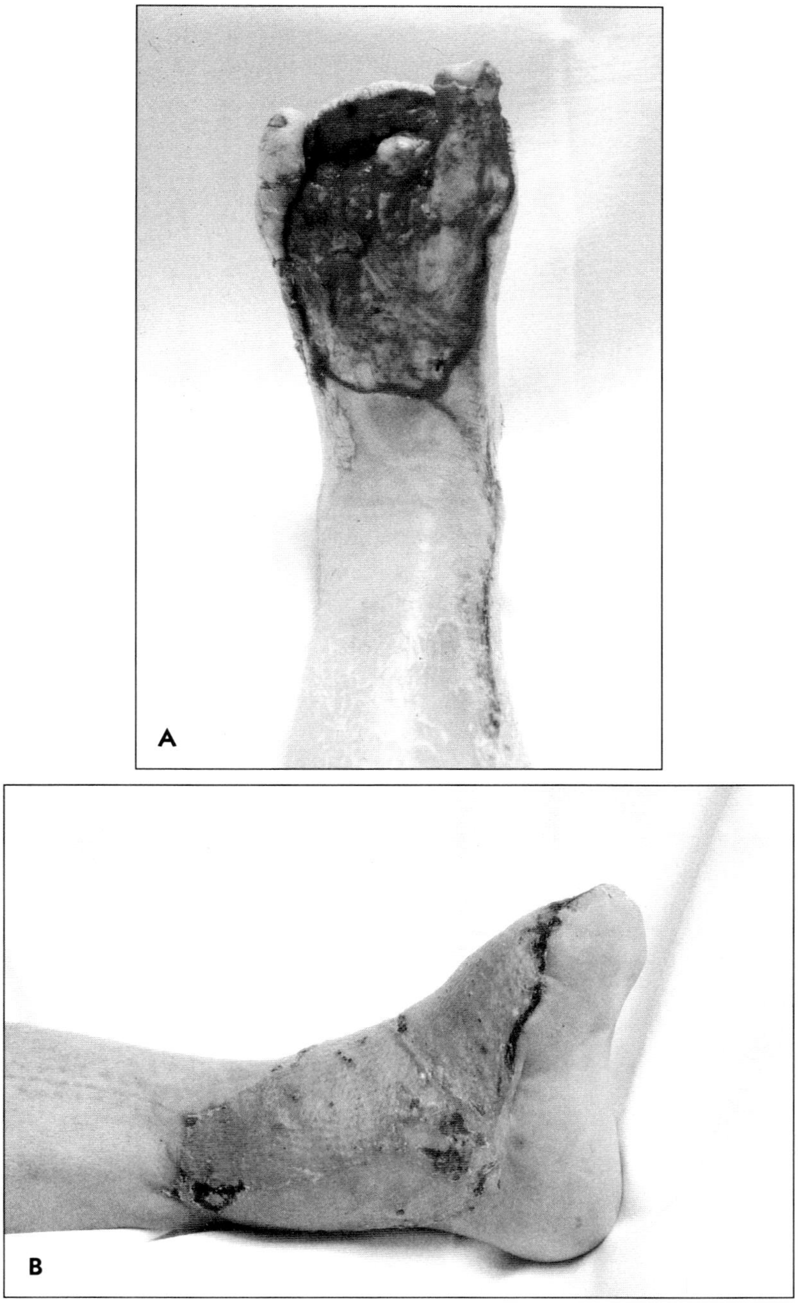

FIG. 7 A - Outcome after failure of an infrainguinal bypass graft.
B - Outcome after redo-bypass grafting and free flap transfer.

flap and graft patency with preserved limb function. Interestingly, three patients among those with an inadequate outflow artery who had a vein graft to the free flap alone had leg salvage without evidence of leg ischemia. Gooden et al. [2] performed 17 LLR-FFTs and ten free flaps alone for limb salvage of major tissue loss of the leg. The leg salvage rate at 2-year follow-up was 70%, and 88% of these patients achieved functional ambulation. However, it is likely that not all patients had critical leg ischemia, particularly those who had free flap transfer alone. Karp et al. [7], Oishi et al. [11], and Ciresi et al. [5] reported similar results with leg salvage rates greater than 70% after a mean follow-up of 10-22 months.

Deep vein graft infection is a somewhat different problem, but can be treated also by microvascular free flap transfer combined with other necessary measures. We have an experience of 14 patients with infrapopliteal autogenous vein graft infection which already caused graft rupture in six cases [14]. The

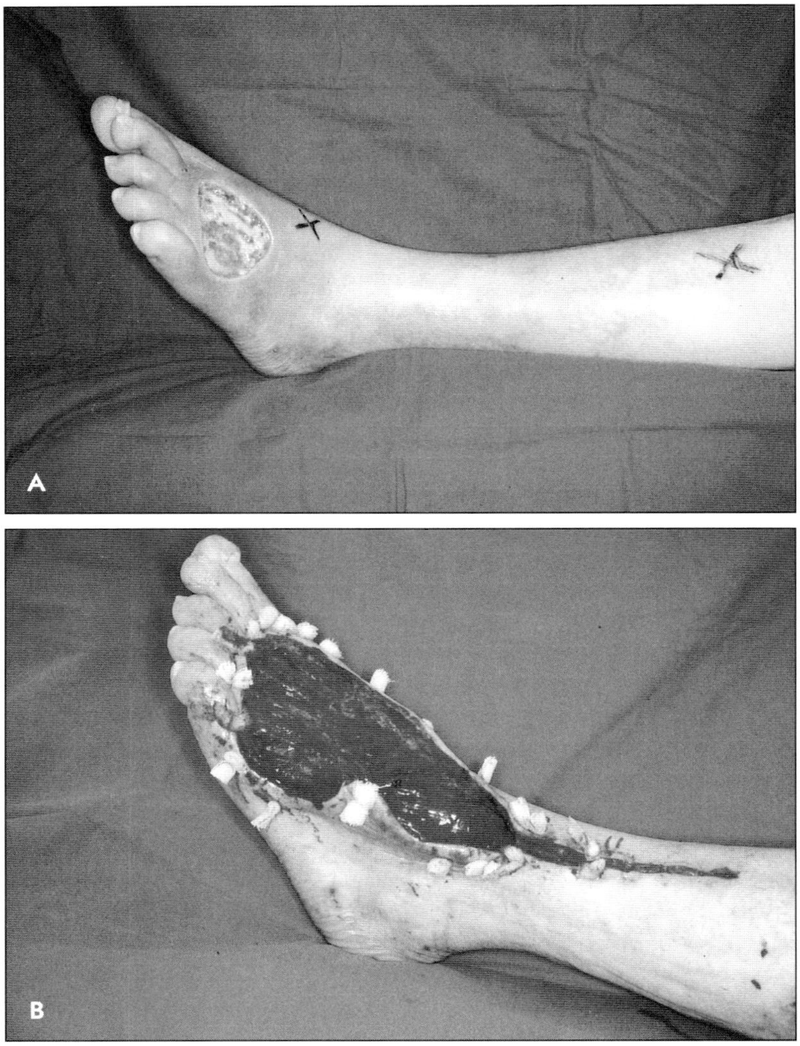

FIG. 8 A - Ulcer of the dorsum of the foot. B - The bypass and free flap transfer have been done.

infected grafts were treated with debridement of all the infected tissue, removal of the infected graft with interposition regrafting in cases of need and immediate muscle flap coverage in ten patients (Figs. 5A, 5B, and 5C). Interposition regrafting plus free muscle was done in four cases. Four patients with poor risk conditions underwent only immediate local muscle coverage [14]. Thirty-day postoperative leg salvage was achieved in all but one patient who underwent local muscle coverage and had above-knee amputation because of infection despite patent bypass graft. At 1-month, 6-month, 1-year and 2-year follow-up, leg salvage rates were 92%, 75%, 55%, and 44%, respectively, while 85%, 68%, 34%, and 34% were alive with legs at the same intervals. Among six patients who presented with rupture of the vein graft, four patients had graft occlusion. These results are acceptable, as the alternative would mostly have been a major amputation.

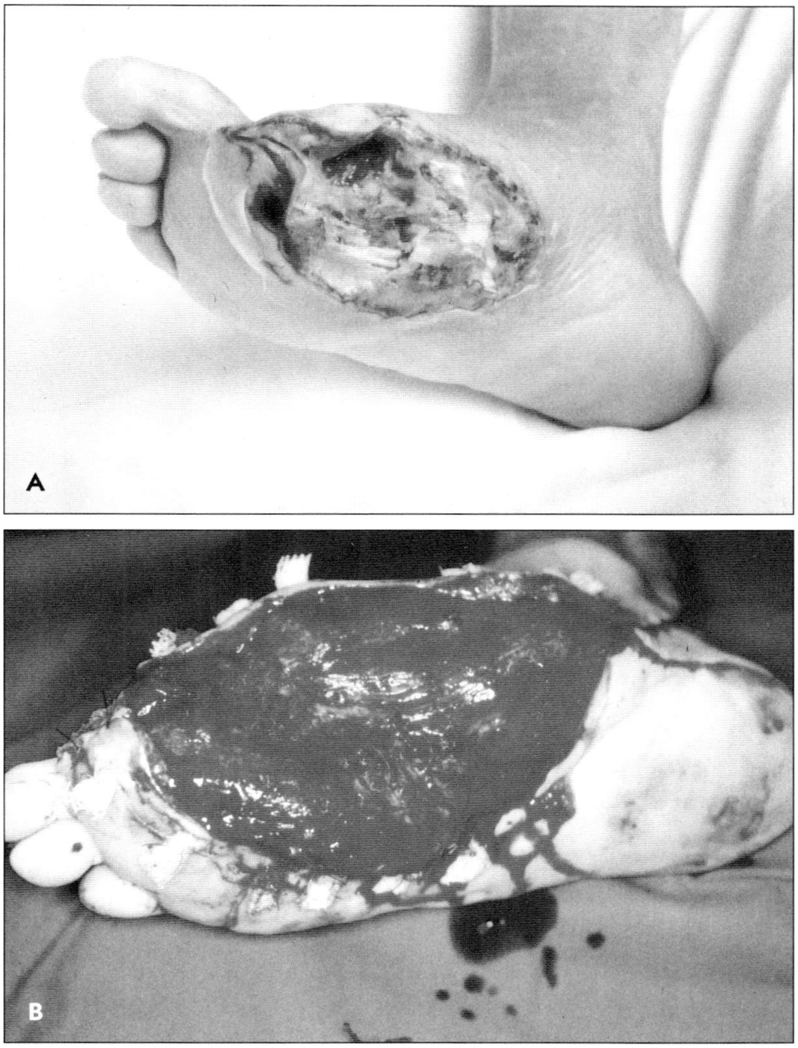

FIG. 9 A - Ulcer of the plantar area of the foot. B - The bypass and free flap transfer have been carried out.

■ DISCUSSION

The final role of combined LLR-FFT procedure is not well established. The data for its support is based mostly on uncontrolled case series without cumulative survival data [5,6,9,11-13]. In any case, combination of these two methods offers an attractive approach to legs previously candidates for amputation.

As to the technique, some authors [5,12] favor a simultaneous over a staged LLR-FFT procedure. Advantages of simultaneous approach are the avoidance of a second operation into scarred tissues, minimizing the anesthetic risk of a multi-staged procedure and of the duration of hospitalization. On the other hand, Cronenwett et al. [6] favor a staged LLR-FFT, the arterial reconstruction and debridement of the wound being performed first, as to avoid a very long initial operation. Early outcome is then evaluated as to decide if local conditions require delayed wound coverage. Our experience shows that in a few patients, good long-term graft patency and leg salvage rate can be achieved despite early free flap failure. This observation suggests that a staged approach consisting of arterial reconstruction with aggressive local wound care may better select those patients who definitely require delayed free flap coverage.

On the contrary, good results are expected with the immediate use of free muscle flap coverage of exposed, intact autogenous vein graft. This aggressive method may successfully treat the serious infectious process and prevent further injury to the functioning graft. According to our experience, this approach is seldom successful in the management of infected, ruptured vein grafts, a condition that should deserve a separate category in the classification of deep wound infections. Nevertheless, an attempt may be acceptable as the alternative would be major amputation, but at present there is data only on six patients [14].

All clinical series dealing with LLR-FFT in critically ischemic extremities reported good results in this patient population at high-risk for leg loss. However, cautious enthusiasms arise from a critical analysis of clinical experience and data reported in the literature. This aggressive approach is performed as the last therapeutic resource in a large number of relatively young patients [2,6,7-12] whose age is often under 50 years and, therefore, great and qualified efforts have been put for leg salvage. Furthermore, such good results have been achieved in a selected patient population who likely had better general conditions to sustain such major operations and further procedures.

Although cost-benefit analysis favors revascularization procedures for critical limb ischemia [1], arterial bypass grafting and free flap coverage of large ischemic wounds seems to require major medical and economic efforts. Simultaneous LLR-FFT procedures may last up to 9 hours with a mean operative time of about 6-7 hours, and the mean postoperative in-hospital stay may be more than 2 months with a mean stay of about 30 days [12]. Even longer operative time and postoperative in-hospital stay are expected for delayed free tissue transfer as well [12]. Moreover, morbidity at the recipient site may highly be expected requiring several early and late additional procedures to achieve both free flap and leg salvage [11]. In fact, thrombosis of the free flap vein or artery, marginal flap necrosis, wound edge necrosis, hematoma, and persistence or recurrence of infection are common complications which require early and long-term continuous medical and surgical surveillance. In our series of 27 combined procedures, 38 new admissions in our hospital for minor and major procedures as to maintain leg salvage have been recorded.

In conclusion, free flap coverage associated with arterial reconstruction seems like an effective procedure for the management of major tissue loss due to critical leg ischemia. However, because it requires major medical and economic efforts, the selection of patients undergoing LLR-FFT is of a major importance.

REFERENCES

1 Luther M. Treatment of chronic critical leg ischaemia. A cost benefit analysis. *Ann Chir Gynaecol* 1997; 86 (suppl. 213): 1-142.

2 Gooden MA, Gentile AT, Mills JL et al. Free tissue transfer to extend the limits of limb salvage for lower extremity tissue loss. *Am J Surg* 1997; 174: 644-649.

3 Calligaro KD, Veith FJ, Dougherty MJ, DeLaurentis DA. Management and outcome of infrapopliteal arterial graft infections with distal graft involvement. *Am J Surg* 1996; 172: 178-180.

4 Swartz WM, Mears DC. The role of free-tissue transfers in lower extremity reconstruction. *Plast Reconstr Surg* 1985; 76: 364-373.

5 Ciresi KF, Anthony JP, Hoffman WY et al. Limb salvage and wound coverage in patients with large ischemic ulcers: a multidisciplinary approach with revascularization and free tissue transfer. *J Vasc Surg* 1993; 18: 648-655.

6 Cronenwett JL, McDaniel MD, Zwolak RM et al. Limb salvage despite extensive tissue loss. Free tissue transfer combined with distal revascularization. *Arch Surg* 1989; 124: 609-615.

7 Karp NS, Kasabian AK, Siebert JW et al. Microvascular free-flap salvage of the diabetic foot:a 5-year experience. *Plast Reconstr Surg* 1994; 94: 834-840.

8 Lepäntalo M, Tukiainen E. Combined vascular reconstruction and microvascular muscle flap transfer for salvage of ischaemic legs with major tissue loss and wound complications. *Eur J Vasc Endovasc Surg* 1996; 12: 65-69.

9 Mimoun M, Hilligot P, Baux S. The nutrient flap: a new concept of the role of the flap and application to the salvage of arteriosclerotic lower limbs. *Plast Reconstr Surg* 1989; 84: 458-467.

10 Mätzke S, Tukiainen E, Lepäntalo M. Survival of a microvascular muscle flap despite the late occlusion of the inflow artery in a neuroischaemic diabetic foot. *Scand J Plast Reconstr Hand Surg* 1997; 31: 71-75.

11 Oishi SN, Levin LS, Pederson WC. Microsurgical management of extremity wounds in diabetics with peripheral vascular disease. *Plast Reconstr Surg* 1993; 92: 485-492.

12 Serletti JM, Deuber MA, Guidera PM et al. Atherosclerosis of the lower extremity and free-tissue reconstruction for limb salvage. *Plast Reconstr Surg* 1995; 96: 1136-1144.

13 Shestak KC, Hendricks DL, Webster MW. Indirect revascularization of the lower extremity by means of microvascular free-muscle flap. A preliminary report. *J Vasc Surg* 1990; 12: 581-585.

14 Tukiainen E, Biancari F, Lepäntalo M. Deep infection of infrapopliteal autogenous vein grafts:immediate use of muscle flaps in leg salvage. *J Vasc Surg* 1998; 28: 611-616.

15 Salmi A, Ahovuo J, Tukiainen E et al. Use of ultrasonography to evaluate muscle thickness and blood flow in free flaps. *Microsurgery* 1995; 16: 601-605.

16 McDaniel MD, Zwolak RM, Schneider JR et al. Indirect revascularization of the lower extremity by means of microvascular free-muscle flap. A preliminary report. *J Vasc Surg* 1991; 12: 829-830.

25

IS THERE A PLACE FOR PRIMARY AMPUTATION?

JOY ROBINSON, JONATHAN D. BEARD

The management of critical lower limb ischemia has changed considerably over recent years with the refinement of distal bypass procedures and the development of endovascular procedures. Some now argue that revascularization should be attempted in all cases and high limb salvage rates have been reported from specialized centers. Economic studies also suggest that successful revascularization is less expensive than amputation and more recent work, looking at the quality of life after surgery for critical limb ischemia, supports the widely held belief that revascularization is better than amputation. However, these quality of life studies are hampered by the use of measures which impose preselected values on the individual rather than seeking to define those factors which motivate each individual enabling them to determine their own quality of life. Surgeons have continued to advocate reconstructive surgery based on the poor mobility achieved by amputees in various retrospective studies [1]. However, the absence of randomized controlled trials of amputation versus reconstruction means that there is no good scientific evidence for the superiority of either treatment group.

An aggressive approach to arterial reconstruction has been advocated for some years [2], even for elderly patients in district general hospitals [3]. However, it has been difficult to prove that an increase in reconstruction rates causes a decrease in amputation rates. In many centers the number of reconstructions has increased, but so has the amputation rate. Over a 10-year period in Scotland the overall incidence of amputations fell, but this was mostly in patients younger than 65; the rate rose in those older [4]. There was a marked increase in the rates of reconstructive surgery, but this was more than compensated for the fall in amputations, and reflected an increase in the number of patients being treated. A retrospective review over 22 years has, however, demonstrated that an increase in arterial reconstruction is associated with a decrease in the amputation rate [5].

■ COST-EFFECTIVENESS

With the increasing financial pressures on health services, attempts are being made to bring cost-effectiveness into consideration as an outcome of surgery. The high cost of arterial reconstruction, including grafts, theatre and nursing time, is offset by more rapid rehabilitation in comparison to the intensive physiotherapy, prosthetic services or home modifications that amputees require. When considering cost comparisons between these groups it has been demonstrated on many occasions that amputation is more costly than successful revascularization [6]. However, the cost of reintervention for graft failure can be substantial, and with prosthetic grafts to crural vessels this can approach the cost of amputation. The majority of patients in one study underwent a single procedure, but almost 20% underwent four or more [7]. Such aggressive reintervention can be justified for the individual by reducing the secondary amputation rate, but it is clear that to maintain economic superiority, the costs of reintervention must be kept to a minimum, and this can undoubtedly be influenced by preoperative selection of patients. The clear short-term cost benefit offered by reconstruction might be lost with longer follow-up as the costs of reintervention accumulate.

Conventionally, the outcome of intervention is measured by morbidity, mortality, limb salvage rates and a return to the patients' pre-morbid state, frequently assessed by mobility and return of the patient to their home. The revascularization of limbs by grafting down to crural vessels has resulted in limb salvage for patients whom only three decades ago would have faced amputation, and has led many to advocate an aggressive policy of revascularization, even with prosthetic material and venous *cuffs* [8]. Moreover, arguments for undertaking primary amputation in patients considered *unfit* for anesthesia are less persuasive with the increasing sophistication of epidural anesthesia. A review of the outcome of surgery in the medically compromised patient (with patient groups comparable for severity of medical illnesses) showed that those undergoing amputation had a higher perioperative mortality rate and shorter long-term survival, leading to the conclusion that revascularization should not to be ruled out on the basis of medical illnesses, and that limb salvage might offer the best chance of survival [9].

A national survey of the management of critical limb ischemia, (defined as Fontaine III or IV) was conducted by *the Vascular Surgical Society of Great Britain and Ireland* in 1993 [10]. The prevalence of critical limb ischemia was found to be 1 in 2500 of the population, and with an increasingly elderly population this is undoubtedly rising. Patients were seen and treated at both teaching and district general hospitals, and the rates of amputation versus revascularization clearly varied between hospitals and surgeons. Some form of revascularization (surgery or percutaneous angioplasty) was offered to 70% of patients, and the overall limb salvage was 75%. The incidence of diabetes was 30% in this study, and diabetics were offered primary amputation more often than revascularization. An increasing incidence of diabetes in the population is likely to effect the amputation versus revascularization rate in the future. Good short-term results can be obtained in diabetics offered tibial arterial reconstruction, although the long-term salvage rates are less encouraging. Primary patency rates of 82% at 1 year can be obtained with a 1-year survival of 90%. The combination of diabetes mellitus with end stage renal failure leads to a decrease in graft patency and a reduced 1-year survival of 52% [11].

When infrainguinal reconstruction is performed for limb salvage, the ideal of uncomplicated surgery, with the relief of pain, swift wound healing and a rapid return to pre-morbid function with no further surgery is rarely achieved. In one retrospective review [12], only 16 of 112 patients had the *ideal* result, and this study goes some way to quantifying the proportion of patients attending clinic and awaiting wound healing. Although persistence of symptoms is often related to the progression of disease in patients with a limited life expectancy, having a functioning graft may be less satisfying for the patient than it is for the surgeon.

Measuring the outcome of surgery with reference to the preservation of pre-morbid residential status is of value to the elderly whose only other option is to move to a residential or nursing home, with major implications in terms of their loss of independence. When the outcome of lower limb revascularization was measured in octogenarians, the primary and secondary graft patency rates were 72% and 87% at 5 years [13]. An arbitrary scoring system measuring ambulatory and residential status showed a reduction in both, although this mainly reflected the fact that more were walking assisted by a cane, crutch or walker. The percentage of non-ambulatory patients remained the same, although the state of some individuals changed, such that a number of previous walkers became confined to bed or a wheelchair, while half of those patients who were previously non-ambulatory became ambulant. The residential status also deteriorated, with less living in their home, although a significant number, none of whom required nursing home care, chose to live with their family. The only predictor of improved postoperative functioning was the individual's baseline perception of their health. By implication, in those patients with a poor preoperative level of functioning, arterial reconstruction as a means to preserving mobility and residential status may well not be justified.

In a review of the treatment options available to those with critical limb ischemia, vascular reconstruction was recognized as the treatment of choice in most patients [14]. However, it was suggested that primary amputation should be considered in cases where the benefits of surgical reconstruction would be negligible along with those in whom the chances of successful reconstruction would be small or in cases of progressive ischemia. Identifying those who might expect *negligible benefit* can be difficult, although focusing on quality of life issues is likely to prove to be a useful method.

■ QUALITY OF LIFE

Recent studies looking at the success of intervention for critical limb ischemia have looked at the quality of life experienced by individuals as an outcome measure. The *Quality Adjusted Life Year (QALY),* which gives the concept of the value of each life year for the individual, provides a single measure of the value that a respondent attaches to their overall health status. This provided one of the earliest tools for assessing quality of life [15] and although it has been developed to make it more applicable to every day clinical practice, including vascular surgery [2], many other tools have superseded it. Health economists have developed the (for many patients) complicated concept of standard gamble and time trade-off which provide a single utility index for cost-benefit analyses. They have also been instrumental in developing standardized generic instruments for describing health-related quality of life from which a single index can be calculated, such as the *EuroQol.* The vast array of scales and profiles used by clinicians and health economists to assess quality of life ranges from domain specific scales, such as the Hospital Anxiety and Depression scale, the Barthel Activities of Daily Living index, to those which measure responses across a range of domains such as emotional reactions and physical energy. *The Nottingham Health Profile (NHP)* and the *Medical Outcomes Short Form 36 item (SF-36)* both fall into the latter category and the individuals' responses are scored for each domain; 6 for the NHP and 8 for the SF-36 (Table) to provide a profile for each individual. The SF-36 is now widely used in many studies, and health economists are also developing a system to provide an index from this profile, thereby allowing data to be used by themselves as well as clinicians. Suggestions have been made to reach a consensus over which is the most reliable of quality of life measure, advocating the use of a single one in an international setting [16]. The SF-36 has been shown to be the most appropriate generic quality of life analysis tool in current use, and it's use has been advocated, not as a *gold standard* but in an attempt to standardize the assessment of outcome from vascular surgery [17].

The limitations of quality of life measurements in general is that they frequently measure an external value system derived from group data, so a total score may bear little relationship to the experience of an individual, akin to the problem of defining *normality.* Events do not necessarily retain the same meaning for individuals over time or the course of an illness, and the impact of any change is difficult to assess with a system that is confined to a few specific domains. Many of the quality of life analysis tools

Table THE DOMAINS OF THE SF-36 QUESTIONNAIRE

Domain	*Number of items*	*Areas covered*
Self-reported general health	5	Rating of own health and in comparison to other people
Physical functioning	10	Ability to walk, dress, kneel
Physical role limitations	4	Limits that health has on daily activities
Bodily pain	2	Severity and impact on daily activities
Mental health	5	Degree of happiness, nervousness or peacefulness
Emotional role limitations	3	Limits that *emotional problems* have on daily activities
Social functioning	2	Impact of health or *emotional problems* on social activities
Energy	4	Incidence of feeling tired or energetic

N.B. An additional item (not a domain) relates to changes in health in comparison to one year ago

also have a *floor* or *ceiling* effect, i.e. limited ability to detect changes in poor or good health, and the SF 36 seems to have a *floor* effect when used on patients with critical limb ischemia. Physical role frequently scores the lowest possible, so any deterioration is undetected, and improvement must be marked to register a different score. Severe quality of life states may thus go undetected. Using any quality of life analysis tool has particular problems in the elderly by virtue of the fact that their perception of their own health is often related to their peer group, and consequently their scores are usually higher than an objective observer might have anticipated. An elderly person with rest pain, diabetes and angina will often not rate their health as worse than others they know, nor be of the opinion that they get unwell more frequently than others.

More recently, tools have been developed in an attempt to assess patient-determined quality of life. These are few, and have been developed by a variety of paramedical groups, including physiotherapists and psychologists. One such tool is the *Schedule for the Evaluation of the Individual Quality of Life (SEIQoL)*, developed by psychologists at the *Royal College of Surgeons in Ireland* [18]. This has previously been evaluated in a number of groups; healthy individuals, those with gastrointestinal disease and patients undergoing hip replacements. It is a system based on judgement analysis, and allows individuals to describe their lives in terms of the factors they consider important, and to measure how they rate those factors and their relative importance in their lives. The technique aims to highlight individuality while describing quality of life in terms which can be critically assessed. It overcomes some limitations of other quality of life tools by not imposing any external value system on the respondent, and also considers their change in priorities. Individuals are asked to nominate five areas (referred to as cues) in their life, which they feel determine their quality of life, whether those cues have a positive or negative effect at that time. They are then asked to rate on a visual analogue scale, which ranges from worst possible to best possible, how well each of those cues are at that time. Finally they are asked to rate the relative importance (referred to as weights) of each cue in their life at that time. The combination of cues, weights, and the score from the visual analogue scale are combined to give a profile for each individual at any given time (Fig. 1). In this figure, the ratings are scored from 0 to 100 according to the visual analogue scale, and the weights are derived as percentages; the total of all weights will always be 100. From this particular example, it can be seen that health was slightly more important than the home, with the latter rated as going well at that time. The ability to garden had obviously been curtailed and this was rated at *very bad*,

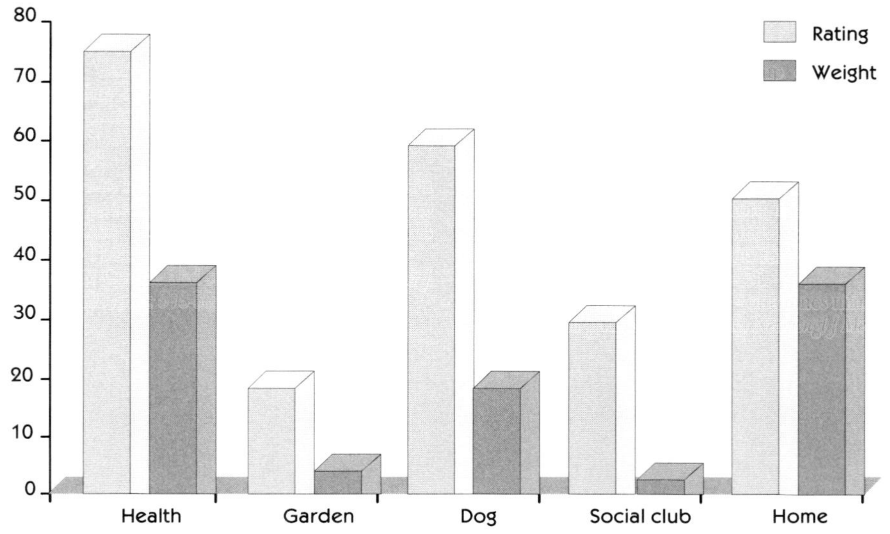

FIG. 1 Example of patient quality of life priorities obtained using the schedule for the evaluation of the individual quality of life.

but since it's relative importance was not great in his life, the impact of the limitation on his quality of life was not great, and overall he rated his quality of life as good.

A number of studies looking specifically at the quality of life in patients with critical limb ischemia have been published over recent years, and some are reviewed below.

It is clear that a successful infrainguinal reconstruction results in an improvement in quality of life, and this has been demonstrated succinctly using the SF-36 in a prospective comparative analysis of 55 patients with critical limb ischemia [19]. All patients with patent grafts had an immediate and lasting improvement in physical functioning, physical role, pain and social functioning, with improved scores for those with an occluded graft in pain, vitality and social functioning but no improvement in physical functioning or role. Those who required secondary amputation had similar improvements in pain, vitality, mental health and social functioning as those with a patent graft, and sometimes demonstrated higher scores for the psychological domains. While this demonstrates that the time and expense incurred in infrainguinal reconstructions is justified for the improvement offered in quality of life, it also highlights the fact that amputees can also benefit from an improvement in their quality of life in all except physical domains.

A prospective study from our institute looked at the outcome in 150 patients admitted with critical limb ischemia and offered radiological intervention, surgical reconstruction or amputation [20]. The quality of life was measured at baseline, 6 and 12 months post-intervention by a number of instruments including the Burford pain thermometer, the *Hospital Anxiety and Depression scale* and the *Barthel Activities of Daily living index*. Postoperatively, patients who had a successful bypass procedure were more anxious than amputees, although less anxious than they were preoperatively. Only those who underwent successful surgical reconstruction had an improvement in their depression scores. The Barthel index was improved in both successful reconstruction and primary amputation groups, whilst those with failed limb salvage and bilateral amputees maintained their preoperative, independent status. All patients had an improvement in their pain scores, and this was statistically significant in all but those with a failed reconstruction. There was little difference in the mobility prior to intervention between amputees or those who underwent revascularization, but mobility increased significantly in the latter. The six-month mortality across the whole study group was 23%. Thus the improved quality of life of those undergoing successful revascularization is largely as a result of the improved mobility that limb salvage offers. Given that on other quality of life scales amputees scored no worse than those who kept their limbs, those in whom mobility is not a priority may be best served by primary amputation. This group might include patients who request amputation, those with concomitant disability such as stroke and those who do not consider mobility to be an important determinant in their quality of life. The assumption of society, and many clinicians, that the only ethical way to treat critical limb ischemia is by revascularization is not borne out on many vascular wards. The perceptions of an individual, their life experience and factors which determine their quality of life cannot be prescribed, and there are patients who adjust to loss of a limb with little apparent concern just as there are those for whom amputation seems incompatible with living. Some amputees do achieve good mobility and many others report an increase in their quality of life because of lack of pain or worries about their graft failing. These patients challenge the assumption that amputation reflects failure for the patient and surgeon. Whilst primary amputation can be proposed on the basis of quality of life assessment alone, it can also be supported by the observation that careful selection of patients for revascularization helps avoid the high morbidity and financial cost associated with revision surgery.

INDIVIDUAL QUALITY OF LIFE

We have been evaluating the SEIQoL within the setting of a vascular surgery ward and patients admitted for intervention for lower limb critical ischemia of longer than two weeks duration. If it is possible to identify those areas of most concern to individuals, then proposed interventions may well be offered based on those quality of life priorities the patient holds. If it were possible to define those factors which motivate patients and use this to predict who would be expected to have a better outcome from revascularization, then identifying patients who would recover well from amputation, whether they are mobilized or not, would also be possible. The SEIQoL has been administered to a number of patients with critical limb ischemia over recent months, and highlights the variety of factors which individuals nominate in determining their quality of life. The most commonly nominated factor is the presence of family and the associated relationships, with a significant number declaring that their quality of life is completely dependent on the relationship with their wife. These individuals enjoy walking and hope for an improvement in their mobility, but do not often look to this to cause a significant improvement in their quality of life.

Case Scenario 1

An 87-year-old gentleman with chronic obstructive airway disease was admitted with ischemic leg ulceration. His angiogram showed vessels suitable for reconstruction, and he underwent femoropopliteal reconstruction. Unfortunately he required two revisions for graft complications and failed to recover, dying on day 12. Such a scenario is not uncommon on a vascular ward, and it is easy to say in retrospect that he may have been best served by a swift amputation. When considering his preoperative quality of life priorities, this gentleman enjoyed his life to the full. His mobility was limited, but he managed to get out of the house on a weekly basis to have a drink with his friends at the local public house. Otherwise he stayed in his daughters' house, his daughter being the most important determinant in his quality of life, closely followed by eating good food (prepared by the daughter), watching the television, getting to the pub and smoking. (Fig. 2) He considered his quality of life was the best it could be, and his expectations of the outcome of surgery were not great.

Case Scenario 2

A 71-year-old lady with a painful gangrenous toe was admitted for femorodistal bypass. She was an insulin dependent diabetic with end stage renal failure and on dialysis. Although she mobilized around the house, she was dependent on her wheelchair outdoors, and was visually impaired. Her surgery was uncomplicated, but she died four months after surgery as a result of her renal failure. Her preoperative

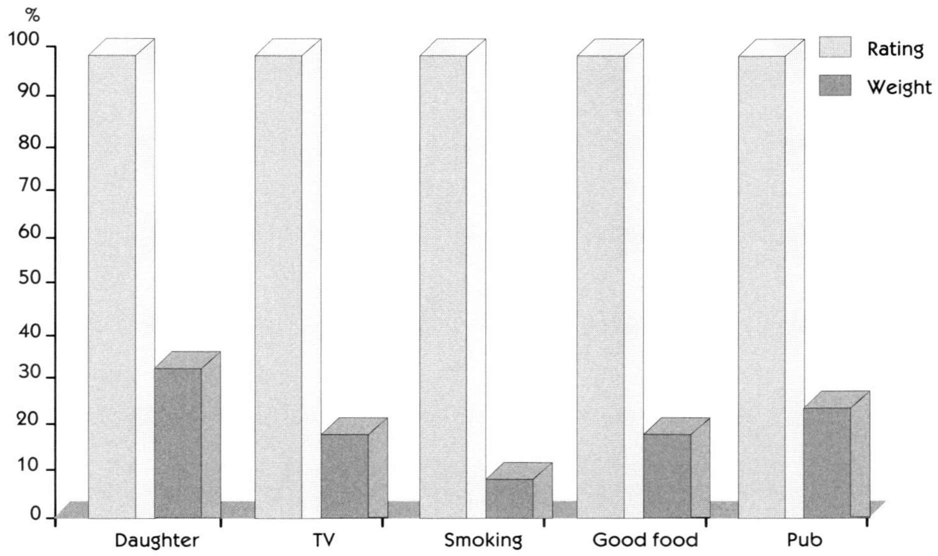

FIG. 2 Profile of patient quality of life priorities.

quality of life priorities mostly revolved around her family; she lived with her daughter and family on whom she was dependent, and although she wanted her health to be better, this primarily related to a desire to be alive for her granddaughter's wedding. The concept of limb salvage to improve her quality of life was not important to her.

Whilst such scenarios are anecdotal, and follow-up application of SEIQoL is presently limited to a few patients at six months, it is clear that it can highlight different patient priorities and their expectations. It has also been useful to demonstrate how these priorities vary with time; even in those who have undergone successful reconstruction, those cues relating to mobility and outside hobbies such as gardening or simply walking the dog, are frequently abandoned with the passage of time. If the patients perspective in determining their quality of life is to be considered, then the results are likely to be more valid in assessing the impact of surgical intervention than the application of questionnaires which have been developed from studying various cross-sections of the population, healthy or otherwise. Those patients who fail to mobilize or who are dependent on support from either family or social services will not necessarily fall into the *poor outcome* group. Using a system which clearly highlights an individuals' perspective can identify those who will not contemplate an amputation, whether it be for mobility or psychological reasons, while identifying those for whom preservation of the limb is not important for their quality of life.

■ CONCLUSION

We hope that in the future, the decision to perform revascularization or primary amputation will be based on an analysis of what is best for the individual patient and not simply on technical feasibility. Such an analysis would be invaluable for identifying those patients in whom the benefit of surgical arterial reconstruction would be negligible, based on their quality of life priorities and not the pre-conceived opinions of others.

REFERENCES

1 McWhinnie DL, Gordon AC, Collin J et al. Rehabilitation outcome 5 years after 100 lower-limb amputations. *Br J Surg* 1994; 81: 1596-1599.
2 Hickey NC, Thomson IA, Shearman CP, Simms MH. Aggressive arterial reconstruction for critical lower limb ischaemia. *Br J Surg* 1991; 78: 1476-1478.
3 Humphreys WV, Evans F, Watkin G, Williams T. Critical limb ischaemia in patients over 80 years of age: options in a district general hospital. *Br J Surg* 1995; 82: 1361-1363.
4 Pell JP, Fowkes FGR, Ruckley CV et al. Declining incidence of amputation for arterial disease in Scotland. *Eur J Vasc Surg* 1994; 8: 602-606.
5 Luther M. The Influence of arterial reconstructive surgery on the outcome of critical leg ischaemia. *Eur J Vasc Surg* 1994; 8: 682-689.
6 Singh S, Evans L, Datta D et al. The costs of managing lower limb-threatening ischaemia. *Eur J Vasc Endovasc Surg* 1996; 12: 359-362.
7 Cheshire NJW, Wolfe JHN, Noone MA et al. The economics of femorocrural reconstruction for critical leg ischaemia with and without autologous vein. *J Vasc Surg* 1992; 15: 167-175.
8 Wolfe JHN, Tyrell MR. Justifying arterial reconstruction to crural vessels even with a prosthetic graft. *Br J Surg* 1991; 78: 897-898.
9 Ouriel K, Fiore WM, Geary JE. Limb-threatening ischemia in the medically compromised patient: amputation or revascularization? *Surgery* 1988; 104: 667-672.
10 Anonymous. Critical limb ischaemia: management and outcome. Report of a National Survey: The Vascular Surgical Society of Great Britain and Ireland. *Eur J Vasc Endovasc Surg* 1995; 10: 108-113.
11 Hakaim AG, Gordon JK, Scott TE. Early outcome of in-situ femorotibial reconstruction among patients with diabetes alone versus diabetes and end-stage renal failure: analysis of 83 limbs. *J Vasc Surg* 1998; 27: 1049-1055.
12 Nicoloff AD, Taylor LM Jr, McLafferty RB et al. Patient recovery after infrainguinal bypass grafting for limb salvage. *J Vasc Surg* 1998; 27: 256-266.
13 Pomposelli FB Jr, Arora S, Gibbons GW et al. Lower extremity arterial reconstruction in the very elderly: successful outcome preserves not only the limb but also residential status and ambulatory function. *J Vasc Surg* 1998; 28: 215-225.
14 Sillesen H. Conservative treatment, amputation or revascularisation for critical limb ischaemia. *Ann Chir Gynaec;* 1998; 87: 159-161.
15 Rosser RM, Watts VC. The management of hospital output. *Int J Epid* 1988; 361-368.
16 Beattie DK, Golledge J, Greenhalgh RM, Davies AH. Quality of life assessment in vascular disease: towards a Consensus. *Eur J Vasc Endovasc Surg* 1997; 13: 9-13.
17 Chetter IC, Spark JI, Dolan P et al. Quality of life analysis in patients with lower limb ischaemia: suggestions for European standardisation. *Eur J Vasc Endovasc Surg* 1997; 13: 597-604.
18 Hickey AM, Bury G, O'Boyle CA et al. A new short form individual quality of life measure (SEIQoL-dw) application in a cohort of individuals with HIV/AIDS. *Br Med J* 1996; 313: 29-33.
19 Chetter IC, Spark JI, Scott DJA et al. Prospective analysis of quality of life in patients following infrainguinal reconstruction for chronic critical ischaemia. *Br J Surg* 1998; 85: 951-955.
20 Johnson BF, Singh S, Evans L et al. A prospective study on the effect of limb-threatening ischaemia and its surgical treatment on the quality of life. *Eur J Vasc Endovasc Surg* 1997; 13: 306-314.

26

INTRAOPERATIVE CONTROL OF ENDOVASCULAR AND SURGICAL RECONSTRUCTIONS FOR CRITICAL LIMB ISCHEMIA

JAN D. BLANKENSTEIJN

The results of limb revascularization are directly related to the quality and initial performance of the reconstruction. This is true for both conventional surgery and endovascular interventions. After endarterectomy and bypass surgery, the first-month failures are attributed mainly to technical errors, intrinsic thrombogenicity, and inadequate outflow tract [1]. Likewise, the results of percutaneous transluminal angioplasty (PTA) are directly proportional to the primary angiographic or intravascular ultrasound (IVUS) appearance of the lesion immediately after balloon dilation.

First-month failure rates of up to 15% after bypass surgery [2,3] and over 30% after PTA have been reported [4,5]. Consequently, the pre- and intraoperative detection of predictive factors for failure have gained substantial interest.

■ RECENT DEVELOPMENTS

Screening for the presence of a hypercoaguable state in vascular surgery patients has been suggested [6], but the yield of such a strategy in all patients scheduled for infrainguinal reconstruction is presently unknown. The identification of grafts at risk by measuring outflow resistance has also been investigated, but results are contradictory [7,8]. With respect to endovascular interventions, IVUS has been reported to be particularly helpful in evaluating the immediate effect of balloon dilation or stent deployment and to indicate the need for stent placement or additional interventions after PTA [9]. As long as it is unclear to which extent problems like incomplete valve lysis, vein wall abnormalities, anastomotic irregularities, and outflow disease increase the risk for graft failure, the reconstruction should be nothing less than perfect. Similarly, IVUS may help to identify plaque rupture, dissection, or media rupture as good or bad prognostic signs after PTA in the future [10]. Currently, however, the most reliable means of getting the best possible results after endovascular intervention is to strive for a perfect angiographic appearance of the lesion treated.

In recent years, minimally invasive techniques have revolutionized vascular surgery. A number of procedures that used to form the main part of conventional vascular surgery, like aortoiliac reconstructions for stenotic disease and above-knee femoropopliteal bypass surgery, have been taken over largely by percutaneous techniques. And more are to follow: aortic aneurysms and carotid artery stenosis. Currently, for the majority of patients with critical limb ischemia, the only hope for long-term limb salvage is bypass surgery. Subintimal angioplasty of the tibial vessels has been reported with acceptable short-term results [11]. With the development of peripheral coaxial systems and smaller catheters, balloons, and stents, minimally invasive procedures will quickly gain significance in this area.

The redistribution of vascular indications between percutaneous techniques and open vascular surgery has led most vascular surgeons to add minimally invasive techniques to their armamentarium. As a consequence, many vascular surgery operating theatres have evolved from being completely deprived of imaging equipment, to being fully-rigged angiography suites. Until recently, the only imaging technique available to most vascular surgeons was the single shot, flush angiogram for bypass and runoff control. With the improved imaging modalities in the operating theatre, digital subtraction angiography (DSA), PTA and stent placement, and fluoroscopically assisted thromboembolectomy [12] have been added to the vascular surgery tools. The vascular surgeon must be familiar with the potential and methodology of these interventional techniques when performing arterial reconstructions for critical limb ischemia.

In this chapter, the available options for intraoperative control of endovascular and conventional vascular procedures are discussed separately.

■ ENDOVASCULAR RECONSTRUCTIONS

Besides PTA and stenting, a variety of minimally invasive procedures for limb ischemia have emerged. Some of these are simply less invasive modifications of conventional vascular reconstructions. The endovascular procedures discussed in this chapter are fluoroscopically assisted thromboembolectomy, remote endarterectomy, and angioscopy assisted bypass procedures. The control modalities for PTA and stenting are not different from those used in the angiography suite (pressure measurements and angiographic appearance) and fall outside the scope of this chapter. The control techniques that will be discussed with respect to the above mentioned procedures are Digital Subtraction Angiography (DSA), Intravascular Ultrasound (IVUS), pressure measurements, and angioscopy.

Fluoroscopically Assisted Thromboembolectomy

With subtraction angiography and *roadmap* facilities in the operating theatre, it is possible to turn the uncontrolled, conventional thrombectomy into a procedure that allows complete removal of thrombus under fluoroscopic guidance and provides complete angiographic control. One of the foremost arguments for treating acute occlusions with thrombolysis first has been the information of what the underlying cause of the occlusion is, in order to proceed with surgery if the problem cannot be solved with PTA [13,14]. With fluoroscopically assisted thromboembolectomy, it is possible to remove the bulk of the clot without the systemic risks of thrombolysis (e.g., cerebral hemorrhage and puncture site bleeding), and to choose further therapy based upon the intraoperative DSA. One big advantage of this approach is that, contrary to initial thrombolytic treatment, all options are available after clot removal (including in- and outflow PTA, surgical revision or PTA of an anastomosis, and the addition of a bypass). Another advantage is that no precious time is lost in case of acute ischemia; this is always a major issue in deciding whether to start thrombolysis or to immediately go to the operating room.

Instead of introducing a single lumen Fogarty thrombectomy catheter, a guidewire is advanced first into the occluded vascular segment. Under fluoroscopy control, the guidewire is passed distally to the occluded segment. Next, a double lumen thrombectomy catheter is passed over the guidewire and again under fluoroscopic control the thrombectomy balloon is inflated (with diluted contrast) and withdrawn. After removal of the bulk of the distal thrombus, a subtraction angiogram is made of the distal segment and the runoff. If thrombus is still present, further runs with the thrombectomy catheter are used. If necessary an adherent clot thrombectomy catheter or a graft thrombectomy catheter is used to achieve complete clearance of thrombus. Remaining thrombus in the runoff vessels can be treated with intraoperative injection of urokinase [15]. After heparinization of the cleared distal segment, the attention is turned to the inflow. It is advised not to start off with opening the inflow, because most of the underlying problems are related to the distal graft or runoff vessels, and because hemostatic control over the inflow, once it is pulsatile, may be cumbersome. A guidewire is passed proximally to the occlusion and the above procedure is repeated. The completion angiograms can be shot through the guidewire lumen of the double lumen thrombectomy catheter. This will give highly selective angiograms with excellent visualization of the proximal and distal anastomoses. Finally, the underlying cause for occlusion is addressed by whatever procedure is deemed necessary. At this stage it is possible to use other modalities to identify the underlying problem. For instance, IVUS is an excellent tool to check for problems at the anastomic areas of a bypass. Pressure measurements can only be done after the arteriotomy has been closed and flow restored; inflow pressure cannot be measured without runoff. Preferably, pressures are measured before and after administering a vasodilator to the runoff arteries to invoke a pressure drop.

Remote Endarterectomy

This procedure is a modification of conventional semi-closed endarterectomy with a proximal and distal incision. In remote endarterectomy, achieving an endpoint of the endarterectomy under fluoroscopic guidance obviates the need for a distal incision. The transition zone is passed with a guidewire and the cut edges of the intima are secured to the wall with the use of a short stent. For a further description of this procedure it is referred to the literature on this subject [16]. For this chapter, it is important to list this less invasive endarterectomy as another option for dealing with critical limb ischemia together with using intraoperative control modalities like DSA and IVUS.

Angioscopy Assisted Bypass Procedures

Angioscopy has been around for a while as an adjunct to in-situ bypass procedures [17]. It allows better control over the quality of the vein and over the completeness of valve lysis. The angioscope can also be used to identify the side branches of the saphenous vein; unfortunately it did not allow for occluding them. As such, the angioscope could be considered an adjunct to conventional in-situ bypass procedures [18]. With the availability of an angioscope that does allow valve lysis under direct vision and side branch coil-occlusion, the in-situ bypass procedure with the *side branch occluder angioscope* (SBO, Baxter Cardiovascular) can justifiably be called an endovascular bypass. Although a large proportion of patients still need several extra incisions to control for big side-branches [19], an in-situ femoropedal bypass graft is possible with only two small incisions at the anastomotic sites. Besides the inherent control modality of this procedure being the angioscope, the additional options being intraoperative DSA and IVUS are discussed together with the conventional reconstructions.

■ CONVENTIONAL RECONSTRUCTIONS

In this paragraph the attention is focussed on bypass surgery. Some control modalities may be used after endarterectomy but their utility can simply be derived from the application to conventional bypass procedures. Before going into details of intraoperative control it must be stressed that preoperative imaging (selective angiography or duplex), good clinical decision-making, and planning must establish adequate bypass in- and outflow tracts.

The control procedures in bypass surgery can be classified into pre-reconstruction studies to evaluate the quality of the venous grafting material to be used and post-reconstruction methods. Pre-reconstruction studies usually involve duplex ultrasound of the long and (if necessary) short saphenous veins. Although some authors claim excellent results of preoperative duplex mapping of venous material [20,21], others considered it inadequate [22]. Certainly, preoperative duplex mapping (especially if not performed by the surgeon who will do the bypass procedure) is rather subjective and operator dependent. There are other ways of judging the quality of the veins intraoperatively, using flebography [21], IVUS [23], or angioscopy [22].

The post-reconstruction methods are used to judge the anastomoses, the quality and the inflow and outflow of the conduit. At first glance, the technical adequacy of an infrainguinal bypass graft can be judged by the external appearance of the bypass and anastomoses, and by the restoration of distal pulses and perfusion. These factors may demonstrate the graft to be patent but they will not guarantee the conduit is perfect in every aspect. Non-invasive (ultrasonic and plethysmographic) instruments and invasive procedures (arteriography and angioscopy) may be added for a more objective evaluation [24-26]. The parameters provided by these methods include blood pressure, flow velocity and volume, pulse waveforms, and anatomic images.

Contrast angiography has two major limitations. First, it requires injection of contrast material that may be detrimental to the graft and distal vessels, or to kidneys already compromised by arteriosclerotic disease. Second, it is unable to grade hemodynamic impairment. With the development of miniaturized fiberoptic catheters, angioscopy became available to the vascular surgeon in the mid-eighties [27]. The angioscope provides a method of controlled guidance of instrumentation and allows immediate assessment of results [25,26]. To date, however, it is a costly and demanding technique with a potential of major complications, including intimal and vessel wall trauma, fluid overload, and thrombosis or embolization. Furthermore, as with arteriography, no information is rendered about the hemodynamic quality of the bypass or about the runoff vessels [28].

Another promising method of intraoperative anatomic imaging is intravascular ultrasound (IVUS), which can be performed with the bypass in circulation [23]. Besides luminal and anastomotic irregularities, vein wall abnormalities and inflow and outflow problems can be detected. However, it is not known whether angioscopy and IVUS are too sensitive which could lead to unnecessary revisions.

The past two decades have shown a remarkable increase in the number of non-invasive modalities. Because these techniques overcome the high costs, discomfort, and risks of angiography, they have become widely used for preliminary evaluation of patients with peripheral vascular disease, intraoperative monitoring bypass grafts, and follow-up. However, the exact value of their utility in the intraoperative evaluation of infrainguinal revascularization has not been the subject of a prospective comparative study.

■ PERSONAL EXPERIENCE

We have compared distal pulse palpation (DPP), qualitative and quantitative continuous wave doppler data (CWD), pulse volume recording (PVR), and graft flow measurement using ultrasonic volume flowmetry (UVF) as primary indicators of intraoperative problems and predictors of graft failure [29]. We also included in the analysis intraoperative angiography and vein diameter. A poor agreement was found among the four intraoperative tests, which confirms that DPP, CWD, PVR, and UVF assess different aspects of the hemodynamic quality of the vascular reconstruction. However, interpretation of audible signals of the CWD had the best likelihood ratios for both a positive and negative test result and thus provided the largest diagnostic gain for the detection of an intraoperative problem. Angiography was only helpful in the correct identification of the problems already detected by CWD and did not detect any additional technical inadequacies. On the other hand, the intraoperative tests were not sensitive for detecting outflow disease.

Angiography may therefore be reserved for cases in which CWD suggests a problem or if preoperative imaging does not allow assessment of the runoff status. Angioscopy or IVUS, however, may be a more accurate method to establish the nature of the problem detected with CWD [23,25]. In predicting early graft failure the results of the intraoperative tests after the necessary corrections of problems were of no value. This corresponds with the assumption that early failure is predominantly related to runoff problems once a perfect vascular conduit is established.

Our data support this theory, as the SVS/ISCVS runoff score [30] was the only significant predictor of early failure. With the use of the SVS/ISCVS runoff score, bypasses with high risk (33%) and low risk (0%) for early failure can be identified. If first-year primary patency is considered, failures appear to be related not only to the SVS/ISCVS runoff score but also to the initial minimal vein graft diameter.

From these results we concluded that, from the intraoperative modalities tested, CWD is the best indicator for technical errors. Routine intraoperative angiography is not necessary and angiography or other means of anatomic imaging may be reserved for situations in which CWD indicates a problem or if runoff information is inadequate. If all technical problems potentially responsible for failure can be eliminated during the initial bypass procedure, the SVS/ISCVS runoff score is a good predictor of early graft failure.

REFERENCES

1 Robinson KD, Sato DT, Gregory RT et al. Long-term outcome after early infrainguinal graft failure. *J Vasc Surg* 1997; 26: 425-437.

2 Craver JM, Ottinger LW, Darling RC et al. Hemorrhage and thrombosis as early complications of femoropopliteal bypass grafts: causes, treatment, and prognostic implications. *Surgery* 1973; 74: 839-845.

3 Whittemore AD, Clowes AW, Couch NP, Mannick JA. Secondary femoropopliteal reconstruction. *Ann Surg* 1981; 193: 35-42.

4 Jeans WD, Armstrong S, Cole SE et al. Fate of patients undergoing transluminal angioplasty for lower-limb ischemia. *Radiology* 1990; 177: 559-564.

5 Matsi PJ, Manninen HI, Suhonen MT et al. Chronic critical lower-limb ischemia: prospective trial of angioplasty with 1-36 months follow-up. *Radiology* 1993; 188: 381-387.

6 Donaldson MC, Weinberg DS, Belkin M et al. Screening for hypercoagulable states in vascular surgical practice: a preliminary study. *J Vasc Surg* 1990; 11: 825-831.

7 Ascer E, Veith FJ, Morin L et al. Components of outflow resistance and their correlation with graft patency in lower extremity arterial reconstructions. *J Vasc Surg* 1984; 1: 817-828.

8 Peterkin GA, LaMorte WW, Menzoian JO. Runoff resistance and early graft failure in infrainguinal bypass surgery. *Arch Surg* 1988; 123: 1199-1201.

9 Van der Lugt A, Gussenhoven EJ, Pasterkamp G et al. Intravascular ultrasound predictors of restenosis after balloon angioplasty of the femoropopliteal artery. *Eur J Vasc Endovasc Surg* 1998; 16: 110-119.

10 Van der Lugt A, Gussenhoven EJ, Mali WP et al. Effect of balloon angioplasty in femoropopliteal arteries assessed by intravascular ultrasound. *Eur J Vasc Endovasc Surg* 1997; 13: 549-556.

11 Bolia A, Bell PR. Femoropopliteal and crural artery recanalization using subintimal angioplasty. *Semin Vasc Surg* 1995; 8: 253-264.

12 Parsons RE, Marin ML, Veith FJ et al. Fluoroscopically assisted thromboembolectomy: an improved method for treating acute arterial occlusions. *Ann Vasc Surg* 1996; 10: 201-210.

13 STILE investigators. Results of a prospective randomized trial evaluating surgery versus thrombolysis for ischemia of the lower extremity. The STILE trial. *Ann Surg* 1994; 220: 251-266.

14 Ouriel K, Veith FJ, Sasahara AA. A comparison of recombinant urokinase with vascular surgery as initial treatment for acute arterial occlusion of the legs. Thrombolysis or Peripheral Arterial Surgery (TOPAS) Investigators. *N Engl J Med* 1998; 338: 1105-1111.

15 Comerota AJ, Rao AK, Throm RC et al. A prospective, randomized, blinded, and placebo-controlled trial of intraoperative intra-arterial urokinase infusion during lower extremity revascularization. Regional and systemic effects. *Ann Surg* 1993; 218: 534-541.

16 Moll FL, Ho GH. Closed superficial femoral artery endarterectomy: a 2-year follow-up. *Cardiovasc Surg* 1997; 5: 398-400.

17 Miller A, Campbell DR, Gibbons GW et al. Routine intraoperative angioscopy in lower extremity revascularization. *Arch Surg* 1989; 124: 604-608.

18 Sales CM, Goldsmith J, Veith FJ. Prospective study of the value of prebypass saphenous vein angioscopy. *Am J Surg* 1995; 170: 106-108.

19 Piano G, Schwartz LB, Foster L et al. Assessing outcomes, costs, and benefits of emerging technology for minimally invasive saphenous vein in-situ distal arterial bypasses. *Arch Surg* 1998; 133: 613-617.

20 McShane MD, Field J, Smallwood J, Chant AD. Early experience with B mode ultrasound mapping of the long saphenous vein prior to femorodistal bypass. *Ann R Coll Surg Engl* 1988; 70: 147-149.

21 Veith FJ, Moss CM, Sprayregen S, Montefusco C. Preoperative saphenous venography in arterial reconstructive surgery of the lower extremity. *Surgery* 1979; 85: 253-256.

22 Wilson YG, Davies AH, Currie IC et al. Angioscopy for quality control of saphenous vein during bypass grafting. *Eur J Vasc Endovasc Surg* 1996; 11: 12-18.

23 Van der Lugt A, Gussenhoven EJ, The SH et al. Femoropopliteal bypass evaluated with intravascular ultrasound. *Eur J Vasc Endovasc Surg* 1995; 9: 394-402.

24 Schmitt DD, Seabrook GR, Bandyk DF et al. Early patency of in-situ saphenous vein bypasses as determined by intraoperative velocity waveform analysis. *Ann Vasc Surg* 1990; 4: 270-275.

25 Gilbertson JJ, Walsh DB, Zwolak RM et al. A blinded comparison of angiography, angioscopy, and duplex scanning in the intraoperative evaluation of in-situ saphenous vein bypass grafts. *J Vasc Surg* 1992; 15: 121-129.

26 Miller A, Marcaccio EJ, Tannenbaum GA et al. Comparison of angioscopy and angiography for monitoring infrainguinal bypass vein grafts: results of a prospective randomized trial. *J Vasc Surg* 1993; 17: 382-398.

27 White GH. Angioscopy. *Surg Clin North Am* 1992; 72: 791-821.

28 Woelfle KD, Kugelmann U, Bruijnen H et al. Intraoperative imaging techniques in infrainguinal arterial bypass grafting: completion angiography versus vascular endoscopy. *Eur J Vasc Surg* 1994; 8: 556-561.

29 Blankensteijn JD, Gertler JP, Brewster DC et al. Intraoperative determinants of infrainguinal bypass graft patency: a prospective study. *Eur J Vasc Endovasc Surg* 1995; 9: 375-382.

30 Rutherford RB, Baker JD, Ernst C et al. Recommended standards for reports dealing with lower extremity ischemia: revised version. *J Vasc Surg* 1997; 26: 517-538.

27

THE CAUSE OF MID-TERM AND LONG-TERM FAILURES OF ARTERIAL RECONSTRUCTIONS: THE ROLE OF SURVEILLANCE WITH SUBSEQUENT ACTION

JAAP BUTH, MIRZA M. IDU

The cause of failures of arterial reconstructions or interventions after the perioperative period can be separated into those that are amenable to limited revisions, and those that can only be resolved by complete redo procedures. Selective use of limited reinterventions, either by an endovascular or a surgical procedure, may be indicated for impending, but not complete, arterial or graft occlusion. Stenosis at graft, anastomotic, or angioplasty sites may become apparent because of the return of symptoms or a decrease of ankle pressures noted at scheduled follow-up visits. In other cases a paucity of symptoms may cause substantially delay, and complete occlusion will have occurred at the time of patient presentation. Both scenarios can be encountered in the practice of follow-up. However, it has become apparent that patterns of failure vary with the different types of reconstruction for lower limb ischemia as will be demonstrated in the present chapter. The role of surveillance will be discussed in the context of infrainguinal vein bypass, femoropopliteal prosthetic bypass, endovascular femoropopliteal interventions, and reconstructions for aortailiac obstructive disease.

■ SURVEILLANCE AFTER INFRAINGUINAL VEIN GRAFTS

BACKGROUND

Autologous vein bypass grafting is considered the best infrainguinal reconstruction for alleviating symptoms of lower limb ischemia. However, there is a high incidence of graft occlusion, resulting in patency rates of 30% to 75% at 5 years [1-3]. While approximately 10% of grafts fail within the first month, most commonly because of operative technical errors, it has long been assumed that later failures were due to progression of atherosclerosis within the inflow and runoff vessels. However, in the recent years it has become apparent that stenotic lesions in the graft (Fig. 1) and at the anastomotic areas are the main cause of graft failure, with 65% to 75% occurring within the first postoperative year [4]. There are compelling reasons to adopt a surveillance program for infrainguinal vein grafts. Because most of these bypasses are performed for critical ischemia, failure often jeopardizes limb survival. Once a vein graft occludes, the results of thrombectomy or thrombolysis are dismal, with a long-term patency of 20% to 28%. At this time it is generally accepted that the follow-up of patients with infrainguinal vein grafts should be directed towards the identification of failing grafts [5,6]. It follows that recognition and revision of a stenosis before occlusion develops, should prolong graft survival [7].

Only one-third of hemodynamically significant graft stenoses can be diagnosed by the return of ischemic symptoms or decreased pulses at physical examination [7,8]. In contrast, it has been shown that periodic surveillance using subtraction angiography or non-invasive methods, such as ankle pressure measurements or duplex scanning are effective in identifying graft stenoses. Using these methods stenotic lesions have been found in 25% to 30% of vein grafts, with the majority identified in the first year [7,9-12].

Which bypasses should be systematically followed? There is no significant difference between the incidence of myointimal hyperplasia or valve site fibrosis in in-situ, non-reversed translocated, reversed, and arm vein grafts [13-15]. Selective repair of vein graft stenosis is most commonly performed by patchplasty, excision with direct repair, and the use of jumpgrafts at the anastomotic sites [16,17]. The results of balloon angioplasty seem less durable than those of surgical revision, as combined recurrence and failure rates of 33%-50% were seen [18-21]. Especially when lesions are longer than one cm, poor results can be expected. It should be considered that myointimal hyperplasia occasionally causes a very diffusely diseased graft, which may be impossible to salvage by a revision.

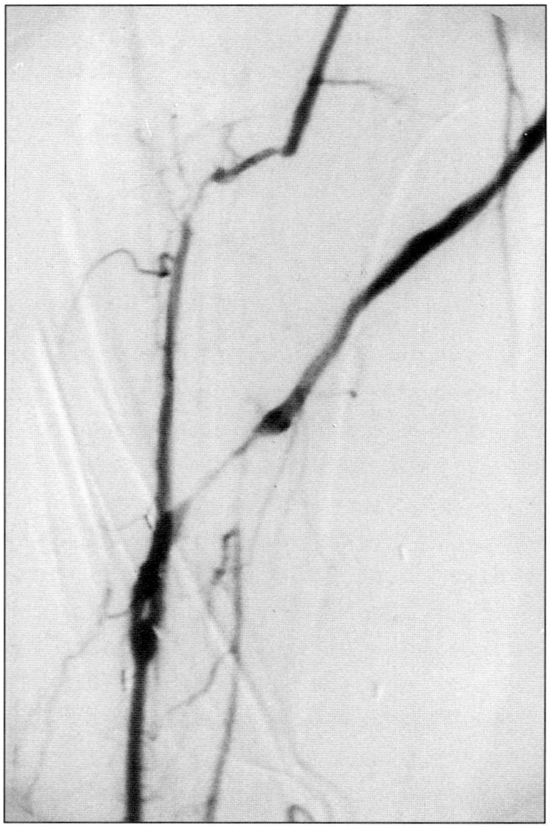

FIG. 1 Long stenotic lesion near the distal anastomosis of a femorotibial vein graft.

ETIOLOGY OF VEIN GRAFT STENOSES

Histologically, the major feature of vein graft stenosis is intimal hyperplasia and the predominant cytological component is the proliferating muscle cell and its secreted matrix. It is likely that much of the research performed on restenosis after balloon angioplasty applies also for vein graft stenosis [22]. There are still many blank areas in the pathogenesis of intimal hyperplasia, several factors were recognized to correlate with its occurrence, such as a high fibrinogen concentration, hyperhomocystinemia, the presence of anticardiolipin antibodies, high platelet concentration in the plasma, pre-existing fibrosis, and small graft diameter [23-25]. Although instrumental manipulation during the vein graft preparation and implantation has been suggested as a cause, Moody et al. found no relation between the sites of potential clamp injury and stenosis development [26].

Smooth muscle cell proliferation is not the only cause of graft stenosis. Another contributing component may be vein graft stricture formation, comparable with constrictive remodeling seen after balloon angioplasty [27]. This process is associated with a decrease in the cross-sectional area of the whole artery. Both phenomena, intimal hyperplasia and constrictive remodeling, play a role in the development of vein graft stenoses.

Although in animal studies stenosis development after angioplasty could be pharmacologically prevented, this has not appeared feasible in human restenosis [22]. There are no established criteria to determine which vein graft is at risk for stenosis development or at which site a lesion may arise. Therefore, preventive measures, if they were available, would be neither practical nor justifiable.

Surveillance Methods, Criteria, and Current Schedule

Most investigators have found the ankle-brachial blood pressure index to be an unreliable indicator of significant stenosis in grafts, and adjacent inflow and runoff arteries [28-31]. The usefulness of duplex scanning, in particular color duplex, for serial follow-up of infrainguinal vein grafts is generally acknowledged (Fig. 2). The superiority of duplex scanning over conventional non-invasive methods was confirmed in a number of studies [8,10,11,32]. Moreover, moderate graft stenosis may also be diagnosed and its progression followed to allow timely intervention.

In the identification of stenotic lesions in vein grafts, several duplex parameters have been correlated with the angiographic severity of the lesion, and with the clinical outcome of the untreated lesion. Of all parameters the ratio of the peak systolic velocity in the stenosis and in a nearby non-stenotic artery segment (PSV-ratio; Fig. 3) had the best correlation [10,11,14,15,31,33,34]. It is our policy to perform a graft revision only if the lesion has a diameter reduction greater than 70% at angiography. If a less severe stenosis is detected, it is followed by duplex at 6-weeks intervals to see if the lesion progresses. The data of a recently published Dutch multicenter study were used to define an algorithm to minimize the use

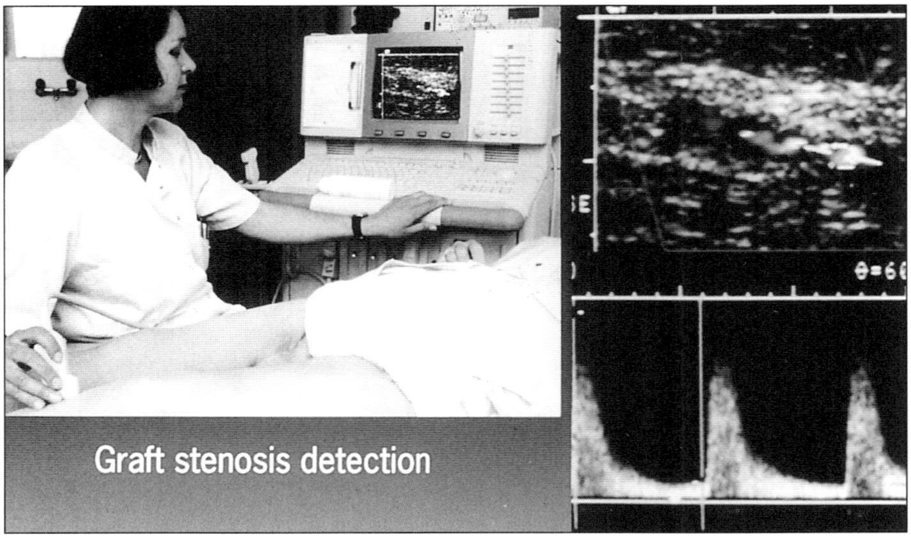

FIG. 2 Color duplex scanning of the entire infrainguinal vein graft. At the site of a stenosis the peak systolic velocity is increased and the doppler-spectrum broadened.

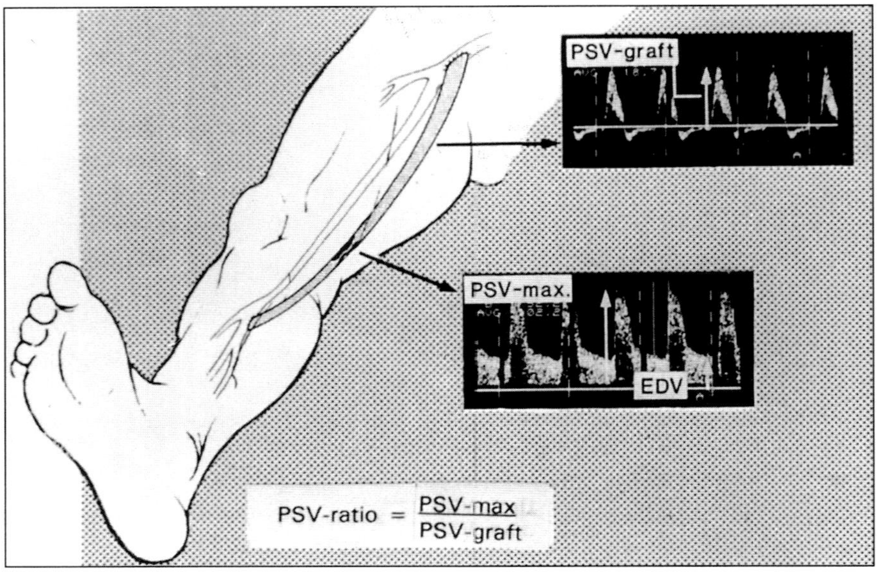

FIG. 3 Schematic illustration how the PSV-ratio is determined.

of angiography during follow-up without compromising the accuracy of indications for graft revision [34]. This assessment was performed by using a cumulative hazard state analysis with transient state method. If a PSV-ratio of < 2.5 was found a conservative approach without angiography was a safe policy. If the PSV-ratio was > 4.0 a vein graft revision without preceding angiography was appropriate, while patients with values between 2.5 and 4.0 constituted a gray zone with regard to duplex determination of stenosis severity. In this latter group an angiogram was performed to decide on clinical management. Using this algorithm a reduction of 50% of the number of angiographies was obtained compared to the previous policy in which angiograms were performed in all patients with a PSV-ratio greater than 2.5.

The optimal surveillance time schedule was assessed in a different study in the same multicenter series [35]. It was made plausible that the duration of the surveillance may be restricted to the first 6 months after the operation, if the patient had no evidence of a graft stenosis within this period. Only 2% of de-novo stenoses would be missed by this policy. Our current surveillance schedule includes complete duplex scanning of the graft and adjacent areas before discharge from the hospital, at 6 weeks, 3 months and 6 months. In case an increased PSV-ratio is found the management schedule described above is followed. If a graft is stenotic but does not require a revision (PSV-ratio < 2.5 or an angiographic stenosis < 70% diameter reduction), duplex examinations are performed at reduced intervals until it is apparent whether the stenosis progresses or is stable. Stenoses that appear stable are followed by duplex for at least another year. It has been noted that stenoses, detected early after the operation, tended to progress to occlusions more frequently than stenoses detected after 3 months [31]. Stenoses identified after 3 months represent more slowly progressive lesions, and stenoses identified after one year imposed the smallest risk of occlusion [31,36,37].

Effects of Surveillance and Stenosis Revision

To determine the potential impact of surveillance several questions need to be answered. What is the natural history of untreated vein graft stenoses? Does surveillance and selective revision contribute significantly to graft patency? Because surveillance is expensive, time consuming, and has the potential to lead to unnecessary surgery, these are important questions. Stenoses develop in 10% to 36% of vein bypass grafts. Approximately two-third are located in the graft body, a quarter at the anastomosis, and approximately 10% in the inflow or runoff arterial segments. Studies on the natural history of stenotic grafts are scant. Szilagyi et al. found in a series of 260 grafts followed by angiography stenoses in 33%, of which 38% occluded [9]. Another angiographic follow-up study with deliberate policy of nonintervention for asymptomatic stenoses in 28% of the grafts, lesions developed of which one quarter occluded [38]. Nielsen found that grafts with duplex-detected stenoses failed within the first year in 32% if the reduction in ankle/brachial pressure index (ABI) between surveillance visits was < 0.15, and in 67% with greater ABI reduction [31].

In other cohort studies 21% to 70% of non-revised stenotic grafts occluded during the study period. In contrast, an average of only 12% of revised grafts occluded. Cumulative assisted primary patency rates of revised grafts compare favorably with primary patency rates of non-revised grafts, exceeding 90% at 2 years [30,33].

To evaluate the impact of surveillance, a comparison of assisted primary or secondary patency rates of grafts followed by conventional clinical follow-up and by duplex surveillance is the best method for evaluation. Lundell et al., in the only prospective randomized study, found that grafts in the surveillance category had a 25% higher secondary patency rate at 2 to 3 years, compared with grafts that were followed clinically [39]. In studies by Idu et al. and Bergamini et al. using contemporary treated control groups with conventional follow-up, it was found that patency rates of grafts in surveillance programs were 91% at 3 years, and 80% at 5 years, which was significantly better than the 78% and 67% patency rates in the control groups (Fig. 4) [40,41].

Although the case of surveillance appeared to be closed with regard to improvement of graft patency, a statistical advantage to safe more limbs requires larger studies than have been performed so far. Only Bergamini et al. [41] found a significant improvement in limb salvage rate for bypasses followed by duplex at five years of follow-up (94% *versus* 73% in clinically followed patients). To demonstrate an unequivocal advantage of surveillance over clinical follow-up in terms of limb preservation some have called for a large randomized prospective trial [42,43].

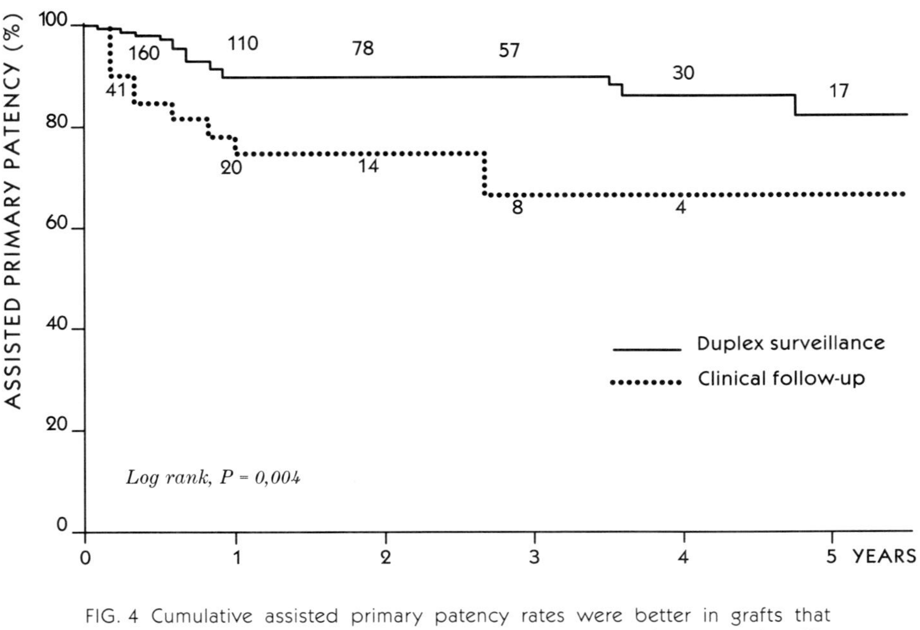

FIG. 4 Cumulative assisted primary patency rates were better in grafts that underwent color-duplex surveillance than in grafts with clinical follow-up (Idu [40]).

We investigated the cost-effectiveness of vein graft surveillance in an unpublished study showing that the greatest advantage of surveillance would be obtained by avoiding amputations in the patient category with severe critical ischemia, while the balance of cost and effectiveness was neutral in patients with intermittent claudication only. In this assessment the financial loss (secondary costs) from recurrent claudication was not taken into account.

Surveillance in Prosthetic Bypass Grafts

The role of regular duplex surveillance was also investigated in a group of patients that had undergone implantation of a femoropopliteal prosthetic bypass. In this unpublished series from our department PTFE-grafts were implanted in 101 patients, of which 45 were followed by clinical examination and 56 by regular duplex scanning, depending of the vascular surgeon in charge. Development of stenoses and patency of grafts in the two groups were compared retrospectively. In the group with clinical follow-up no stenoses were detected in patent grafts and only one was detected after thrombectomy and revised during the same procedure. In patients with duplex surveillance two stenoses were revised, one in a patent graft and one after a thrombectomy. There was no significant difference in assisted primary patency between the patient groups with different follow-up schedules. This made us conclude that, in contrast to vein grafts, stenosis development is not a major cause of failure in prosthetic bypasses. This renders regular duplex surveillance ineffective. The mechanism of graft failure is different in PTFE-grafts than in vein grafts. Thrombogenicity of the prosthesis material may be a more important factor. Our experience was similar to the outcome of a cohort study from Lalak et al. [44]. These investigators found stenoses amenable to intervention in only 6% of grafts during a 4-year follow-up. They concluded that the low yield of remediable disease does not justify the cost of duplex surveillance after this type of procedure.

Surveillance Following Endovascular Intervention in Femoropopliteal Obstructive Disease

In endovascular treatment of femoropopliteal disease, similarly as in vein grafts, stenotic lesions are easier to treat and have better results than occlusions. Therefore duplex surveillance may be potentially advantageous after balloon angioplasty or other endovascular therapies in the femoropopliteal segment. In a prospective study we have compared: (1) conventional clinical follow-up, and (2) serial duplex follow-up, as diagnostic schedules, with regard to the ability to indicate impending failure [45]. In a group of 124 patients with successful endovascular treatment recurrent preocclusive lesions were identified in 52 patients. A non-interventional policy for asymptomatic patients was followed. Failure was defined as the development of occlusion or the necessity of an intervention. Conventional clinical follow-up including regular ankle pressure measurements and treadmill exercise testing performed better than duplex testing in predicting failure, with a sensitivity of 93% and a specificity of 90%. Impending failure was predicted by duplex with a sensitivity of 86% and a specificity of 95%. On the basis of these observations we concluded that clinical/conventional non-invasive assessment was comparable, if not superior, to duplex surveillance after endovascular femoropopliteal treatment. It appeared that many patients with a restenosis develop symptoms before occlusion occurred, most notably intermittent claudication. Compared to the pattern in vein graft failure a sudden occlusion of a known restenotic segment infrequently occurred. Moreover, in a limited number of patients an occlusion developed without preceding stenosis. Clinical assessment and conventional noninvasive measurements appeared quite sufficient for monitoring patients after endovascular interventions.

Stenosis Development after Prosthetic Bypass for Aortoiliac Occlusive Disease

Since the report of Imparato et al. in 1972, stenosis development at the distal anastomosis has been recognized as a cause of graft limb occlusion following prosthetic bypass procedures for aortoiliac occlusive disease [46]. However, few studies on the long-term results of aortofemoral bypass have addressed the development of perianastomotic stenotic lesions. In a retrospective study we investigated the occurrence of abnormalities at the distal anastomosis in a series of 103 bypasses implanted for aortoiliac occlusive disease, using digital subtraction angiography (DSA) and color-flow duplex [47]. Stenotic lesions at or just beyond the distal anastomosis were identified by DSA in 30 patients with a follow-up of one to nine years. The incidence of stenoses was 33% in the first 3 years following the operation, 20% in the period of 3 to 6 years, 32% in the period of 6 to 9 years, and 50% in bypasses followed for longer than 9 years. Color-flow imaging had a 89% sensitivity in identifying distal anastomotic stenosis and a 85% specificity in ruling out significant lesions. Prophylactic repair of a high-grade stenosis (> 70% diameter reduction) was performed in only two patients. All other identified stenoses were of mild to moderate severity, not requiring any treatment. We concluded that color-flow duplex was accurate in identifying distal anastomotic stenosis. However, our data suggested a low yield of cases requiring prophylactic repair. Therefore, routine surveillance of bypass grafts for aortoiliac occlusions can not be recommended.

CONCLUSION

Serial duplex scanning was used in the follow-up of various types of reconstructions or interventions for lower limb arterial insufficiency. Accumulated data only support entering infrainguinal vein grafts in a surveillance program. The duration of surveillance in this category should be at least 6 months. Vein grafts with high-grade stenoses should be revised before they occlude, while it appeared safe to continue duplex follow-up for low-grade stenoses. The relative benefit of surveillance compared to clinical follow-up alone appears substantial with regard to maintenance of vein graft patency. Whether limb salvage is improved may be addressed in a larger scale prospective randomized trial.

REFERENCES

1 Darling RC, Linton RR. Durability of femoropopliteal reconstruction. Endarterectomy versus vein bypass grafts. *Am J Surg* 1972; 123: 472-479.
2 Veith FJ, Gupta SK, Ascer E et al. Six-year prospective multicenter randomized comparison of autologous saphenous vein and expanded polytetrafluoroethylene grafts in infrainguinal arterial reconstructions. *J Vasc Surg* 1986; 3: 104-114.
3 Taylor LM Jr, Edwards JM, Porter JM. Present status of reversed vein bypass grafting. Five-year results of a modern series. *J Vasc Surg* 1990; 11: 193-206.
4 Brewster DC, La Salle AJ, Robinson JG et al. Femoropopliteal graft failures. Clinical consequences and success of secondary reconstructions. *Arch Surg* 1983; 118: 1043-1047.
5 Cohen JR, Mannick JA, Couch NP, Whittemore AD. Recognition and management of impending vein-graft failure. Importance for long-term patency. *Arch Surg* 1986; 121: 758-759.
6 Belkin M, Donaldson MC, Whittemore AD et al. Observations of the use of thrombolytic agents for thrombotic occlusions of infrainguinal vein grafts. *J Vasc Surg* 1990; 11: 289-296.
7 Moody P, Gould DA, Harris PL. Vein graft surveillance improves patency in femoropopliteal bypass. *Eur J Vasc Surg* 1990; 4: 117-121.
8 Disselhoff B, Buth J, Jakimowicz J. Early detection of stenosis of femoro-distal grafts. A surveillance study using colour-duplex scanning. *Eur J Vasc Surg* 1989; 3: 43-48.
9 Szilagyi DE, Elliot JP, Haglesman JH et al. Biologic fate of autologous vein implants as arterial substitutes: clinical, angiographic and histopathologic observations in femoropopliteal operations for atherosclerosis. *Ann Surg* 1973; 178: 232-246.
10 Grigg MJ, Nicolaides AN, Wolfe JHM. Femorodistal vein bypass graft stenosis. *Br J Surg* 1988; 75: 737-740.
11 Sladen JG, Gilmour JL. Vein graft stenosis: characteristics and effect of treatment. *Am J Surg* 1981; 141: 549-553.
12 Whittemore AD, Clowes AW, Couch NP, Mannick JA. Secondary femoropopliteal reconstruction. *Ann Surg* 1981; 193: 35-42.
13 Harris PL, Decossart LM, Moody P, Douglas H. How can we detect and manage fibrous structures within new grafts? Are these more problematic with reversed than in-situ bypass. In Greenhalgh RM, Jamieson CW, Nicolaides AN, (eds.). *Limb salvage and amputation for vascular disease.* London, WB Saunders, 1988, pp 221-231.
14 Bandyk DF, Johnson BL, Gupta AK, Esses GE. Nature and management of duplex abnormalities encountered during infrainguinal bypass grafting. *J Vasc Surg* 1996; 24: 430-438.
15 Chalmers RTA, Hoballah JJ, Kresowik TF et al. The impact of color duplex surveillance on the outcome of lower limb bypass with segments of arm veins. *J Vasc Surg* 1994; 19: 279-288.
16 Breslau RC, DeWeese JA. Successful endophlebectomy of autogenous venous bypass grafts. *Ann Surg* 1965; 162: 251-254.

17 McNamara JJ, Darling RC, Linton RR. Segmental stenosis of saphenous-vein autografts: preventable cause of late occlusion in arterial reconstruction. *N Engl J Med* 1967; 277: 290-292.

18 Bandyk DF, Bergamini TM, Towne JB et al. Durability of vein graft revision: the outcome of secondary procedures. *J Vasc Surg* 1991; 13: 200-210.

19 Perler BA, Osterman FA, Mitchell SE et al. Balloon dilatation versus surgical revision of infra-inguinal autogenous vein graft stenosis long-term follow-up. *J Cardiovasc Surg* 1990; 31: 656-667.

20 Whittemore AD, Donaldson MC, Polak JF, Mannick JA. Limitations of balloon angioplasty for vein graft stenosis. *J Vasc Surg* 1991; 14: 340-345.

21 Sanchez LA, Suggs WD, Marin ML et al. Is percutaneous balloon angioplasty appropriate in the treatment of graft and anastomotic lesions responsible for failing vein bypasses. *Am J Surg* 1994; 168: 97-101.

22 Chan P. Prospects for prevention of graft stenosis and angioplasty restenosis. *Eur J Vasc Endovasc Surg* 1997; 13: 429-431.

23 Panetta TF, Marin ML, Veith FJ et al. Unsuspected preexisting saphenous vein disease. An unrecognized cause of vein bypass failure. *J Vasc Surg* 1992; 15: 102-112.

24 Cheshire NJW, Wolfe JHN, Baradas MA et al. Smoking and plasma fibrinogen, lipoprotein-a, and serotonin are markers for postoperative infrainguinal graft stenosis. *Eur J Vasc Endovasc Surg* 1996; 11: 479-486.

25 Idu MM, Buth J, Hop WCJ et al. Factors influencing the development of vein-graft stenosis and their significance for clinical management. *Eur J Vasc Endovasc Surg* 1999; 17: 15-21.

26 Moody AP, Edwards PR, Harris PL. The aetiology of vein graft strictures; a prospective marker study. *Eur J Vasc Surg* 1992; 6: 509-511.

27 Mintz GS, Popma JJ, Pichard AD et al. Arterial remodeling after coronary angioplasty: a serial intravascular ultrasound study. *Circulation* 1996; 94: 35-43.

28 Wolfe JHN, Thomas ML, Jamieson CW et al. Early diagnosis of femorodistal graft stenoses. *Br J Surg* 1987; 74: 268-270.

29 Mills JL, Harris JE, Taylor LM et al. The importance of routine surveillance of distal bypass grafts with duplex scanning: a study of 379 reversed vein grafts. *J Vasc Surg* 1990; 12: 379-386.

30 Green RM, McNamara J, Ouriel K, DeWeese JA. Comparison of infrainguinal graft surveillance techniques. *J Vasc Surg* 1990; 11: 207-215.

31 Nielsen TG. Natural history of infrainguinal vein bypass stenosis: early lesions increase the risk of thrombosis. *Eur J Vasc Endovasc Surg* 1996; 12: 60-64.

32 Davies AH, Magee TR, Tennant SGW et al. Criteria for the identification of the *at risk* infrainguinal bypass graft. *Eur J Vasc Surg* 1994; 8: 315-319.

33 Mattos MA, Van Bemmelen PS, Hodgson KJ et al. Does correction of stenoses identified with color-duplex scanning improve infrainguinal graft patency? *J Vasc Surg* 1993; 17: 54-66.

34 Idu MM, Buth J, Hop WCJ et al. Vein graft surveillance: is graft revision without angiography justified and what criteria should be used? *J Vasc Surg* 1998; 27: 399-413.

35 Idu MM, Buth J, Cuypers P et al. Economising vein graft surveillance programs. *Eur J Vasc Endovasc Surg* 1998; 15: 432-438.

36 Harris PL. Vein graft surveillance. All part of the service. *Br J Surg* 1992; 79: 97-99.

37 Mills JL, Bandyk DF, Gahtan V, Esses GE. The origin of infrainguinal vein graft stenosis. A prospective study based on duplex surveillance. *J Vasc Surg* 1995; 21: 16-25.

38 Moody P, De Cossart LM, Douglas HM, Harris PL. Asymptomatic strictures in femoro-popliteal vein grafts. *Eur J Vasc Surg* 1989; 3: 389-392.

39 Lundell A, Lindblad B, Bergqvist D, Hansen F. Femoro-popliteal-crural graft patency is improved by an intensive surveillance program: a prospective randomized study. *J Vasc Surg* 1995; 21: 26-34.

40 Idu MM, Blankensteijn JD, de Gier P et al. Impact of a color-flow duplex surveillance program on infrainguinal vein graft patency. A five-year experience. *J Vasc Surg* 1993; 17: 42-53.

41 Bergamini TM, George SM Jr, Massey HT et al. Intensive surveillance of femoropopliteal-tibial autogenous vein bypasses improves long-term graft patency and limb salvage. *Ann Surg* 1995; 221: 507-516.

42 Golledge J, Beattie DK, Greenhalgh RM, Davies AH. Have the results of infrainguinal bypass improved with the wide-spread utilisation of postoperative surveillance? *Eur J Vasc Endovasc Surg* 1996; 11: 388-392.

43 Ihlberg L, Luther M, Tierala E, Lepäntalo M. The utility of duplex scanning in infrainguinal vein graft surveillance: results from a randomised controlled study. *Eur J Vasc Endovasc Surg* 1998; 16: 19-27.

44 Lalak NJ, Hanel KC, Hunt J, Morgan A. Duplex scan surveillance of infrainguinal prosthetic bypass grafts. *J Vasc Surg* 1994; 20: 637-641.

45 Tielbeek AV, Rietjens E, Buth J et al. The value of duplex surveillance after endovascular intervention for femoro-popliteal obstructive disease. *Eur J Vasc Endovasc Surg* 1996; 12: 145-150.

46 Imparato AM, Bracco A, Kim GE, Zeff R. Intima and neointimal fibrous proliferation causing failure of arterial reconstructions. *Surgery* 1972; 72: 1007-1017.

47 De Gier P, Sommeling C, Van Dulken E et al. Stenosis development at the distal anastomosis of prosthetic bypasses for aortoiliac occlusive disease. Incidence and accuracy of colour-flow duplex in the diagnosis. *Eur J Vasc Surg* 1993; 7: 237-244.

28

LONG-TERM MORBIDITY AND MORTALITY OF PATIENTS TREATED FOR CRITICAL LIMB ISCHEMIA

THOMAS TROËNG

Most patients suffering from critical limb ischemia have well passed the retirement age. They often have complicating diseases that, together with their advanced vascular disease, lead to a limited expected survival even after successful treatment. The balance between the risks of active intervention and the benefits of a functional extremity is difficult to decide upon, especially in the individual patient. An observational follow-up study of consecutively treated patients may show within what limits treatment outcome is to be expected.

Thus, the aim of this study was to investigate what postoperative course patients can expect after open surgery or angioplasty for critical limb ischemia in routine health care.

■ MATERIAL AND METHODS

The *Swedish Vascular Registry* started to register peripheral vascular operations and angioplasties in southern Sweden in 1987. It records preoperative risk factors, details of procedures performed and 30 days and 1-year clinical outcome. Details of the registry have been described elsewhere [1]. By 1994 most Swedish vascular surgeons and all major centers participated in the registry [2]. During the first five years 2828 consecutive patients treated for critical limb ischemia were registered. They constituted the study population and the characteristics are described in Table I. As the 644 patients treated by angioplasty had virtually identical overall 1-year clinical outcome compared to the 2184 patients treated with open surgery, they were analyzed together as one single group. Patient survival can be monitored indefinitely through the national population registry, in this study until the end of 1997.

In-hospital stays for these patients were identified until the end of 1995 in the *National In-Patient Registry (NIR)* kept by the *National Board for Health and Social Welfare.* All Swedish hospitals report every episode of in-hospital care to this registry including dates of admission and discharge, diagnostic codes, procedure codes and possible in-hospital death. Reporting is compulsory and quality of data is regularly controlled. Thus, all patients had a 1-year follow-up of clinical outcome and a minimal 4-year follow-up concerning possible in-hospital care, amputations or other operations and a minimum of six years as to survival. Comparisons between groups were performed using the chi-square test, risk factors for death by logistic regression, and survival by Kaplan-Meier analysis. The SPSS for Windows program package version 6.1 was used for all analyses.

Table I CHARACTERISTICS OF PATIENT POPULATION — 1388 MEN MEDIAN AGE 74 YEARS; 1440 WOMEN MEDIAN AGE 77 YEARS

Risk factors (%)	*Men*	*Women*	*Difference*
Cerebrovascular	16.5	11.6	$p = 0.0002$
Diabetes	33.3	35.7	NS
Hypertension	34.8	44.0	$p < 0.0001$
Heart disease	54.2	48.8	$p = 0.004$
Pulmonary disease	13.1	7.3	$p < 0.0001$
Renal disease	7.1	5.6	NS
Smoking	52.4	23.5	$p < 0.0001$
Previous vascular surgery	25.6	19.4	$p < 0.0001$
Degree of ischemia			
Rest pain	49.1	46.6	
Ischemic ulcer	32.1	39.8	
Gangrene	18.9	13.6	$p < 0.0001$
Level of reconstruction (%)	*Surgery*	*Angioplasty*	
Above groin	14.3	22.0	
Thigh level	30.1	61.3	
Below knee	55.6	16.6	

NS: not significant

■ RESULTS

The 1-year clinical outcome is presented in Table II. The 30-day results with 77.8% of the patients improved and/or with a patent reconstruction are modified considerably during the first postoperative year by subsequent amputations and deaths. It was possible to identify 2744 (97%) of the 2828 patients in the NIR. There was a median number of four stays during the first 4 postoperative years, and at least one more hospital stay after the primary stay was experienced by 2530 (92%) patients. The cause of this stay was circulatory disease in 59.6%, diabetes in 8.7%, femoral neck fracture in 3.5%, respiratory disease in 3.1 %, a chronic ulcer in 2%, malignant disease in 1.9%, and various others in 22.8%. In 66.4% of these first postoperative stays no surgery was performed, while a vascular procedure was done in 12.9%, an orthopedic procedure, including amputations, in 8.7%, and some other surgery in 11.9%. The median time to the first next hospital stay was 55 days. The time distribution of the occurrence of postoperative stays is shown in Figure 1. During the first postoperative year another 6214 hospital stays were necessary.

Table II CLINICAL OUTCOME BY 30 DAYS AND AT 1 YEAR (%)

	30 days	*1 year*
Improved/ patent reconstruction	77.8	55.9
Not improved	11.9	8.0
Amputated	6.7	14.0
Dead	3.6	22.1

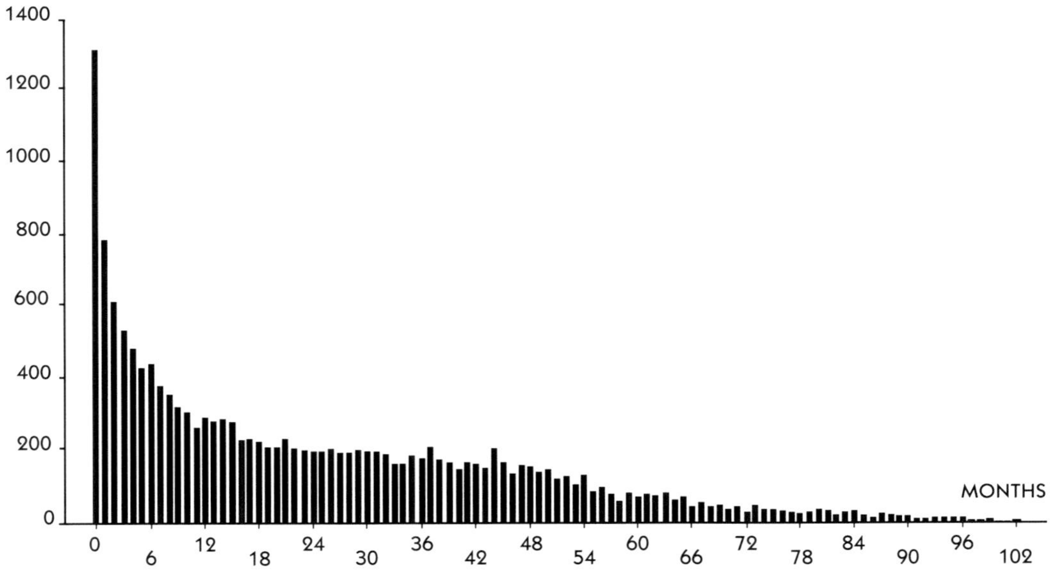

FIG. 1 Number of postoperative hospital stays for each month after intervention for critical limb ischemia in 2744 patients.

Figure 2 shows the distribution of groups of diagnoses during the 15 first postperative hospital stays. Some circulatory disease was the primary diagnosis in 62% of the first postoperative stay, which decreased to 34% by the 15th stay.

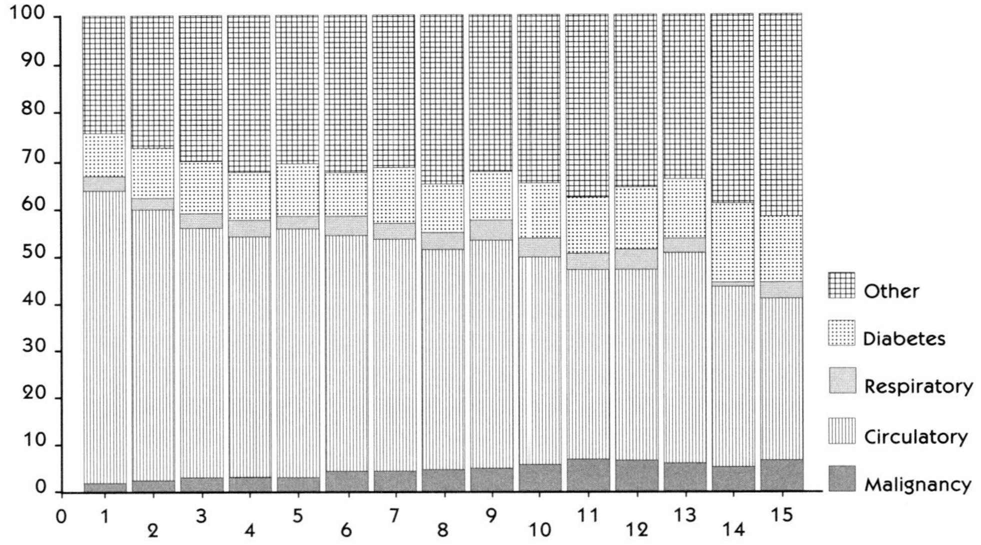

FIG. 2 Primary diagnosis at 15 first postoperative hospital stays which represent 97.5% of all postoperative stays during 4 years after intervention for critical limb ischemia.

The type of postoperative hospital stays is seen in Figure 3. Some surgical treatment was given to 33.5% of patients during their first postoperative hospital stay and in decreasing proportion during later stays. Patients admitted to a tenth stay were treated surgically in only 22.3% and those few with a 15th stay had surgery in only 15.3%. The primary vascular procedure needed a mean of 16.3 days in hospital for a total of 44,887 days. Subsequent stays during the next 4 years added to 15,052 days for repeated vascular

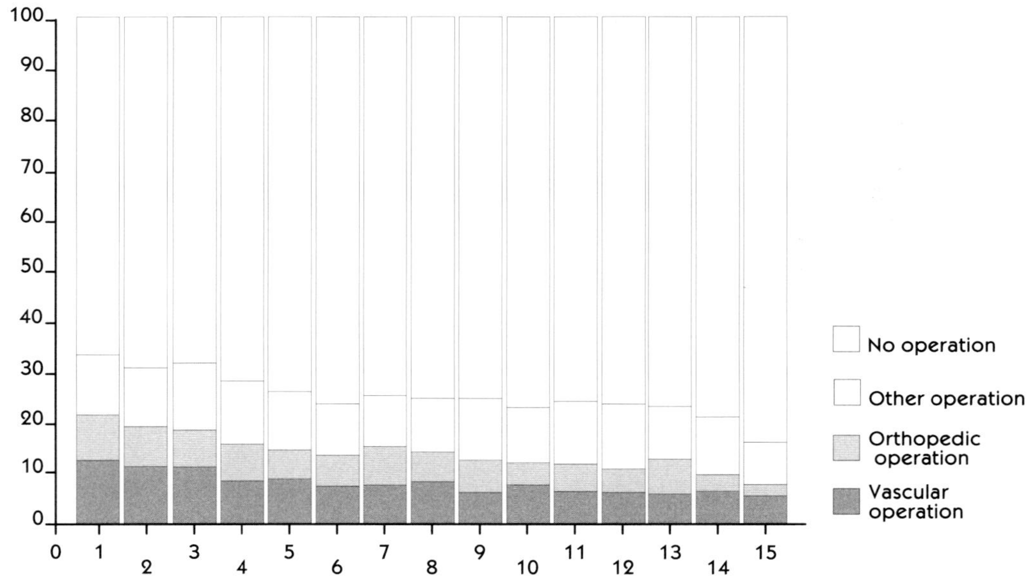

FIG. 3 Proportions of modes of treatment during 15 postoperative hospital stays.

surgery, 24,210 days for orthopedic procedures, 23,465 days for other surgery and 199,761 days for non-operative treatment. The primary procedure generated only 14.6% of all days in hospital and the patients stayed an average of 28 days in hospital per year during these 4 years, corresponding to 10.7% of their days alive.

Patient survival at 4 years was 52%. It was modified by a number of factors. Cerebrovascular, heart, pulmonary, renal disease and diabetes, as well as degree of ischemia, were all independent risk factors for death within 4 years. The small group of patients younger than 60 years survived 4 years in 75.6% while only 40.4% of patients 80 years of age or older were alive 4 years after the primary vascular procedure. As an example, Figure 4 shows the significantly inferior survival of patients with diabetes compared to non-diabetics.

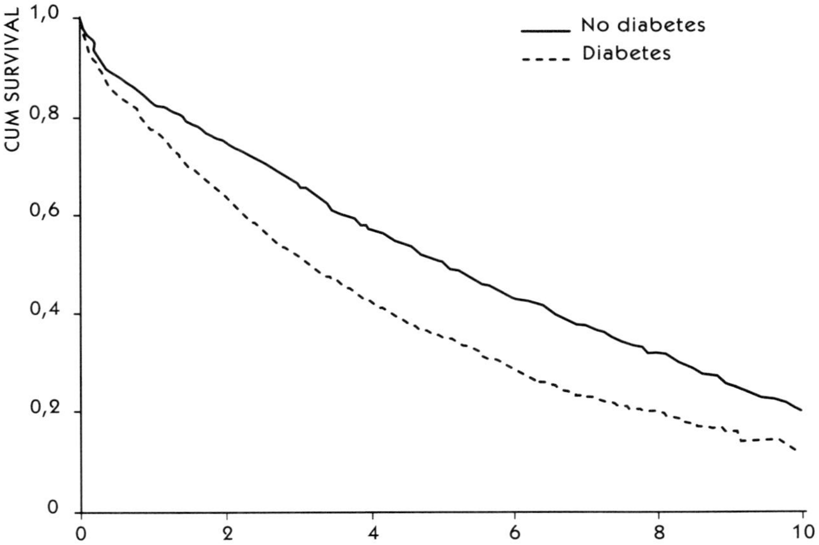

FIG. 4 Patient survival (Kaplan-Meier, proportion surviving by postoperative years) after intervention for critical limb ischemia.

■ DISCUSSION

The clinical outcome after reconstructive procedures for chronic limb ischemia is commonly expressed in terms of graft patency, incidence of amputations and short-time survival [3]. Such data are highly relevant as quality indicators for surgical and endovascular treatment techniques. The role of vascular surgery in a wider perspective and the long-term outcome of patients may, however, provide complementary data. This study combined data on unselected consecutive vascular patients who were prospectively registered by their surgeons, with an administrative database containing national data on all hospital stays, surgical and non-surgical. As it is highly unlikely that any significant number of inpatient treatment episodes are missed, the resulting picture of the postoperative course for patients after treatment for critical limb ischemia will be very close to reality.

The amputation rate and short-term survival in this study was similar to what has been described for a high-risk group of patients with critical ischemia in other studies [4-6]. Such clinical outcome may be inferior to what is reported from specialized centers treating selected patients but represents a more realistic picture of what these patients may expect.

We have previously analyzed the *Swedish Vascular Registry* data and reported that the incidence of reoperations, future redo procedures, and other-site vascular interventions is high after different types of vascular procedures [7]. To our knowledge, no one has previously described the total postoperative morbidity in patients undergoing vascular reconstructions. The occurrence of in-hospital treatment, possible operations, and the diagnosis are quite valid indicators of morbidity. Virtually every vascular patient needs future hospital care, in fact they need an average of almost a month per year during the following few years. It is more common that hospital care is needed for non-operative treatment. The cause for rehospitalization is circulatory disease in almost two-thirds of the first postoperative stays but only in one-third by the 15th stay. Diabetes is the cause in approximately 10% of all subsequent stays. Malignant disease increases from 1.8% to 6.4% from the first to the 15th stay. Vascular surgeons ought to consider this future need for treatment before they decide to embark on open or endovascular reconstructive surgery.

The selection of patients for active surgery in the first place might be improved. Even simple vascular registry data can be used to identify high-risk groups. An example is that only 37% of women older than 75, with peripheral gangrene are alive and improved while men 75 years old or younger with rest pain only have this outcome in 65%. Another example, focussing social outcome, is that 82% of patients with rest pain only and no cardiac or cerebrovascular risk factors live in their homes 1 year after intervention. Corresponding figures for patients with gangrene and cardiac or cerebrovascular disease is only 45%. Such differences clearly ought to influence decisions on treatment, although it is not easy to state how or to what extent.

High risk may imply an increased risk for complications like myocardial infarction or cerebrovascular lesions in connection with the selected procedure. It may also mean an increased risk for inferior treatment outcome in the sense of not having the peripheral ischemia relieved or even deteriorated. Should all patients with a chance of a favorable outcome of, say, no more than 25% or 30% actually have active treatment?

Other medical specialties have managed to reduce the need for in-hospital care considerably by active home care and other efforts. This should not be impossible also for vascular patients. Home care aimed at preventing the need for hospital care and vigorous risk factor modification in cooperation with angiologists, cardiologists, and others are some measures to consider.

In conclusion, open or endovascular surgery for critical limb ischemia is not only a one-shot treatment attempt to solve a localized vascular problem, but also a step in a series of, sometimes, repeated more or less palliative hospital stays involving both vascular and non-vascular surgery as well as non-surgical treatment. The patient's future experience [8] may be an important factor to consider in the surgeon's decision.

ACKNOWLEDGMENTS

Without the continued support and report of data from virtually all Swedish vascular surgeons in more than 50 centers, this presentation would not have been possible. The support from the Regional Oncological Register, Lund University, in the acquisition and management of data, by grants from the "National Dagmar Fund" and from The Epidemiological Center of The National Board for Health and Social Welfare is gratefully acknowledged.

REFERENCES

1 Anonymous. The Vascular Registry - A responsibility for all vascular surgeons? Vascular Registry in Southern Sweden (VRISS) *Eur J Vasc Surg* 1987; 1: 219-226.

2 Auditing surgical outcome. Ten years with The Swedish Vascular Registry - Swedvasc. *Eur J Vasc Endovasc Surg* 1998; 164 (Suppl 581): 1-48.

3 Dormandy JA, Thomas PRS. What is the natural history of a critically ischaemic patient with and without his leg? In *Limb salvage and amputation for vascular disease.* London, WB Saunders, 1988, pp 11-26.

4 Wolfe JH, Wyatt MG. Critical and subcritical ischaemia. *Eur J Vasc Endovasc Surg* 1997; 13: 578-582.

5 Luther M. Surgical treatment of chronic critical leg ischaemia. A five-year follow-up of survival, mobility and treatment level. *Eur J Vasc Endovasc Surg* 1998; 164: 35-43.

6 Holdsworth J. District hospital management and outcome of critical lower limb ischaemia: comparison with national figures. *Eur J Vasc Endovasc Surg* 1997; 13: 159-163.

7 Anonymous. Reoperations, Redo surgery and other interventions constitute more than one-third of vascular surgery. A study from Swedvasc (the Swedish Vascular Registry). The Swedish Society for Vascular Surgery. *Eur J Vasc Endovasc Surg* 1997; 14: 244-251.

8 Ellwood P. Shattuck lecture - outcomes management - a technology of patient experience. *New Engl J Med* 1988; 318: 1549-1556.

Conception et réalisation
ODIM, Z.A. La Carretière
04130 VOLX

Achevé d'imprimer sur ses presses
en mars 1999
N° d'imprimeur : 92.775
Dépôt légal : 1er trimestre 1999

Imprimé en FRANCE